Jim Janley

LINEAR SYSTEMS AND DIGITAL SIGNAL PROCESSING

THOMAS YOUNG

PRENTICE-HALL, INC., Englewood Cliffs, NJ 07632

Library of Congress Cataloging in Publication Data

Young, Thomas,
  Linear systems and digital signal processing.

  Includes bibliographies and index.
  1. Signal processing—Digital techniques.
  2. Analog-to-digital converters.   I. Title.
  TK5102.5.Y68   1985   621.38'043   84-18375
  ISBN 0-13-537366-2

Editorial/production supervision and
  interior design: *Ellen Denning*
Cover design: *Ben Santora*
Manufacturing buyer: *Gordon Osbourne*

© 1985 by Prentice-Hall, Inc., Englewood Cliffs, New Jersey 07632

All rights reserved. No part of this book may be
reproduced, in any form or by any means,
without permission in writing from the publisher.

Printed in the United States of America

10 9 8 7 6 5 4 3 2 1

ISBN 0-13-537366-2 01

PRENTICE-HALL INTERNATIONAL, INC., *London*
PRENTICE-HALL OF AUSTRALIA PTY. LIMITED, *Sydney*
EDITORA PRENTICE-HALL DO BRASIL, LTDA., *Rio de Janeiro*
PRENTICE-HALL CANADA INC., *Toronto*
PRENTICE-HALL HISPANOAMERICANA, S.A., *Mexico*
PRENTICE-HALL OF INDIA PRIVATE LIMITED, *New Delhi*
PRENTICE-HALL OF JAPAN, INC., *Tokyo*
PRENTICE-HALL OF SOUTHEAST ASIA PTE. LTD., *Singapore*
WHITEHALL BOOKS LIMITED, *Wellington, New Zealand*

This book is dedicated to:

        my wife, Nancy, and
        my first grandchild, Erik Edward.

CONTENTS

PREFACE ix

1   LINEAR SYSTEMS   1

    1.1   Definition of a Function, 2
    1.2   Linearity, 7
    1.3   Time Invariance, 8
    1.4   Causality, 10
    1.5   Delta Function: $\delta(t)$, 11
    1.6   Time-Shifted Delta Function, 13
    1.7   Impulse Response, 14
    1.8   Time-Shifting Property of the Delta Function, 14
    1.9   System Output: Convolution, 16
    1.10   Evaluation of the Convolution Integral, 18
    1.11   Fourier Transform, 25
    1.12   Convolution in the Frequency Domain, 35
    1.13   Correlation, 39

1.14 Differential Equation Representation of a Linear System, 47
1.15 Fourier and Laplace Transfer Functions, 48
1.16 Summary, 49
Problems, 50

2 AN OVERVIEW OF A DIGITAL SIGNAL PROCESSING SYSTEM 57

2.1 Definition of Signal Processing, 58
2.2 Analog Signal Processing, 60
2.3 Digital Signal Processing, 61
2.4 Description of a Digital Processing System, 62
2.5 Summary, 65
Problems, 66

3 ANALOG SIGNAL PROCESSING 67

3.1 Analog Filters, 67
3.2 Filter Characteristics, 70
3.3 Amplifiers, 76
3.4 Track-and-Hold Circuits, 85
3.5 Practical Sampling Theory, 96
3.6 Summary, 104
Problems, 104

4 ANALOG-TO-DIGITAL CONVERTERS 106

4.1 Characteristics of Conversion, 107
4.2 ADC Types, 115
4.3 IC ADCs, 117
4.4 Data Acquisition ICs, 135
4.5 Summary, 147
Problems, 147

5 DISCRETE-TIME SIGNALS AND THE z TRANSFORM 149

5.1 Discrete-Time Signals, 150
5.2 Impulse Response, 151
5.3 Input Sequence, 152

- 5.4 Discrete Convolution Example, 154
- 5.5 Transform Concepts, 156
- 5.6 Frequency Response of a Discrete-Time System, 157
- 5.7 Convolution in the Frequency Domain, 160
- 5.8 z Transform, 161
- 5.9 Role of the z Plane, 162
- 5.10 Examples of the z Transform, 163
- 5.11 Fourier Transform of the DTS from the z Transform, 166
- 5.12 z Transform Properties, 166
- 5.13 Summary, 168

Problems, 169

6 DIGITAL SYSTEMS 173

- 6.1 Discrete Linear System Fundamentals, 174
- 6.2 z Transform of a Difference Equation, 182
- 6.3 System Transfer Function, $H(z)$, 182
- 6.4 Block Diagram Representation of a Digital System, 189
- 6.5 Discrete System Circuit Analysis, 194
- 6.6 Digital Filter Synthesis, 196
- 6.7 Frequency Transformation, 234
- 6.8 Summary, 238

Problems, 238

7 THE DISCRETE AND THE FAST FOURIER TRANSFORMS 244

- 7.1 Discrete Fourier Transform, 245
- 7.2 Convolution Using the DFT, 250
- 7.3 Correlation Using the DFT, 254
- 7.4 Filtering Long Sequences of Samples, 254
- 7.5 Fast Fourier Transform, 257
- 7.6 Application of the FFT, 268
- 7.7 Summary, 274

Problems, 275

8 PROCESSOR INTEGRATED CIRCUITS 277

- 8.1 AMD 29500 Family, 277
- 8.2 Intel 2920 Signal Processor, 283

8.3 NEC μP7720, 287
8.4 TRW 10xx Family, 291
8.5 TMS32010 Digital Signal Processor, 293
8.6 Summary, 296

9 DIGITAL-TO-ANALOG CONVERTER 297

9.1 DAC Circuits, 298
9.2 Commercial DACs, 301
9.3 DAC Terminology, 323
9.4 DAC Output, 325
9.5 Output Processing, 327
9.6 Summary, 327

APPENDIX 328

A ANALOG FILTER CHARACTERISTICS 328

B PASCAL PROGRAM FOR THE DFT 344

C FORTRAN PROGRAM FOR THE FFT 350

INDEX   352

PREFACE

Courses in Digital Signal Processing (DSP) have been offered at the graduate and senior level by Electrical Engineering departments for several years. As they require a course in linear systems as a prerequisite, many Electrical and Computer Technology students are precluded from taking DSP as linear systems courses are not usually offered in their programs.

This text is based upon the lectures that presented the concepts of DSP to the Bachelor of Engineering Technology students at Rochester Institute of Technology, and it reflects the concept that DSP should be presented as a regular part of a technology program. It begins with a chapter that serves to introduce the reader to linear systems and to provide the background for the DSP material of the later chapters.

This is followed by chapters that introduce sampling concepts, the representation of a discrete signal using the z transform, digital systems, and the Fast Fourier Transform (FFT). The material on analog processing and interfacing the analog and digital worlds was kept brief as there are many excellent texts and articles on these topics. This allowed more space to be dedicated to the main ideas of linear systems and DSP. An appendix on analog filters is included to serve as a refresher or as a brief introduction. This provides the reader with some background when analog filters are encountered in the text.

Due to the rapid development in the hardware used to process the digital signals, it was decided to concentrate on some of the ICs that are available and to present only a brief look at the devices presently in use. The large number of journals and magazines along with the literature provided by the manufacturers should be consulted to keep abreast of the introduction of more powerful and faster devices.

In its present form, the text is suitable for people in Electrical Engineering who would like an introduction to DSP. It can also be used as a primer before reading the classical texts on the topic. In addition, the large number of people in other areas of Engineering, Computer Science, and Computer Technology will also find it valuable as an introductory source of information.

I would like to take this opportunity to thank the manufacturers who allowed me to use their data sheets, the students who provided feedback on the material of the text, and the people at Prentice-Hall whose work helped bring this text into the world.

THOMAS YOUNG

LINEAR SYSTEMS

1

The main purpose of this text is to introduce the reader to *digital signal processing* (DSP); however, it is advisable to present the concepts and terminology of linear systems first, as many of the techniques and applications of DSP require an understanding of them. In this chapter we consider the linear system in sufficient detail to provide the background required for later chapters.

When applied to a system, the concept of linearity allows one to use relatively simple analytical techniques to determine the input and output (I/O) relations of the system or its transfer function. These techniques will be examined from the time- and frequency-domain points of view. This beginning will lead us to the concepts of convolution (filtering) and correlation, preparing us for their use in DSP.

An *electrical system* is defined as any combination of electrical components. The system can be as simple as an RC circuit or as complex as a computer-controlled space shuttle. Regardless of a system's actual composition, we will assume that the relationship between its output and input signal is linear to simplify our analysis. Figure 1-1 will be used to represent an arbitrary system.

As the first step in establishing the concept of linearity, the definition of a function is presented. This will be useful as the values assumed by the input

Figure 1-1 Block representation of an arbitrary system.

and output signals of the system are determined by their functional relationships with time and frequency.

1.1 DEFINITION OF A FUNCTION

A function is defined when the following three items are provided: (1) a collection (set) of numbers, real or complex, called the *domain*; (2) a second set of numbers, also real or complex, called the *range*; and (3) a rule that relates the numbers in the range to numbers in the domain.

1.1.1 Domain

In electronics, the domains of interest usually consist of time or frequency values. Once a domain is decided upon (e.g., time), we can arbitrarily select any one of its members. As the values in the domain are chosen at our convenience, the domain is termed the *independent variable*.

In specifying a domain, maximum and minimum values are indicated which form the upper and lower limits of the domain. If all numerical values between the limits, possibly including them, are elements of the domain, it is termed *continuous*; if only a finite number of values are allowed, the domain is *discrete*. As an example, if all values of time between, and including, the limits of 0 and 10 seconds form the domain, it is continuous; if only integer values of time (e.g., 0s, 1s, 2s, . . . , 10s) form the domain, it is discrete.

If either or both of the limits of the domain are infinity, the domain, whether discrete or continuous, is said to be *infinite*. As an example, it is possible to consider all values of frequency from $-\infty$ to $+\infty$ as the domain, making it continuous and infinite in extent. Had we chosen a discrete set of frequency values as our domain, it would be discrete and infinite in extent. When representing functions in two- or three-dimensional graphs, the domain is the *abscissa*.

1.1.2 Range

The range usually consists of a collection of real or complex numbers that represent either voltage, current, power, the amplitude of a frequency component, or a phase-shift angle. As with the domain, the range can be continuous (all values allowed) or discrete (a finite amount of values allowed), finite or infinite in extent. As the numbers that form the range are determined by the choice of elements in the domain, the former are termed *dependent variables*. Even though a dependency exists between the values of the elements of a domain and its range, the range can be continuous or discrete, independent of the domain: a continuous domain, time, may give rise to either a continuous or a discrete range. Examples of this are the voltage output of a sine wave generator (continuous domain-continuous range) or the voltage of a digital signal (continuous domain-discrete range). A discrete domain may also give rise to either a continuous or a discrete range, as with the amplitudes of frequency components obtained by a Fourier analysis (discrete domain-continuous range) or the output of an analog-to-digital converter (ADC) (discrete domain-discrete range) in which the output levels can assume only discrete values at discrete sampling times. These relationships are displayed in Figures 1–2 and 1–3. In a graphical representation of the range, it is termed the *ordinate*.

1.1.3 Rule

The rule that states how to relate numbers in the domain (e.g., time, t, or frequency, f) with numbers in the range (e.g., voltage, v, power, p, or spectral component amplitude) completes the definition of a function. As an example, consider the voltage function, $v = v(\cdot)$, where $\cdot$ represents an arbitrary set of numbers forming the domain of interest, v represents the set of numbers forming the range, and $v(\cdot)$ represents the rule that specifies the way to relate values in the domain (t or f) to those of the range (e.g., v). Another name for the term inside the parentheses is the *argument*.

Choosing the time domain and selecting an element, t_1, the rule will identify a particular value of the range, v_1, associated with t_1. This is shown in Equation 1–1.

$$v(\cdot) = 10 \sin 2\pi f \cdot \quad (1\text{–}1a)$$

$$v = v(t) = 10 \sin 2\pi ft \quad (1\text{–}1b)$$

$$v_1 = v(t_1) = 10 \sin 2\pi ft_1 \quad (1\text{–}1c)$$

where Equation 1–1a specifies the rule, Equation 1–1b indicates the range and domain, and Equation 1–1c identifies the two numbers, v_1 and t_1, that are related by the rule. For any value of t (continuous and infinite domain), the values the voltage can assume are limited to all values between and including +10 and −10 V (continuous and finite range). What mathematicians call a

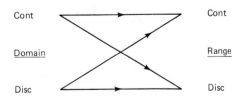

Figure 1–2 Representation of the domain/range relationship.

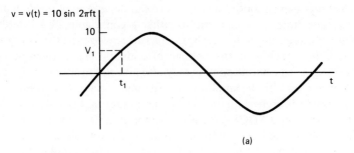

(a)

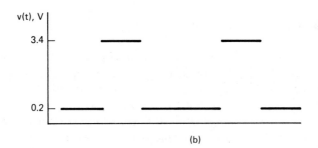

(b)

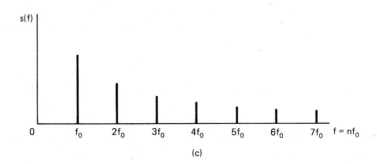

(c)

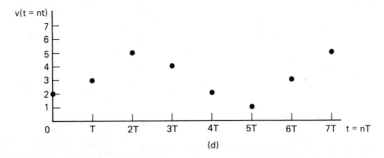

(d)

Figure 1-3 Graphical representation of a function; (a) continuous domain, continuous range; (b) continuous domain, discrete range; (c) discrete domain, continuous range; (d) discrete domain, discrete range.

Sec. 1.1   Definition of a Function   5

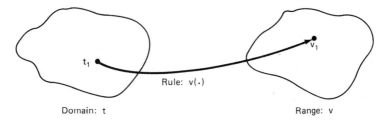

Figure 1–4 Pictorial representation of a time function.

function, we in the area of electronics call a signal: a voltage or current whose value (the range) can be measured or observed at different times (the domain). We will use the terms signal and function interchangably. A signal whose domain and range are continuous is termed *analog*.

In order to have our results relate to practical experience, the functions we will use in this text are absolutely integrable [i.e., the integral of the function's magnitude over all time must be finite]. This is expressed in Equation 1–2.

$$\int_{-\infty}^{\infty} |f(t)| \, dt \leq M < \infty \tag{1-2}$$

where $|\,|$ indicates magnitude and M is a finite constant. If the function represents voltage or current, this is another way of stating that the signal contains a finite amount of energy. Figures 1–3, 1–4, and 1–5 are graphical and pictorial representations of a function.

1.1.4 Implicit Function

If the domain of a given function is itself the range of another function, an implicit relationship exists between the functions. This occurs, for example, when we specify the domain in radians, ω, in the function $v = v(\omega)$. ω is also the range for the function $\omega = \omega(f) = 2\pi f$. This yields the implicit function $v = v(\omega(f)) = v'(f)$. Either by specifying $f = f_1$ for $v'(f)$ or $\omega = \omega_1$ for $v(\omega)$, we will obtain $v = v_1$. This is shown pictorially in Figure 1–6.

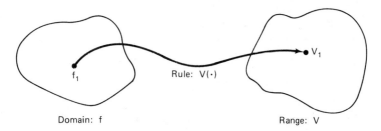

Figure 1–5 Pictorial representation of a frequency function.

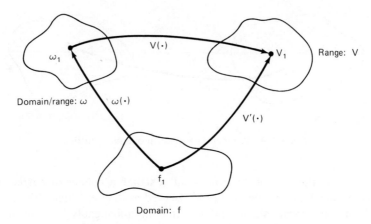

Figure 1–6 Pictorial representation of an implicit function.

By the foregoing definition of a function, $v(\omega) \neq v'(f)$; however, as the arguments are functionally related [$\omega = \omega(f)$ in this situation] and they have the same range, we will consider $v(\omega)$ and $v'(f)$ as being equivalent: $v(\omega) \Longleftrightarrow v'(f)$. From a graphical point of view, this implies that we are using a different abscissa to display the function.

Example 1–1

Represent the magnitude of the frequency components of a periodic signal (e.g., a square wave) using radians (ω) and hertz (f).

Solution This is given in Figure 1–7, where $|C_n|$ has been plotted against both domains, ω and $f = \omega/2\pi$.

In Example 1–1, both domains were discrete, while the range was continuous.
When dealing with implicit functions, we will make use of their equivalency in choosing a domain that will provide a different quantity of information.

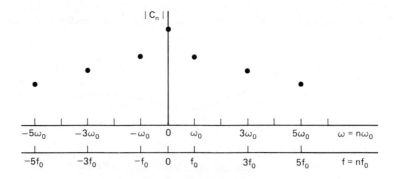

Figure 1–7 Solution to Example 1–1.

1.2 LINEARITY

Linearity can be defined using the arithmetical concepts of addition and multiplication as applied to functions.

1.2.1 Addition

Given a system with an input that consists of a sum of signals, if the system is linear, its output is obtained by summing the individual outputs for each of the input signals. This can be shown by considering the following. Given two system inputs, $x_1(t)$, and $x_2(t)$, that produce the outputs $y_1(t)$ and $y_2(t)$, respectively, $(x_i(t) \rightarrow y_i(t))$, an input $x(t) = x_1(t) + x_2(t)$ will produce an output $y(t) = y_1(t) + y_2(t)$. This is shown in Figure 1-8. For N inputs, we can generalize the definition of linearity as in Equation 1-3.

$$\sum_{i=1}^{N} x_i(t) = x(t) \longrightarrow y(t) = \sum_{i=1}^{N} y_i(t) \qquad (1\text{-}3)$$

where $y_i(t)$ is the output due to the input $x_i(t)$.

Example 1-2

$x_1(t) = 3t + 2$ and $x_2(t) = 2 - t/2$. If a linear system produces $y_1(t) = (t/4) - 1$ and $y_2(t) = t$ for the inputs above, what is the system output if its input is $x(t) = x_1(t) + x_2(t) = (3t + 2) + (2 - t/2)$?

Solution Using Equation 1-3, we have

$$y(t) = y_1(t) + y_2(t) \qquad (1\text{-}3)$$
$$= \left(\frac{t}{4} - 1\right) + t$$
$$= \frac{5t}{4} - 1$$

1.2.2 Multiplication

If a system is linear, multiplying its input by a (complex) constant will cause the output to be multiplied by the same constant. If an input signal

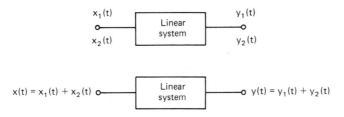

Figure 1-8 Representation of a linear system: addition.

$x(t)$ produces an output $y(t)$, then $cx(t)$ applied at the input will produce the output $cy(t)$, where c can be real or complex. This is given in Equation 1–4 for N inputs, each multiplied by a (possibly) different constant, c_j.

$$\sum_{j=1}^{N} c_j x_j(t) = x(t) \longrightarrow y(t) = \sum_{j=1}^{N} c_j y_j(t) \qquad (1\text{–}4)$$

where $y_i(t)$ is the output due to $x_i(t)$ and c_i may be real or complex. In this text we assume that all systems discussed are linear.

1.3 TIME INVARIANCE

If the components that form the system are constant in value (i.e., they do not change with time) the system is said to be *time invariant*. To define time invariance mathematically, we introduce the time-shifted, or time-delayed function. If $s(t)$ is a given function, then a time-delayed version of this function is $s(t - \tau)$, where τ is the amount of the delay. This is shown in Figure 1–9.

The effect of changing the domain (argument) from t to $t - \tau$ is to shift the graphical representation of the function to the right on the time axis for values of $\tau > 0$. This can be stated as follows: The time-shifted function will provide the same set of values (range) for $s(t - \tau)$ as $s(t)$; that is, their ranges are identical when their arguments assume the same value. This is shown in Equation 1–5, which is the functional form of the waveform in Figure 1–9.

$$s(\cdot) = \begin{cases} \dfrac{2\,V}{T_1}(\cdot), & 0 \le \cdot \le \dfrac{T_1}{2} \\ \dfrac{-2\,V}{T_1}(\cdot) + 2\,V, & \dfrac{T_1}{2} \le \cdot \le T_1 \\ 0, & \text{otherwise} \end{cases} \qquad (1\text{–}5a)$$

$$s(t) = \begin{cases} \dfrac{2\,V}{T_1}t, & 0 \le t \le \dfrac{T_1}{2} \\ \dfrac{-2\,V}{T_1}t + 2\,V, & \dfrac{T_1}{2} \le t \le T_1 \\ 0, & \text{otherwise} \end{cases} \qquad (1\text{–}5b)$$

$$s(t - \tau) = \begin{cases} \dfrac{2\,V}{T_1}(t - \tau), & 0 \le t - \tau \le \dfrac{T_1}{2} \\ \dfrac{-2\,V}{T_1}(t - \tau) + 2\,V, & \dfrac{T_1}{2} \le t - \tau \le T_1 \\ 0, & \text{otherwise} \end{cases} \qquad (1\text{–}5c)$$

Sec. 1.3 Time Invariance

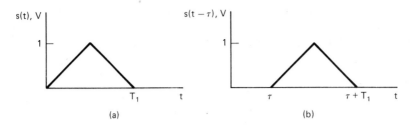

Figure 1-9 A signal and its time delayed form.

Equation 1-5b is the nonshifted waveform (Figure 1-9a) and Equation 1-5c is the time-shifted waveform (Figure 1-9b). Equation 1-5c is obtained from Equation 1-5a by replacing ($\cdot$) with ($t - \tau$), where τ is a parameter [i.e., a "constant" variable], that indicates the amount of the shift or delay. For negative values of τ, the shift would be to the left, representing an advanced signal. To observe the values of t for which $s(t - \tau)$ has a nonzero value, we add τ to both sides of the inequality in Equation 1-5c to obtain Equation 1-5d.

$$s(t-\tau) = \begin{cases} \dfrac{2V}{T_1}(t-\tau), & \tau \leq t \leq \dfrac{T_1}{2}+\tau \\ \dfrac{-2V}{T_1}(t-\tau), & \tau + \dfrac{T_1}{2} \leq t \leq T_1+\tau \\ 0, & \text{otherwise} \end{cases} \quad (1\text{-}5d)$$

The effect of time invariance on a linear system can be stated as follows: If an input signal $s(t)$ produces an output $y(t)$, a time-invariant (TI) system will produce $y(t - \tau)$ if the input is $s(t - \tau)$. This is shown in Figure 1-10. For a time-invariant system, there is no change in the shape of the output

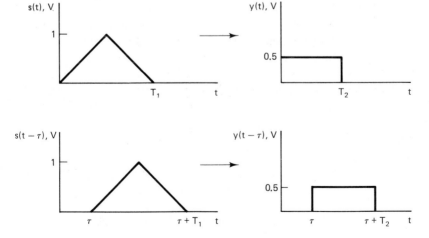

Figure 1-10 Effect of time invariance on system output.

waveform if identical inputs are applied at different times. As the system does not change, the resulting output will have the same waveform in each instance, but shifted in time. A linear time-invariant system (LTIS) will be assumed for our investigations.

Example 1–3

If a 1-V, 3-ms pulse applied to a LTIS at $t = 0$ s produces a 0.5-V, 2-ms pulse, what is the output if the same pulse is applied 10 ms later?

Solution As the system is TI, the output for the second input pulse will have the same waveform as for the first, but it will appear 10 ms later. This is shown in Figure 1–11.

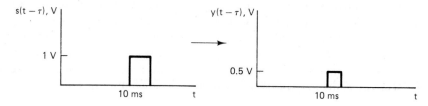

Figure 1–11 Solution to Example 1–3.

1.4 CAUSALITY

A system is assumed to be unable to predict which inputs will be applied to it; that is, an output signal cannot be generated until it is caused to do so by a signal at the system's input. The relation between the cause, which must come first, and the effect, which is a result of the cause, is termed *causality*. A system is causal if it requires an input to produce an output: $s(t)$ is required to produce $y(t)$. This can be stated as: If $s(t)$ is zero for all $t < 0$, then $y(t)$ is also zero for all $t < 0$. This is shown in Figure 1–12.

If the system is noncausal, an output will appear without, or prior to, an input signal. This is shown in Figure 1–13, where it is assumed that $s(t) \longrightarrow y(t)$. A noncausal system is physically unrealizable and is considered only from a theoretical point of view.

If a causal system is also time-invariant, we can modify the definition of causality by stating: If $s(t - \tau)$ is zero for all $t - \tau < 0$ (i.e., $t < \tau$), then

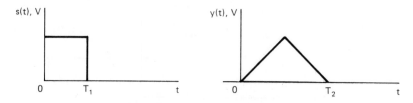

Figure 1–12 Output of a causal system.

Sec. 1.5   Delta Function: δ(t)   11

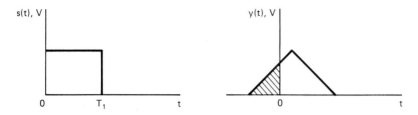

Figure 1–13   Output of a noncausal system.

$y(t - \tau)$ is also zero for all $t < \tau$. In any reference to a system, it is assumed to be linear, time invariant, and causal.

1.5 DELTA FUNCTION: δ(t)

To determine the I/O relationship of a system, we must know how it will respond to a special type of input called the *impulse*. The impulse is represented by the delta function, $\delta(t)$. Its domain is all t, and its range contains only two values, zero and infinity. The rule that relates the range and the domain is

$$\delta(\cdot) = \begin{cases} 0, & \cdot \neq 0 \\ \infty, & \cdot = 0 \end{cases} \quad (1\text{–}6a)$$

$$\delta(t) = \begin{cases} 0, & t \neq 0 \\ \infty, & t = 0 \end{cases} \quad (1\text{–}6b)$$

that is, the range is zero for all values of time except when $t = 0$, at which time the function has the value of infinity (i.e., it "blows up"). The graphical representation of the delta function is given in Figure 1–14, where the arrow represents a signal of zero width and infinite height.

To obtain some insight into the delta function, we can consider the function $\Delta(t)$ as shown in Figure 1–15. This function is defined in Equation 1–7.

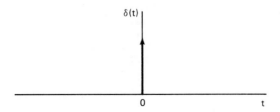

Figure 1–14   Graphical representation of the δ function.

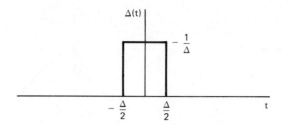

Figure 1–15 Graphical representation of $\Delta(t)$.

$$\Delta(t) = \begin{cases} \dfrac{1}{\Delta}, & -\dfrac{\Delta}{2} \le t \le \dfrac{\Delta}{2} \\ 0, & \text{otherwise} \end{cases} \qquad (1\text{-}7)$$

The function has a height of $1/\Delta$ and a width of Δ, resulting in an area of 1. This can be shown by integrating $\Delta(t)$ over all time.

$$\int_{-\infty}^{\infty} \Delta(t)\, dt = \int_{-\frac{\Delta}{2}}^{\frac{\Delta}{2}} \frac{1}{\Delta}\, dt = \frac{1}{\Delta}\left[\frac{\Delta}{2} - \left(-\frac{\Delta}{2}\right)\right] = 1 \qquad (1\text{-}8)$$

The δ function is obtained from $\Delta(t)$ by the limiting process given in Equation 1-9.

$$\delta(t) \equiv \lim_{\Delta \to 0} \Delta(t) \qquad (1\text{-}9)$$

The width of the function goes to zero as its amplitude goes to infinity in a manner that causes the area under the curve to remain constant at unity. The limiting process is shown in Figure 1–16. As Δ decreases in value, the function becomes narrower and higher. In the limit, as $\Delta \longrightarrow 0$, $\Delta(t) \longrightarrow \delta(t)$.

As the integral of $\Delta(t)$ is unity for any value of Δ, it follows that the integral of the delta function is also unity provided that the origin is included in the limits of integration; otherwise, it is zero. This is given in Equation 1–10.

$$\int_a^b \delta(t)\, dt = \begin{cases} 1, & a \le 0 \le b \\ 0, & \text{otherwise} \end{cases} \qquad (1\text{-}10)$$

Figure 1–16 Limiting process for $\Delta(t)$.

Sec. 1.6   Time-Shifted Delta Function

If we multiply the delta function by a constant, k, and then integrate the product over all time, we obtain Equation 1-11.

$$\int_{-\infty}^{\infty} k\delta(t)\, dt = k \int_{-\infty}^{\infty} \delta(t)\, dt = k \qquad (1\text{-}11)$$

Multiplication of $\delta(t)$ by a constant is equivalent to choosing $\Delta(t)$ with a height of k/Δ. This results in an area equal to k.

Example 1-4

Evaluate the following integrals.

(a) $\displaystyle\int_{-1}^{3} \delta(t)\, dt$ (b) $\displaystyle\int_{-26}^{-15} \delta(t)\, dt$

(c) $\displaystyle\int_{-10}^{100} 3\delta(t)\, dt$ (d) $\displaystyle\int_{40}^{97} 7\delta(t)\, dt$

Solution   Using Equations 1-10 and 1-11, we have

(a) 1   (b) 0

(c) 3   (d) 0

where the limits of integration for parts (b) and (d) did not include the origin, resulting in the zero value.

1.6 TIME-SHIFTED DELTA FUNCTION

If the argument of the δ function is replaced with $t - \tau$, we see that the impulse still occurs when the argument takes on the value of zero (i.e., when $t - \tau = 0$ or $t = \tau$). This is given in Equation 1-12 and shown in Figure 1-17.

$$\delta(t - \tau) = \begin{cases} 0, & t - \tau \neq 0 \\ \infty & t - \tau = 0 \end{cases} \qquad (1\text{-}12)$$

As in Equation 1-10, the integral of the time-shifted δ function is given as

$$\int_{a}^{b} \delta(t - \tau)\, dt = \begin{cases} 1, & a \leq \tau \leq b \\ 0, & \text{otherwise} \end{cases} \qquad (1\text{-}13)$$

Figure 1-17   Time shifted δ function.

If the time-shifted δ function is multiplied by a constant k and integrated over all time, we obtain

$$\int_{-\infty}^{\infty} k\delta(t - \tau)\, dt = k \tag{1-14}$$

1.7 IMPULSE RESPONSE

If the input to a system is the impulse, $\delta(t)$, the output is termed the *impulse response* and designated $h(t)$. This is shown in Figure 1–18, where the input signal is $\delta(t)$ and the output is defined as $h(t)$.

x(t) = δ(t) ○———[LTICS]———○ y(t) ≡ h(t)   **Figure 1–18** Impulse response of a LTIC system.

A system can be identified by its impulse response. Several techniques are available to determine $h(t)$, the most common being the inverse Laplace transform of the output/input ratio, where both factors are expressed as functions of the complex variable s.

If the input to the system is the time-shifted version of the δ function, the output is a time-shifted version of the impulse response. This is shown in Figure 1–19. This result is due to the time invariance of the system. As the system is linear, applying $c\delta(t - \tau)$ at the input will produce $ch(t - \tau)$ at the output.

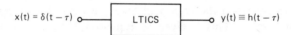

Figure 1–19 Time shifted impulse response.

1.8 TIME-SIFTING PROPERTY OF THE DELTA FUNCTION

If we integrate the product of the delta function and any time function, we can observe the *sifting property* of the δ function.

$$\int_{-\infty}^{\infty} f(t)\delta(t)\, dt = \int_{-\infty}^{-\epsilon} f(t)\delta(t)\, dt + \int_{-\epsilon}^{\epsilon} f(t)\delta(t)\, dt + \int_{\epsilon}^{\infty} f(t)\delta(t)\, dt \tag{1-15}$$

where ϵ is arbitrarily small.

The first and last integrals on the right-hand side are zero, as the limits of integration do not include the origin. If $f(t)$ is "reasonably well behaved" at the origin (i.e., no jumps), as $\epsilon \to 0$, $f(t)$ can be approximated by $f(0)$. Since $f(0)$ is a constant, it can be taken outside the integral, as shown in Equation 1–16. Passing to the limit as $\epsilon \to 0$, we have

Sec. 1.8 Time-Sifting Property of the Delta Function

$$\lim_{\epsilon \to 0} \int_{-\epsilon}^{\epsilon} f(t)\delta(t)\, dt \simeq \int_{-\epsilon}^{\epsilon} f(0)\delta(t)\, dt = f(0) \int_{-\epsilon}^{\epsilon} \delta(t)\, dt = f(0) \quad (1\text{-}16)$$

The integral has a value of 1, as the variable of integration includes the origin. Equations 1-15 and 1-16 are combined to yield Equation 1-17.

$$\int_{-\infty}^{\infty} f(t)\delta(t)\, dt = f(0) \quad (1\text{-}17)$$

From all possible values of $f(t)$, the δ function has sifted out the value $f(0)$. This is shown in Figure 1-20.

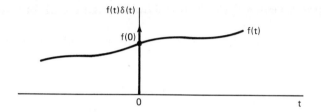

Figure 1-20 Sifting property of the δ function.

If we use the time-shifted δ function in Equation 1-17, we obtain

$$f(\tau) = \int_{-\infty}^{\infty} f(t)\delta(t - \tau)\, dt \quad (1\text{-}18)$$

As in Equation 1-15, the integrand is zero for all $t \neq \tau$. As $t \to \tau$, $f(t)$ can be approximated by $f(\tau)$, a constant value which is taken outside the integral. As the variable of integration includes τ, the integral will have a value of unity.

Example 1-5
Evaluate the following integrals.

$$\text{(a)} \int_{-\infty}^{\infty} (3t + 2)\delta(t)\, dt \quad (1\text{-}19a)$$

$$\text{(b)} \int_{-\infty}^{\infty} (2 \cos \omega t)\delta(t)\, dt \quad (1\text{-}19b)$$

$$\text{(c)} \int_{-\infty}^{\infty} (\sqrt{t} + 6)\delta(t - 3)\, dt \quad (1\text{-}19c)$$

$$\text{(d)} \int_{-\infty}^{\infty} (t^2 + 7)\delta(t + 4)\, dt \quad (1\text{-}19d)$$

Solution   Using Equation 1-18, we obtain

(a) $(3 \cdot 0 + 2) = 2$ (b) $2 \cos \omega \cdot 0 = 2$
(c) $\sqrt{3} + 6 = 7.732$ (d) $(-4)^2 + 7 = 23$

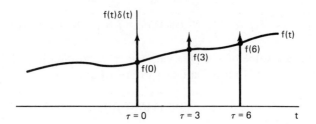

Figure 1-21 Sifting property of the time shifted δ function.

In this manner, all values of a function (its range) can be represented by the sifting integral, Equation 1-18. This is shown in Figure 1-21 for several values of τ.

As $\delta(t - \tau) = \delta(\tau - t)$, we can interchange t with τ in Equation 1-18 and also obtain

$$f(t) = \int_{-\infty}^{\infty} f(\tau)\delta(t - \tau)\, d\tau \tag{1-20}$$

This allows us to represent any function of time by the sifting integral treating τ as the variable of integration and t as a parameter: To find the value of $f(t)$ for a particular (fixed) value of t, the integral of Equation 1-20 is evaluated. Interchanging t and τ is a technique we will use when the convolution equation is developed.

1.9 SYSTEM OUTPUT: CONVOLUTION

With the information given above, we are able to establish a procedure to obtain the system output for any input using the system impulse response. Although this process of analysis is quite useful, we will find that it is more useful to be able to determine the system impulse response given the system input and its desired output. The ability to find a system's impulse response allows one to take a given signal and with the desired output in mind, determine the required system response (i.e., we can design a circuit to modify a signal to suit our needs). In later chapters we combine the concepts of linear systems with those of digital signal processing to aid in our analysis and design of a digital system.

To develop the system response, we refer to Figure 1-19, in which the time-shifted δ function produced a time-shifted impulse response. This is given in Equation 1-21.

$$\delta(t - \tau) \longrightarrow h(t - \tau) \tag{1-21}$$

where τ is a specific value of the variable t.

It was also mentioned that, because of linearity, a constant times the δ function would produce the impulse response multiplied by the same constant.

Sec. 1.9 System Output: Convolution

If we evaluate a function of time at a given value of τ, we obtain $f(t = \tau) = f(\tau)$, which is a constant. Multiplying the δ function of Equation 1–21 by the constant value $f(\tau)$, we obtain

$$f(\tau)\delta(t - \tau) \longrightarrow f(\tau)h(t - \tau) \qquad (1\text{–}22)$$

As the system is linear, if we compose an input created from several inputs of the form $f(\tau)\delta(t - \tau)$, each with a different value of τ, τ_j, the output will be composed of the sum of the individual outputs, one for each of the inputs above. This is shown in Equation 1–23.

$$\sum_{j=1}^{N} f(\tau_j)\delta(t - \tau_j) = f(t) \longrightarrow y(t) = \sum_{j=1}^{N} f(\tau_j)h(t - \tau_j) \qquad (1\text{–}23)$$

In this situation, we are using τ as a parameter; that is, τ represents specific values of the time, t, and $f(\tau_j)\delta(t - \tau_j) \longrightarrow f(\tau_j)h(t - \tau_j)$.

To obtain the desired result, we must realize that since τ can take on any value that t does, we can think of τ as the variable and t as the parameter. This allows us to consider an infinite sum of inputs in Equation 1–23, one for each value of τ. As τ can take on any value that t does, it must also be continuous. This implies that the domain τ equals the domain t, and as τ is continuous, we can represent the sums in Equation 1–23 as integrals.

$$\sum_{j=1}^{N} f(\tau_j)\delta(t - \tau_j) \longrightarrow \int_{-\infty}^{\infty} f(\tau)\delta(t - \tau)\, d\tau \qquad (1\text{–}24\text{a})$$

$$\sum_{j=1}^{N} f(\tau_j)h(t - \tau_j) \longrightarrow \int_{-\infty}^{\infty} f(\tau)h(t - \tau)\, d\tau \qquad (1\text{–}24\text{b})$$

where the integration is over τ and t is serving as the parameter. The integral in Equation 1–24a is the same as in Equation 1–20, the sifting integral, and must represent $f(t)$. Keep in mind that we are presently treating t as a parameter and that evaluating the sifting integral yields a single value of $f(t)$. To obtain all values of $f(t)$, we would have to evaluate the integral for all values of t; however, we will not do this but will concern ourselves only with the integral on the right-hand side of Equation 1–24b. As this represents the system output for the arbitrary input $f(t)$, we have

$$y(t) = \int_{-\infty}^{\infty} f(\tau)h(t - \tau)\, d\tau \qquad (1\text{–}25)$$

This is the convolution equation and is the desired result that allows us to find the output of a system whose impulse response is $h(t)$ for an arbitrary input $f(t)$. Again, t is being treated as the parameter (or given value) and τ as the variable of integration. The integral in Equation 1–25 is one of the major results of linear system theory. Equation 1–25 is represented in shortened form as

$$y(t) = f(t) * h(t) = \int_{-\infty}^{\infty} f(\tau)h(t-\tau)\, d\tau \tag{1–26}$$

Evaluating the integral provides us with the value $y(t)$ for a single value of time. At this point, it is convenient again for us to consider t as the independent variable. Evaluating the integral provides us with the functional form of $y(t)$.

In the following section, we present an approach that will allow us readily to evaluate the convolution integral.

1.10 EVALUATION OF THE CONVOLUTION INTEGRAL

The standard approach to evaluate the convolution integral uses a graphical procedure to determine the limits of integration and is called the "fold, displace to the left, and slide to the right" technique. This is quite helpful, as the integrand is a product of two terms and it is not always clear exactly what the limits are from the functional form of the input signal and impulse response. As t is once again considered to be a parameter for this technique, we will look for values of t that result in a nonzero product of $f(t)$ and $h(t-\tau)$. This occurs whenever the graphical representation of the functions overlap, as shown below.

To demonstrate the procedure used for evaluating the convolution integral we will consider the functions shown in Figure 1–22, where $f(t)$ is the input signal to a system whose impulse response is $h(t)$. If $f(t)$ is in volts, $y(t)$ will also be in volts if $h(t-\tau)$ is in units of reciprocal seconds: 1/s.

The first step in the evaluation of the integral is to replace t with τ. This is done in Figure 1–23. Next, τ is replaced with $-\tau$ in one of the functions.

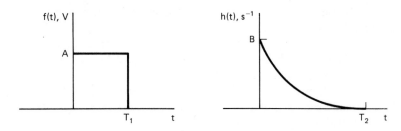

Figure 1–22 Graphical form of input and transfer function.

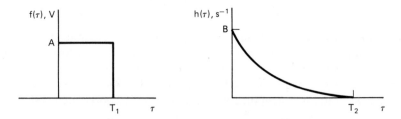

Figure 1–23 Change of domain for Figure 1–22.

Sec. 1.10   Evaluation of the Convolution Integral   19

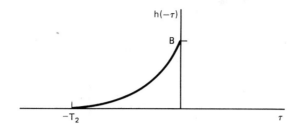

Figure 1–24   Replacement of τ with $-\tau$

For this example, $h(t)$ was arbitrarily chosen and is shown in Figure 1–24. This is the "folding" part of the technique: the curve is folded about the vertical axis.

To displace the curve to the left, we add a "constant" to the argument. For our purposes, the constant will be the parameter t. This yields Figure 1–25a. The position of the curve along the axis is determined by the value of t we choose. This is shown in Figure 1–25b, where the curve location is shown for values of t equal to t_1, t_2, t_3: as t becomes more positive, the curve "slides" to the right.

The last step is to show the product of $f(\tau)$ and $h(t - \tau)$ (i.e., the integrand), as in Figure 1–26. For the chosen value of t, the curves do not overlap; that is, they have no common values of τ that produce a nonzero product. In fact, for any value of t such that $-\infty < t < 0$, there is no overlap, causing

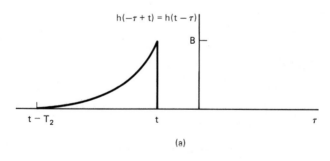

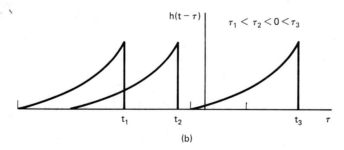

Figure 1–25(a)   Replacement of $-\tau$ with $-\tau + t$; (b) $h(t - \tau)$ for several values of t.

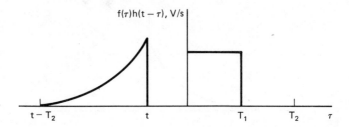

Figure 1–26 Graph of the integrand for $t < 0$.

the product and its integral to be zero. This indicates that $y(t)$ must also be zero for all values of t less than zero: $y(t) = 0$, $t < 0$. This is what we expect for a causal system: no output prior to the application of an input signal.

For causal systems, we can modify the procedure for evaluating the convolution integral. When $t < 0$, the integrand is also zero as there is no overlap of the product functions: One of the functions is always zero as shown in Figure 1–26. Therefore $y(t) = 0$ for all $t < 0$. In addition, the integral is always zero for $-\infty < \tau < 0$. Equation 1-26 can be written as

$$y(t) = 0, \quad t < 0 \tag{1-27a}$$

$$y(t) = \int_0^\infty f(\tau)h(t-\tau)\,d\tau \tag{1-27b}$$

As t is allowed to become more positive, the curve of $h(t-\tau)$ moves (slides) to the right and will overlap $f(\tau)$ when $t \geq 0$, resulting in a nonzero product. This is shown in Figure 1–27. As the overlap (nonzero product) occurs only for values of τ greater than zero and less than t, the limits of the convolution integral become

$$y_1(t) = \int_0^t f(\tau)h(t-\tau)\,d\tau, \qquad 0 \leq t \leq T_1 \tag{1-28}$$

This integral is valid for any value of time between 0 and T_1, yielding the output $y_1(t)$ for $0 \leq t \leq T_1$.

For all values of t greater than T_1 and less than T_2, the product of the

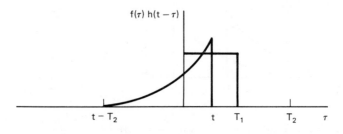

Figure 1–27 Graph of the integrand for $0 \leq t \leq T_1$.

Sec. 1.10   Evaluation of the Convolution Integral

functions is zero for values of $\tau < 0$ and greater than T_1. This is shown in Figure 1–28. Overlap occurs for any value of τ between 0 and T_1, resulting in an integral

$$y_2(t) = \int_0^{T_1} f(\tau) h(t - \tau) \, d\tau, \qquad T_1 \leq t \leq T_2 \qquad (1\text{–}29)$$

This integral is valid for any value of t greater than T_1, but as we will see, less than T_2. It provides us with $y_2(t)$ for $T_1 \leq t \leq T_2$.

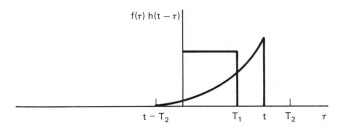

Figure 1–28 Graph of the integrand for $T_1 \leq t \leq T_2$.

For values of t greater than T_2 and less than $T_2 + T_1$, the curve of $h(t - \tau)$ will overlap $f(\tau)$ only from $t - T_2$ to T_1, as shown in Figure 1–29. The output is determined for this value of t as

$$y_3(t) = \int_{t-T_2}^{T_1} f(\tau) h(t - \tau) \, d\tau, \qquad T_2 \leq t \leq T_1 + T_2 \qquad (1\text{–}30)$$

We now have $y_3(t)$ for $T_2 \leq t \leq T_1 + T_2$. Finally, when t is greater than $T_1 + T_2$, there is no longer any overlap and the integrand, integral, and $y(t)$ are zero.

To obtain $y(t)$ for all t, the values of $y(t)$ for each value of t that provides a nonzero integral are pieced together.

$$y(t) = \begin{cases} y_1(t), & 0 \leq t \leq T_1 \\ y_2(t), & T_1 \leq t \leq T_2 \\ y_3(t), & T_2 \leq t \leq T_1 + T_2 \\ 0, & \text{otherwise} \end{cases} \qquad (1\text{–}31)$$

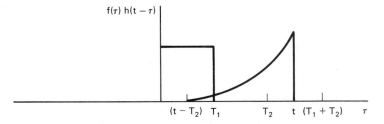

Figure 1–29 Graph of the integrand, $T_2 \leq t$.

From a mathematical point of view, we see that a change of limits is required each time one curve crosses a point in which the slope (derivative) of the second curve has a discontinuity, e.g., at $\tau = 0$ and T_1. Observing where these points occur will make it somewhat easier to evaluate the convolution integral as they indicate when we are to select new limits.

In the following example, the foregoing approach is demonstrated: The output of a system with a known impulse response is determined by convolving it with a given input signal, $s(t)$.

Example 1–6

Determine the output of the system whose input and impulse response are shown in Figure 1–30.

Solution We first express $s(t)$ and $h(t)$ in functional form where $h(t)$ has units of reciprocal seconds, $1/s$.

$$s(t) = \begin{cases} 0.6 \text{ V}, & 0 \leq t \leq T_1 = 50 \text{ ms} \\ 0, & \text{otherwise} \end{cases} \quad (1\text{--}32\text{a})$$

$$h(t) = \begin{cases} \left(\dfrac{-1}{0.1 \ s^2}\right) t + \dfrac{1}{s}, & 0 \leq t \leq T_2 = 100 \text{ ms} \\ 0, & \text{otherwise} \end{cases} \quad (1\text{--}32\text{b})$$

Next, $h(t)$ is converted to $h(t - \tau)$.

$$h(\tau) = \begin{cases} \left(\dfrac{-1}{0.1 \ s^2}\right) \tau + \dfrac{1}{s}, & 0 \leq \tau \leq T_2 = 100 \text{ ms} \\ 0, & \text{otherwise} \end{cases} \quad (1\text{--}33\text{a})$$

$$h(-\tau) = \begin{cases} \left(\dfrac{-1}{0.1 \ s^2}\right)(-\tau) + \dfrac{1}{s}, & 0 \leq -\tau \leq T_2 = 100 \text{ ms} \\ 0, & \text{otherwise} \end{cases} \quad (1\text{--}33\text{b})$$

$$h(t - \tau) = \begin{cases} \left(\dfrac{-1}{0.1 \ s^2}\right)(t - \tau) + \dfrac{1}{s}, & 0 \leq t - \tau \leq T_2 = 100 \text{ ms} \\ 0, & \text{otherwise} \end{cases} \quad (1\text{--}33\text{c})$$

The transition to $h(t - \tau)$ can be accomplished directly by replacing $(\cdot)$ with $(t - \tau)$ in the functional representation, $h(\cdot)$. Having made the transition from t to $t - \tau$,

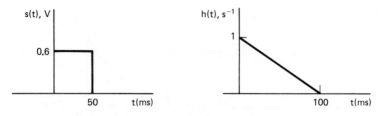

Figure 1–30 Functions for Example 1–6.

Sec. 1.10 Evaluation of the Convolution Integral

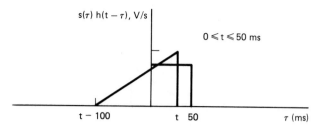

Figure 1-31 Graph of the integrand for $0 \leq t \leq 50$ ms.

we can display $s(\tau)h(t - \tau)$ for the first region of overlap $0 \leq t \leq 50$ ms, as shown in Figure 1-31. We have ignored values of $t < 0$, as they produce $y(t) = 0$.

As the overlap occurs only for $0 \leq \tau \leq t$, Equation 1-28 yields

$$y_1(t) = \int_0^t s(\tau)h(t - \tau)d\tau$$

$$= \int_0^t (0.6 \text{ V})\left\{\left(\frac{-1}{0.1 \text{ s}^2}\right)(t - \tau) + \frac{1}{s}\right\} d\tau \qquad (1\text{-}28)$$

Expanding the integrand, we obtain

$$y_1(t) = \int_0^t \left(\frac{-0.6 \text{ V}}{0.1 \text{ s}^2}\right) t \, d\tau + \int_0^t \left(\frac{0.6 \text{ V}}{0.1 \text{ s}^2}\right) \tau \, d\tau + \int_0^t \frac{0.6 \text{ V} \, d\tau}{s} \qquad (1\text{-}34\text{a})$$

$$y_1(t) = \left(\frac{-0.6 \text{ V}}{0.1 \text{ s}^2}\right) t\tau \bigg|_0^t + \left(\frac{0.6 \text{ V}}{0.1 \text{ s}^2}\right) \frac{\tau^2}{2} \bigg|_0^t + \frac{(0.6 \text{ V})}{s} \tau \bigg|_0^t \qquad (1\text{-}34\text{b})$$

$$= \frac{-0.6 \text{ V}}{0.1 \text{ s}^2} t^2 + \frac{0.6 \text{ V}}{0.1 \text{ s}^2} \frac{t^2}{2} + \frac{0.6 \text{ V}}{s} t \qquad (1\text{-}34\text{c})$$

$$= \frac{-6 \text{ V}}{s^2} t^2 + \frac{6 \text{ V}}{2 s^2} t^2 + \frac{0.6 \text{ V}}{s} t \qquad (1\text{-}34\text{d})$$

$$= \frac{-3 \text{ V}}{s^2} t^2 + \frac{0.6 \text{ V}}{s} t, \qquad 0 \leq t \leq T_1 = 50 \text{ ms} \qquad (1\text{-}34\text{e})$$

where $y_1(t)$ is in volts.

As $h(t - \tau)$ crosses the discontinuity in $s(\tau)$ at $\tau = 50$ ms, a new region of overlap occurs for $50 \text{ ms} \leq t \leq 100 \text{ ms}$ and a new set of limits must be chosen. This is shown in Figure 1-32. Overlap of the curves occurs for $0 \leq \tau \leq T_1 = 50$ ms.

$$y_2(t) = \int_0^{50 \text{ ms}} s(\tau)h(t - \tau) \, d\tau \qquad (1\text{-}35\text{a})$$

$$= \int_0^{50 \text{ ms}} \left(\frac{-6 \text{ V}}{s^2}\right) t \, d\tau + \int_0^{50 \text{ ms}} \left(\frac{6 \text{ V}}{s^2}\right) \tau \, d\tau + \int_0^{50 \text{ ms}} \frac{0.6 \text{ V}}{s} d\tau \qquad (1\text{-}35\text{b})$$

$$= \left(\frac{-6 \text{ V}}{s^2}\right) t\tau \bigg|_0^{50 \text{ ms}} + \left(\frac{6 \text{ V}}{s^2}\right) \frac{\tau^2}{2} \bigg|_0^{50 \text{ ms}} + \left(\frac{0.6 \text{ V}}{s}\right) \tau \bigg|_0^{50 \text{ ms}} \qquad (1\text{-}35\text{c})$$

$$= -\left(\frac{6 \text{ V}}{s^2}\right) t \, 50 \text{ ms} + \left(\frac{6 \text{ V}}{s^2}\right) \frac{(50 \text{ ms})^2}{2} + \left(\frac{0.6 \text{ V}}{s}\right) 50 \text{ ms} \qquad (1\text{-}35\text{d})$$

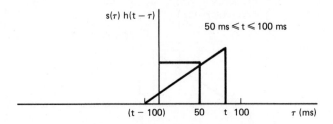

Figure 1–32 Graph of the integrand for 50 ms ≤ t ≤ 100 ms.

This becomes

$$y_2(t) = \frac{-0.3 \, Vt}{s} + 0.0375 V, \qquad T_1 \le t \le T_2 = 100 \text{ ms} \qquad (1\text{-}35e)$$

The last overlap region occurs for $T_2 \le t \le T_1 + T_2$ and is shown in Figure 1–33. Overlap occurs for $t - T_2 < \tau < T_1$.

$$y_3(t) = \int_{t-T_2}^{T_1} s(\tau) h(t-\tau) \, d\tau \qquad (1\text{-}36a)$$

$$= \int_{t-T_2}^{T_1} \left(\frac{-6 \, V}{s^2}\right) t \, d\tau + \int_{t-T_2}^{T_1} \left(\frac{6 \, V}{s^2}\right) \tau \, d\tau + \int_{t-T_2}^{T_1} \left(\frac{0.6 \, V}{s}\right) d\tau \qquad (1\text{-}36b)$$

$$= \left(\frac{-6 \, V}{s^2}\right) t\tau \bigg|_{t-T_2}^{T_1} + \left(\frac{6 \, V}{s^2}\right) \frac{\tau^2}{2} \bigg|_{t-T_2}^{T_1} + \left(\frac{0.6 \, V}{s}\right) \tau \bigg|_{t-T_2}^{T_1} \qquad (1\text{-}36c)$$

$$= \left(\frac{-6 \, V}{s^2}\right) t \, [T_1 - (t - T_2)] + \left(\frac{6 \, V}{s^2}\right) \frac{[T_1^2 - (t - T_2)^2]}{2}$$

$$+ \frac{0.6 \, V}{s} [T_1 - (t - T_2)] \qquad (1\text{-}36d)$$

After expanding the terms in brackets and combining constant terms, we obtain

$$y_3(t) = \frac{3 \, Vt^2}{s^2} - \frac{0.9 \, Vt}{s} + 0.0675 \, V, \qquad T_2 \le t \le T_1 + T_2 \qquad (1\text{-}36e)$$

For $t > T_1 + T_2$, there is no overlap of the curves and $y(t)$ is zero.

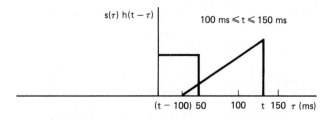

Figure 1–33 Graph of the integrand for 100 ms ≤ t.

The results above are combined to obtain $y(t)$ for all values of time.

$$y(t) = \begin{cases} \left(\dfrac{-3t^2}{s^2} + \dfrac{0.6t}{s}\right) \text{V}, & 0 \leq t \leq 50 \text{ ms} \\ \left(\dfrac{-0.3t}{s} + 0.0375\text{V}\right) \text{V}, & 50 \text{ ms} \leq t \leq 100 \text{ ms} \\ \left(\dfrac{3t^2}{s^2} - \dfrac{0.9t}{s} + 0.0675\right) \text{V}, & 100 \text{ ms} \leq t \leq 150 \text{ ms} \\ 0, & \text{otherwise} \end{cases} \quad (1\text{–}37)$$

Figure 1–34 shows $y(t)$ for all time.

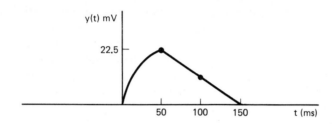

Figure 1–34 Graph of output for Example 1–6.

One of the purposes of presenting a (relatively) simple example is to show the reader the amount of "bookkeeping" and computational labor involved in evaluating the convolution integral to obtain the system output for a given input. We will see that convolution can be more readily performed using the frequency-domain representations of the input signal and impulse response of the system, and it will be this approach that is emphasized, both in the analog and discrete systems. In addition, when the discrete form of convolution and its use in digital signal processing is discussed, the reader will have some background material to which he or she can refer.

Prior to leaving convolution theory, it should be mentioned that either function can be converted from t to $t - \tau$. If, in Equation 1–23, we replace t with $t - \tau$, we obtain

$$y(t) = f(t) * h(t) = \int_{-\infty}^{\infty} f(t - \tau) h(\tau) \, d\tau \quad (1\text{–}38)$$

Both forms of the convolution integral are valid, providing a choice as to which function remains "fixed" in position.

1.11 FOURIER TRANSFORM

In Figures 1–4 and 1–5, a pictorial representation of a function was presented: Figure 1–4 showed a function of time, and Figure 1–5 showed a function of frequency. In many situations, the signals we see on an oscilliscope (a time function) provide us with a limited amount of information. If we want more information, we must look at the signal from a different point of view. The viewpoint that provides additional information is the signal's frequency domain representation (i.e., the frequencies present, their amplitude, and phase). This requires that we make a transition from the time to the frequency domain.

To find the frequency domain representation, $F(f)$, of a time domain signal, $f(t)$, we must be able to take any function of time and convert it to a function of frequency such that there is a unique relationship (one to one) between $f(t)$ and $F(f)$. The procedure that enables one to determine the unique function $F(f)$ that is related to a given $f(t)$ is called a Fourier *transform*.

A transform is a (mathematical) technique that takes a function with a given range, domain, and rule and provides a new function with a different range, domain, and rule. Although there are many transforms available, we will concern ourselves with the ones used mostly in electronics. Foremost among these is the *Fourier transform* (FT). The FT takes a time-domain function whose range is voltage or current and generates the amplitude and phase, which is the range, of the frequency domain representation of the time-domain function. Figure 1–35 combines Figures 1–4 and 1–5 to show a pictoral representation of the FT.

All the transforms we will consider can be put in the general form of Equation 1–39.

$$T\{f(t)\} = \int_a^b f(t)K(t, \lambda)dt = F(\lambda) \qquad (1\text{-}39)$$

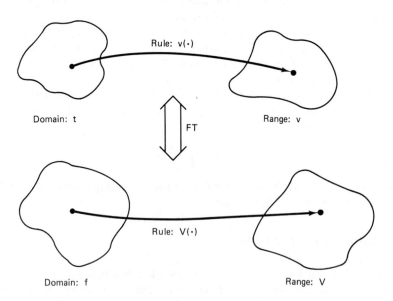

Figure 1–35 Pictorial representation of a Fourier transform.

Sec. 1.11   Fourier Transform   27

where $K(t, \lambda)$ is called the *kernel* and $f(t)$ is an arbitrary function of time. Each transform has a different kernel and definition of λ. In addition, the limits of integration, a, b, will differ for different transforms. As an example, the FT will have $\lambda = j\omega = j2\pi f$ and doubly infinite limits: $a = -\infty$, $b = +\infty$; the Laplace transform will have $\lambda = \sigma + j\omega$, and semi-infinite limits: $a = 0$, $b = +\infty$. In both transforms, the kernel is a complex periodic exponential function of t and λ: $K(t, \lambda) = \exp(\lambda t)$. Also, as the integration is over t, the transform is a function of λ, as indicated in Equation 1–39. To obtain the Fourier transform, we require the Fourier series. As an introduction to the Fourier series, we will provide some background and show that it is an extension of vector concepts.

1.11.1 Vector Concepts

We will begin our discussion of the Fourier series by recalling that a vector (having magnitude and direction) can be represented in three dimensions by three other vectors termed *basis vectors*. This is given in Equation 1–40.

$$\bar{E} = E_1\bar{a}_1 + E_2\bar{a}_2 + E_3\bar{a}_3 \tag{1-40}$$

where $\bar{E}$ is the vector, $\{\bar{a}_i\}$ is the set of basis vectors, and $\{E_i\}$ is the set of components in this basis. (A different basis will result in a different set of components.) This is shown in Figure 1–36 for the rectangular (Cartesian) set of coordinates.

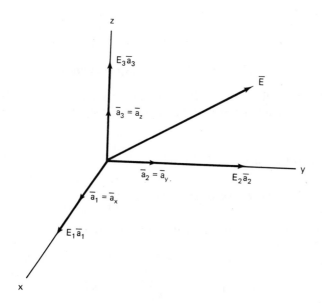

Figure 1–36   Representation of $\bar{E}$ in three dimensions.

The set of basis vectors in Figure 1–36 has the following characteristics:

1. They are orthogonal. In two or three dimensions, this means that they form 90° angles with each other. Mathematically, this implies that their inner (dot) product is zero. This is represented in Equation 1–41.

$$\bar{a}_i \cdot \bar{a}_j = (\bar{a}_i \cdot \bar{a}_j) = \begin{cases} 0, & i \neq j \\ 1, & i = j \end{cases} \quad (1\text{–}41)$$

2. The magnitude (length) of a basis vector is unity, as given in Equation 1–42.

$$|\bar{a}_i| = 1 \quad (1\text{–}42)$$

3. As all three-dimensional vectors can be represented by the orthonormal basis of Figure 1–36, the set $\{\bar{a}_i\}$ is said to span the space. That is, they form a complete set: Any three-dimensional vector can be uniquely represented by them.

Making a transition from the specific three-dimensional example to a more general one, we can state that an n-dimensional space ($n > 3$) will require n-orthonormal basis vectors to span the space with each vector satisfying the conditions of Equations 1–41 and 1–42. The number of dimensions is not limited and, in most instances, will become infinite requiring an infinite number of basis vectors to span the space. The functions $K(t, \lambda) = \exp(j2\pi ft)$ can be used as a basis, as they can be shown to satisfy conditions 1 through 3: they are orthonormal and span a space. Since they are periodic, time-varying functions, the space they span consists of all periodic, time-varying functions. The form of $K(t, \lambda)$ for the Fourier series is given in Equation 1–43.

$$K(t, \lambda) = K(t, 2\pi n f_0) = \exp(j2\pi n f_0 t) \quad (1\text{–}43)$$

where n is the set of all positive and negative integers and f_0 is a fixed (fundamental) frequency. These functions can be shown to satisfy the three characteristics given above:

1. They are orthogonal:

$$(K(t, 2\pi n f_0), K(t, 2\pi m f_0)) = \begin{cases} 0, & m \neq n \\ 1, & m = n \end{cases} \quad (1\text{–}44)$$

where we define orthogonality for these functions in Equations 1–45.

$$\frac{1}{T} \int_{-T/2}^{T/2} (K(t, 2\pi n f_0), K(t, 2\pi m f_0)) \, dt = \frac{1}{T} \int_{-T/2}^{T/2} e^{j2\pi n f_0 t} \cdot e^{j2\pi m f_0 t} \, dt$$

$$= \begin{cases} 0, & m \neq n \\ 1, & m = n \end{cases} \quad (1\text{–}45)$$

where T is the period of $K(t, \lambda)$.

Sec. 1.11 Fourier Transform

2. Their magnitude is unity:

$$|K(t, 2\pi n f_0)| = |\exp(j2\pi n f_0 t)| = 1 \qquad (1\text{--}46a)$$

This is demonstrated by expanding the exponential term.

$$|\exp(j2\pi n f_0 t)| = |\cos 2\pi n f_0 t + j \sin 2\pi n f_0 t|$$
$$= \sqrt{\cos^2 2\pi n f_0 t + \sin^2 2\pi n f_0 t} \qquad (1\text{--}46b)$$
$$= 1$$

3. They span the space of real-valued, periodic time functions: Any function of this form can be represented as a (linear) combination of these functions.

1.11.2 Fourier Series: Periodic Signals

The comments above are an attempt to justify the presentation of the Fourier series in standard exponential form:

$$f(t) = \sum_{n=-\infty}^{\infty} C_n \exp(j2\pi n f_0 t) \qquad (1\text{--}47)$$

where $f(t)$ is any periodic function with period $T = 2\pi/\omega_0 = 1/f_0$, extending from $-\infty$ to $+\infty$, and the C_n are complex numbers. We see the similarity between Equations 1–40 and 1–47: they are both sums of basis vectors, each with a number (real or complex) multiplying it. From previous study of Fourier series, we know that nf_0 represents the discrete frequency components that represent $f(t)$ and that C_n represents the amplitude and phase of those frequency components. This is shown in Figure 1–37, where the (continuous) amplitude of C_n is presented as a function of the (discrete) frequency, f, for an arbitrary function, $f(t)$. The + and − frequency components reflect the fact that in the real world, we work with sines and cosines. Both positive and negative exponential components are required to obtain the frequency component, as shown in Equation 1–48.

$$|C_n|\exp(jnx) + |C_{-n}|\exp(-jnx) \equiv 2|C_n|\cos nx \qquad (1\text{--}48a)$$

$$|C_n|\exp(jnx) - |C_{-n}|\exp(-jnx) \equiv 2j|C_n|\sin nx \qquad (1\text{--}48b)$$

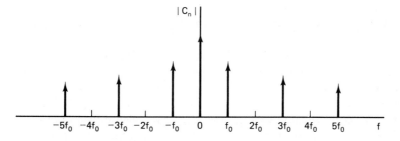

Figure 1–37 $|C_n|$ for an arbitrary function, $f(t)$.

Both components of a given frequency combine to yield a single sine wave, whose amplitude is twice that of its exponential components, C_n as $|C_n| = |C_{-n}|$.

The Fourier series representation of $f(t)$ is obtained once the complex constants, C_n, are determined. Multiplying both sides of Equation 1–47 by $\exp(-j2\pi mf_0 t)$, we can obtain their values.

$$\exp(-j2\pi mf_0 t) f(t) = \exp(-j2\pi mf_0 t) \sum_{n=-\infty}^{\infty} C_n \exp(j2\pi nf_0 t) \quad (1\text{–}49)$$

As the sum is over n, we can bring the exponential term inside the summation and integrate over t for one period.

$$\exp(-j2\pi mf_0 t) f(t) = \sum_{n=-\infty}^{\infty} C_n \exp(-j2\pi mf_0 t) \exp(j2\pi nf_0 t) \quad (1\text{–}50\text{a})$$

$$\frac{1}{T} \int_{-T/2}^{T/2} \exp(-j2\pi mf_0 t) f(t) \, dt$$

$$= \sum_{n=-\infty}^{\infty} \frac{C_n}{T} \int_{-T/2}^{T/2} \exp(-j2\pi mf_0 t) \exp(-j2\pi nf_0 t) \, dt \quad (1\text{–}50\text{b})$$

Based on the orthogonality condition (Equation 1–45), the integrals on the right-hand side are zero for all n, except when $n = m$ at which time the integral equals T.

$$\frac{1}{T} \int_{-T/2}^{T/2} f(t) \exp(-j2\pi mf_0 t) \, dt = C_m \cdot 1 \quad (1\text{–}50\text{c})$$

The C_m are found to be

$$C_m \equiv \frac{1}{T} \int_{-T/2}^{T/2} f(t) \exp(-j2\pi mf_0 t) \, dt \quad (1\text{–}50\text{d})$$

Comparing Equation 1–50d with Equation 1–39, we see that C_m is the (Fourier) transform of $f(t)$ and is a function of frequency. As C_m has a value only for integer multiples of the fundamental frequency f_0, it is a continuous range, discrete domain function. C_m is a complex number, representing the amplitude and phase of the mth harmonic component of $f(t)$.

1.11.3 Fourier Transform for Aperiodic Signals

As only ideal signals can satisfy the conditions for having a unique representation by a Fourier series (FS), we must find a new approach for practical signals that are nonperiodic and finite in extent. For a nonperiodic function, transition from the FS to the FT is brought about by treating $f(t)$ as periodic, but with a period equal to infinity. This is accomplished by a limiting process in which we allow T, the portion of the function for which the range is nonzero, to pass to infinity in the limit: $T \longrightarrow \infty$. Note that this does not mean that

T becomes infinite; rather, it is allowed to approach infinity, a very important distinction. Since f_0, the fundamental frequency of the signal's spectrum, is also the spacing between frequency components, the limiting process implies that as T goes toward ∞, f_0 must go toward zero in a manner that maintains the reciprocal relation between period and frequency. This is shown in Equation 1–51.

$$T \longrightarrow \infty, \qquad f_0 \longrightarrow 0 \Longrightarrow Tf_0 = 1 \qquad (1\text{-}51)$$

As the spacing between components goes to zero, the discrete set of frequency components becomes a continuous set. From this we see that the FT of an aperiodic time-varying signal is a continuous function (i.e., all frequency components are present forming the domain). The discrete integer n, which was used as the summing index in Equation 1–47, cannot be used due to the continuous nature of the spectrum. To make the transition from a discrete to a continuous representation, we can consider what happens to the nth harmonic of the series, nf_0, as $T \longrightarrow \infty$ and $f_0 \longrightarrow 0$. To represent the particular frequency $f = nf_0$ as $f_0 \longrightarrow 0$, n must go toward ∞ such that the product remains constant at f. This is indicated in Equation 1–52.

$$f_0 \longrightarrow 0, \qquad n \longrightarrow \infty \Longrightarrow nf_0 = f \qquad (1\text{-}52)$$

In this manner, the transition is made from the discrete frequency domain nf_0, to the continuous frequency domain, f.

To determine the effect of $T \longrightarrow \infty$ on C_n we can take the magnitude of Equation 1–50d after we form the product TC_n. This is given in Equation 1–53.

$$|TC_n| = \left| \int_{-T/2}^{T/2} f(t) \exp(-j2\pi nf_0 t)\, dt \right| \qquad (1\text{-}53a)$$

$$\leq \int_{-T/2}^{T/2} |f(t) \exp(-j2\pi nf_0 t)\, dt| \qquad (1\text{-}53b)$$

$$\leq \int_{-T/2}^{T/2} |f(t)||\exp(-j2\pi nf_0 t)|\, dt \qquad (1\text{-}53c)$$

$$= \int_{-T/2}^{T/2} |f(t)|\, dt$$

Based on Equation 1–2, we have

$$TC_n \leq |TC_n| \leq \int_{-T/2}^{T/2} |f(t)|\, dt \leq M < \infty \qquad (1\text{-}54)$$

Letting T go to infinity, Equation 1–54 implies that C_n must decrease such that the integral in Equation 1–54 remains finite (i.e., it exists). This is shown in Equation 1–55.

$$\lim_{T \to \infty} TC_n = \int_{-\infty}^{\infty} f(t) \exp(-j2\pi f_0 t n)\, dt \leq M < \infty \qquad (1\text{-}55)$$

As the integration is over t, the integral must be a continuous function of f. This is shown in Equation 1–56 which presents the FT of $f(t)$.

$$\lim_{T \to \infty} TC_n = F(2\pi f) = \int_{-\infty}^{\infty} f(t) \exp(-j2\pi ft) \, dt \qquad (1\text{–}56)$$

The inverse FT (IFT) is given by

$$\text{IFT}\{F(2\pi f)\} = f(t) = \int_{-\infty}^{\infty} F(2\pi f) \exp(j2\pi ft) \, df \qquad (1\text{–}57)$$

The integrand is usually complex and requires theorems from complex variables for its evaluation. As $F(f)$ is a complex number, it can be expressed as

$$F(2\pi f) \Longleftrightarrow F(f) = F_r(f) + jF_I(f) = |F(f)| \angle \theta(f) \qquad (1\text{–}58)$$

where

$$|F(f)| = \sqrt{F_r^2 + F_I^2}, \qquad \theta(f) = \tan^{-1} \frac{F_I}{F_R} \qquad (1\text{–}59)$$

As an example of the FT, consider the following.

Example 1–7

Find the FT of the pulse function defined below

$$f(t) = \begin{cases} A, & |t| \leq \dfrac{T_0}{2} \\ 0, & \text{otherwise} \end{cases}$$

Solution

$$\begin{aligned} F(2\pi f) &= \int_{-\infty}^{\infty} f(t) \exp(-j2\pi ft) \, dt \\ &= \int_{-T_0/2}^{T_0/2} A \exp(-j2\pi ft) \, dt \\ &= \frac{A}{-j2\pi f} \exp(-j2\pi ft) \Big|_{-T_0/2}^{T_0/2} \qquad (1\text{–}56) \\ &= \frac{A}{-j2\pi f} \left[\exp\left(-j2\pi f \frac{T_0}{2}\right) - \exp\left(+j2\pi f \frac{T_0}{2}\right) \right] \\ &= \frac{A}{\pi f} \sin \pi f T_0 = A T_0 \frac{\sin \pi f T_0}{\pi f T_0} \\ &= A T_0 \text{Sa}(\pi f T_0) \end{aligned}$$

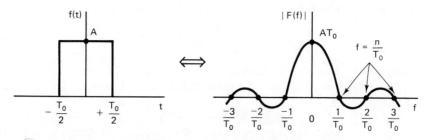

Figure 1–38 FT pair of Example 1–7.

Sec. 1.11 Fourier Transform

where the zeros of $F(f)$ occur when $f = \pm(n/T_0)$. The relation is shown in Figure 1–38.

Figure 1–39 shows several FT pairs.

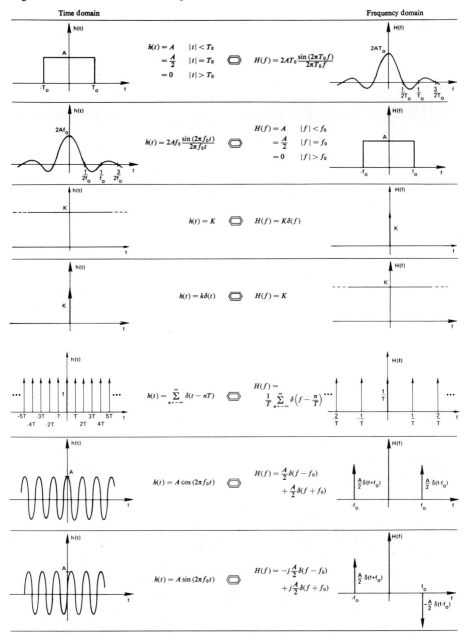

Figure 1–39 FT pairs. (From *The Fast Fourier Transform*, E. Oran Brigham. Copyright © 1974, Prentice-Hall, Inc. Reprinted by permission of Prentice-Hall, Inc.)

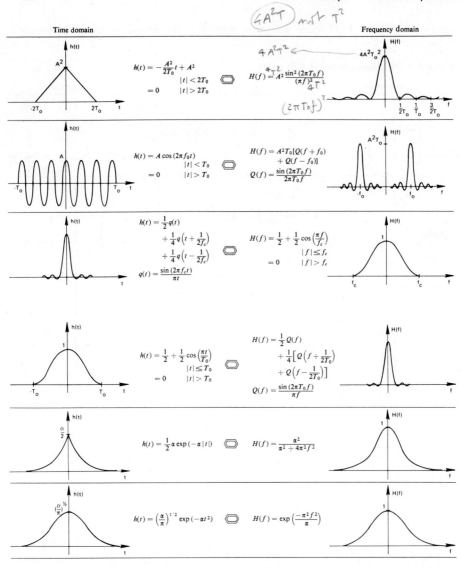

Figure 1-39 (cont.)

1.11.4 Fourier Transform Properties

Several important properties of the FT can be obtained by manipulating Equations 1-56 and 1-57. These properties are summarized in Table 1-1 and will be used throughout the text as needed.

TABLE 1-1 Properties of the FT

Operation	$f(t)$	$F(\omega)$
1. Scaling	$f(at)$	$\dfrac{1}{|a|} F\left(\dfrac{\omega}{a}\right)$
2. Time shifting	$f(t - t_0)$	$F(\omega)e^{-j\omega t_0}$
3. Frequency shifting	$f(t)e^{j\omega_0 t}$	$F(\omega - \omega_0)$
4. Time differentiation	$\dfrac{d^n f}{dt^n}$	$(j\omega)^n F(\omega)$
5. Frequency differentiation	$(-jt)^n f(t)$	$\dfrac{d^n F}{d\omega^n}$
6. Time integration	$\displaystyle\int_{-\infty}^{t} f(\tau)\, d\tau$	$\dfrac{1}{(j\omega)} F(\omega)$
7. Time convolution	$f_1(t) * f_2(t)$	$F_1(\omega) F_2(\omega)$
8. Frequency convolution	$f_1(t) f_2(t)$	$\dfrac{1}{2\pi}[F_1(\omega) * F_2(\omega)]$

From *Signals, Systems, and Communications*, B.P. Lathi. Copyright © 1966, John Wiley and Sons, Inc. Reprinted by permission of John Wiley and Sons, Inc.

1.12 CONVOLUTION IN THE FREQUENCY DOMAIN

We can take the FT of Equation 1–26.

$$\text{FT}\left\{ r(t) = \int_{-\infty}^{\infty} s(\tau) h(t - \tau)\, d\tau \right\} \qquad (1\text{–}60)$$

which can be expressed as

$$r(t) \iff R(f) = \int_{-\infty}^{\infty} \text{FT}\{s(\tau) h(t - \tau)\}\, d\tau \qquad (1\text{–}61)$$

$$= \int_{-\infty}^{\infty} s(\tau) \text{FT}\{h(t - \tau)\}\, d\tau \qquad (1\text{–}62)$$

The change from Equation 1–60 to Equation 1–61 can be performed, as the FT operation is linear. Equation 1–62 follows as h is the only function that involves t.

From the time-shifting property of the FT (operation 2, Table 1–1), we obtain

$$\text{FT}\{h(t)\} = H(f) \iff \text{FT}\{h(t - \tau)\} = H(f) \exp(-j2\pi f \tau) \qquad (1\text{–}63)$$

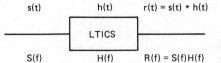

Figure 1–40 Convolution in the time/frequency domains.

Equation 1–62 becomes

$$\int_{-\infty}^{\infty} s(t) \text{FT}\{h(t-\tau)\} \, d\tau = \int_{-\infty}^{\infty} s(\tau) H(f) \exp(-j2\pi f \tau) \, d\tau \quad (1\text{--}64\text{a})$$

$$= H(f) \int_{-\infty}^{\infty} s(\tau) \exp(-j2\pi f \tau) \, d\tau \quad (1\text{--}64\text{b})$$

The integral in Equation 1–64b is recognized as $S(f)$. Equation 1–62 can now be written as

$$R(f) = S(f) H(f) \quad (1\text{--}65)$$

which states that convolution in the frequency domain is obtained by the product of the FTs of the time-domain functions. This is indicated in Figure 1–40.

Example 1–8
Using both techniques for convolution, find the output of a LTIC system with an impulse response given by Figure 1–41 and whose input is given by Figure 1–42.

Solution

1. *Convolution integral.* As the integral is over τ, $h(t-\tau)$ is the reversed and shifted version of Figure 1–41. This is shown in Figure 1–43. In this figure, τ is the variable and t is the fixed parameter.

To evaluate the integral, it is helpful to show the product of both functions on the same axis for several values of t. Figure 1–44 shows an example of this presentation

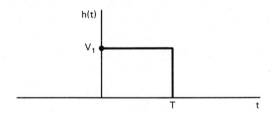

Figure 1–41 System impulse response of Example 1–8.

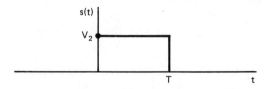

Figure 1–42 System input for Example 1–8.

Sec. 1.12 Convolution in the Frequency Domain

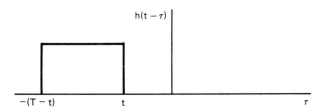

Figure 1–43 $h(t - \tau)$ for Example 1–8.

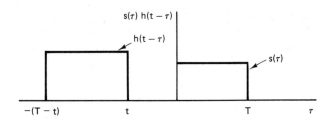

Figure 1–44 Integrand of Example 1–8, $t < 0$.

for a particular value of $t < 0$. For choices of $t < 0$, there is no overlap of the curves. That is, their product is zero for all values of τ: At least one of the factors is zero.

$$s(\tau) \, h \, (t - \tau) = 0 \tag{1-66}$$

This causes the value of the integral to be zero and is true for all values of t that do not produce an overlap of the curves: $r(t)$, the system output, is zero when there is no overlap. This is expressed in Equation 1–67.

$$r(t) = \int_{-\infty}^{\infty} s(\tau) \, h \, (t - \tau) \, d\tau = 0, \qquad t < 0 \tag{1-67}$$

Another condition for which there is no overlap occurs for values of t greater than $2T$. This is shown in Figure 1–45. Overlap occurs only if we have

$$0 \leq t \leq 2T \tag{1-68}$$

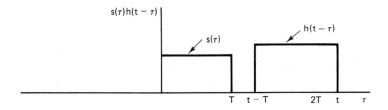

Figure 1–45 Integrand of Example 1–8, $t \geq 2T$.

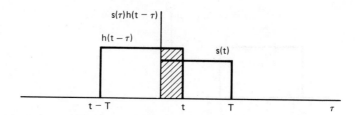

Figure 1-46 Integrand of Example 1-8, $0 \leq t \leq T$.

and is shown in Figure 1-46 for $0 \leq t \leq T$. There is an overlap of curves for the conditions of Equation 1-68, and as the product of the curves is nonzero for these values of t, the integrand is also nonzero.

$$s(\tau)h(t-\tau) \neq 0 \tag{1-69a}$$

$$r(t) = \int_{-\infty}^{\infty} s(\tau)h(t-\tau)\, d\tau$$
$$= \int_{0}^{t} V_1 V_2 d\tau = V_1 V_2 t, \qquad 0 \leq t \leq T \tag{1-69b}$$

$r(t)$ in Equation 1-69b will be an increasing function of t until $t = T$, as shown in Figure 1-47. When $t = T$, we have the maximum amount of overlap, and the value of $r(t)$ becomes its maximum value $r(t = T) = V_1 V_2 T$.

For values of t greater than T, the overlap decreases. Equation 1-69b becomes

$$r(t) = \int_{t-T}^{T} V_1 V_2 \, d\tau = V_1 V_2 \tau \Big|_{t-T}^{T} \tag{1-70a}$$

$$= V_1 V_2 (2T - t) \tag{1-70b}$$

which is a decreasing function of t. When $t = 2T$, the overlap ceases and $r(t) = 0$. Figure 1-48 shows the complete value of $r(t)$.
The output can be expressed as

$$r(t) = \begin{cases} V_1 V_2 t, & 0 \leq t \leq T \\ V_1 V_2 (2T - t), & T \leq t \leq 2T \\ 0, & \text{otherwise} \end{cases} \tag{1-71}$$

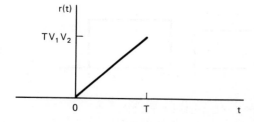

Figure 1-47 System output, r(t), for $0 \leq t \leq T$.

Sec. 1.13   Correlation 39

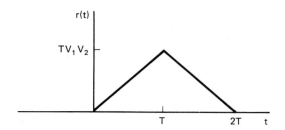

Figure 1-48 Result of convolution, Example 1-8.

2. *Fourier transform method.* Using the time-shifting property of the FT (Table 1-1) and the results of Example 1-7, we obtain

$$\text{FT}\{s(t)\} = V_1 T \,\text{Sa}(\pi fT) \exp(-j\pi fT) = S(f) \tag{1-72a}$$

$$\text{FT}\{h(t)\} = V_2 T \,\text{Sa}(\pi fT) \exp(-j\pi fT) = H(f) \tag{1-72b}$$

$$R(f) = H(f)S(f) = V_1 V_2 T^2 \,\text{Sa}^2(\pi fT) \exp(-j2\pi fT) \tag{1-72c}$$

$$r(t) = \text{IFT}\{R(f)\} \tag{1-72d}$$

Using the results of the first FT pair in Figure 1-39 and the FT time-shifting property, the results of Equation 1-71 are obtained.

Although we did not evaluate the IFT due to its complex form, the important thing to remember is that both techniques give the same results for $r(t)$ and are, therefore, equivalent. This allows us to use either the time- or frequency-domain formulation for convolution. DSP techniques will also use this approach to solve convolution problems.

1.13 CORRELATION

In several instances, we would like to compare two signals to see how much alike they are. Three examples of this are (1) a communication system, where it is desired to determine which one of several known signals was received, (2) a radar system, which must determine whether a transmitted signal was returned and what frequency changes (i.e., Doppler shift) had occurred, and (3) a binary pseudo-random noise signal used in space shuttle communications. This comparison of two signals can be performed by the process of correlation.

1.13.1 Definition of Correlation

Given two signals, $s_1(t)$ and $s_2(t)$, which are real, aperiodic waveforms, the correlation function is given by

$$R_{12}(\tau) \equiv \int_{-\infty}^{\infty} s_1(t) s_2(t - \tau) \, dt \qquad (1\text{--}73)$$

where $R_{12}(\tau)$ is termed the correlation function and τ is termed the searching parameter. The larger the value of $R_{12}(\tau)$, the more the signals agree.

1.13.2 Evaluation of the Correlation Integral

The correlation integral is evaluated in a manner similar to the convolution integral; however, as the variable of integration is t and not τ, we do not have to "fold" any of the functions. For correlation, we choose τ to displace the second waveform to the right and slide it to the left by varying τ, looking for overlap of the curves.

1.13.3 Autocorrelation

Equation 1–73 assumes $s_1(t)$ and $s_2(t)$ to be different signals. This yields the cross-correlation function, $R_{12}(\tau)$; if $s_1(t)$ equals $s_2(t)$, we have the autocorrelation function, as given in Equation 1–74.

$$R_s(\tau) \equiv \int_{-\infty}^{\infty} s(t) s(t - \tau) \, dt \qquad (1\text{--}74)$$

where $R_s(\tau)$ is the autocorrelation function and τ is the searching parameter. It is assumed that $s(t)$ satisfies Equation 1–2. To see how the autocorrelation integral is evaluated, we consider the following example.

Example 1–9

What is the autocorrelation function of the signal shown in Figure 1–49?

Solution We establish the product $s(t)s(t - \tau)$ by displacing a replica of $s(t)$, τ units to the right, as shown in Figure 1–50. The displaced figure is moved to the left by causing τ to become less positive, and a nonzero product is sought. In this example, there is no overlap for $\tau < -1$ s and $\tau > 1$ s; that is, overlap occurs only for -1 s $\leq \tau \leq 1$ s. This is shown in Figure 1–51. Therefore, the autocorrelation function will have a nonzero value for -1 s $\leq \tau \leq 1$ s.

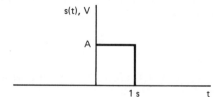

Figure 1–49 Signal for Example 1–9.

Sec. 1.13 Correlation

Figure 1–50 Autocorrelation integrand, no overlap.

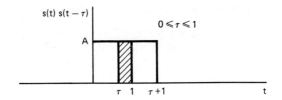

Figure 1–51 Autocorrelation integrand, with overlap.

The integral (Equation 1–74) becomes

$$R_s(\tau) = \int_\tau^{1s} A \cdot A \, dt \tag{1-74}$$

for $0 \leq \tau \leq 1$ s. This is evaluated as

$$R_s(\tau) = \int_\tau^{1s} A^2 \, dt \tag{1-75a}$$

$$= A^2 T \bigg|_\tau^{1s} \tag{1-75b}$$

$$= A^2(1\,\text{s} - \tau) \tag{1-75c}$$

Maximum overlap occurs for $\tau = 0$, which yields

$$R_s(0) = A^2(1\,\text{s}) \tag{1-76}$$

This equation will be useful when we wish to determine the energy content of a signal, and we will return to it in Section 1.13.5.

For values of $-1\,\text{s} \leq \tau \leq 0$, the amount of overlap decreases and we have

$$R_s(\tau) = \int_0^{\tau+1s} s(t)s(t-\tau) \, dt \tag{1-74}$$

as shown in Figure 1–52. Equation 1–74 becomes

$$R_s(\tau) = \int_0^{\tau+1s} A^2 \, dt \tag{1-77a}$$

$$= A^2 t \bigg|_0^{\tau+1s} \tag{1-77b}$$

$$= A^2(\tau + 1\text{s}) \tag{1-77c}$$

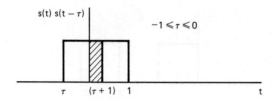

Figure 1–52 Autocorrelation integrand, overlap, $\tau < 0$.

The autocorrelation function is then defined by combining Equations 1–75c and 1–77c:

$$R_s(\tau) = \begin{cases} A^2(1\,\text{s} - \tau), & 0 \le \tau \le 1\text{s} \\ A^2(\tau + 1\text{s}), & -1\,\text{s} \le \tau \le 0 \\ 0, & \text{otherwise} \end{cases}$$

and is shown in Figure 1–53.

The large value of $R_s(\tau)$ at $\tau = 0$ indicates that the correlation between the two waveforms is high, as we would expect since they are identical. If the waveforms were dissimilar, we would evaluate their cross-correlation function in a similar manner; however, the value of $R_{12}(\tau)$ would be small for all τ.

1.13.4 Change of Variable

If we let $t = t + \tau$ in Equations 1–73 and 1–74, we obtain

$$R_{12}(\tau) = \int_{-\infty}^{\infty} s_1(t + \tau) s_2(t)\, dt \qquad (1\text{–}78\text{a})$$

$$R_s(\tau) = \int_{-\infty}^{\infty} s(t + \tau) s(t)\, dt \qquad (1.78\text{b})$$

Either form can be used to evaluate the correlation functions.

1.13.5 Fourier Transform of the Autocorrelation Function

If $s(t)$ is a voltage or current waveform and contains finite energy, we can obtain its FT. This condition is satisfied if $s(t)$ is square integrable, as shown in Equation 1–79.

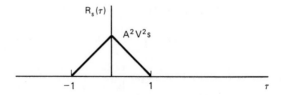

Figure 1–53 Autocorrelation function of Example 1–9.

Sec. 1.13 Correlation

$$\int_{-\infty}^{\infty} |s(t)|^2 dt \leq M < \infty \tag{1-79}$$

where we are considering the power dissipated in a 1-Ω resistor given by $|s(t)|^2/(1\ \Omega)$ if $s(t)$ is a voltage waveform, and by $|s(t)|^2\ (1\ \Omega)$ if $s(t)$ is a current waveform. Taking the FT of Equation 1–74, we have

$$\text{FT}\{R_s(\tau)\} = \text{FT}\left\{\int_{-\infty}^{\infty} s(t)s(t+\tau)\,dt\right\} \tag{1-80}$$

As the FT and the integration are linear operations, they can be interchanged:

$$\text{FT}\int = \int \text{FT} \tag{1-81}$$

Equation 1–80 becomes

$$\text{FT}\{R_s(\tau)\} = \int_{-\infty}^{\infty} \text{FT}\{s(t)\,s(t+\tau)\}\,dt \tag{1-82}$$

Using the properties of the FT of a product and the FT of a time-shifted function, we obtain

$$\text{FT}\{R_s(\tau)\} = S(f)S^*(f) = |S(f)|^2 \tag{1-83}$$

where the * indicates the complex conjugate. The term on the right is the energy density spectrum of the signal $s(t)$, and is defined as

$$W_s(f) \equiv |S(f)|^2 \tag{1-84}$$

$W_s(f)$ indicates the amount of energy contained in a range of frequencies between f and $f + df$. Therefore, the integral of $W_s(f)$ is the total energy content of $s(t)$:

$$\int_{-\infty}^{\infty} W_s(f)\,df = \int_{-\infty}^{\infty} |S(f)|^2\,df = \int_{-\infty}^{\infty} s^2(t)\,dt \tag{1-85}$$

where the last term on the right-hand side is obtained from Parseval's relation and can be recognized as $R_s(0)$. We then have

$$R_s(0) = \int_{-\infty}^{\infty} W_s(f)\,df \tag{1-86}$$

which states that the evaluation of the autocorrelation function for $\tau = 0$ provides the total energy content of the signal $s(t)$ for all frequencies. Referring back to Equation 1–76, we see that the energy stored in the pulse A volts high and $T = 1$ s wide as A^2V^2 (1 s). If we assume that this is dissipated in a 1-Ω resistor, the total energy becomes

$$\frac{A^2V^2(1\text{ s})}{1\ \Omega} = A^2\text{ (watts)}(1\text{ s}) = A^2\text{ joules} \tag{1-87}$$

If we wish to determine the energy contained in a band of frequencies that represent $s(t)$, we integrate $W_s(f)$ over the desired band. This is given in Equation 1–88 and Figure 1–54.

$$E_s(f_2 - f_1) = \int_{f_1}^{f_2} W_s(f)\, df + \int_{-f_2}^{-f_2} W_s(f)\, df \tag{1-88}$$

where the integration occurs over both positive and negative frequency values of interest.

Example 1–10

Find the FT of the autocorrelation function obtained in Example 1–9.

Solution Using Figure 1–39, we see that the FT of the triangular pulse is

$$W_s(f) = A^2 \text{Sa}^2(2\pi T_0 f) \tag{1-89}$$

where $T_0 = 0.5$ s. The total area under the curve is the the energy content of $s(t)$, which can also be found by evaluation of $R_s(\tau)$ for $\tau = 0$.

1.13.6 Uses of Autocorrelation

Although there are many uses of cross- and autocorrelation given in the literature, we will consider only two of them.

Linear system output energy. If we consider a linear system with impulse response $h(t)$ and an input $s_i(t)$, we can express the output $s_o(t)$ in the frequency domain as

$$S_o(f) = S_i(f) H(f) \tag{1-90}$$

Multiplying Equation 1–90 by its complex conjugate, we obtain

$$W_{s_o}(f) = |S_o(f)|^2 = |H(f)|^2 |S_i(f)|^2 = |H(f)|^2\, W_{s_i}^2(f) \tag{1-91}$$

which states that the output energy density of a signal is obtained from the product of the input energy density and the square of the FT of the impulse response.

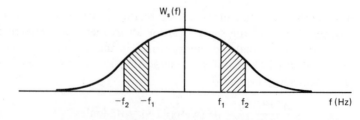

Figure 1–54 Energy content in the frequency band $f_1 - f_2$.

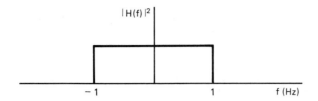

Figure 1–55 Transfer function of an ideal LP filter.

Example 1–11

The pulse of Example 1–9 is fed to an ideal low-pass (LP) filter whose cutoff frequency is 1 Hz. What is the energy density of the output signal?

Solution The magnitude squared of the ideal LP filter is shown in Figure 1–55. If this is multiplied by $|S_i(f)|$ (from Figure 1–30) we see that all components of $S_o(f)$ for $f > 1$ Hz are removed (i.e., all high-frequency components). $W_{s_o}(f)$ is given by

$$W_{s_o}(f) = \begin{cases} W_{s_i}(f), & -1 \text{ Hz} \leq f \leq 1 \text{ Hz} \\ 0, & \text{otherwise} \end{cases}$$

and shown in Figure 1–56.

Matched (Doppler) filter. To determine if one of a predetermined set of finite-length signals $\{s_i(t)\}$ has been received, communications and radar systems make use of a matched filter. The I/O relation of a filter is obtained by convolution: $y(t) = s(t) * h(t)$. In the matched filter, its impulse response is chosen as the reversed-time, complex conjugate of the signal to be identified. This is given in Equation 1–92.

$$h(t) \equiv s_i^*(-t) \quad (1\text{–}92)$$

where $s_i(t)$ is the desired signal of the set $s(t)$. As $h(t)$ is noncausal, it must be delayed by the length of s_i to make it realizable. This allows us to write the convolution equation as

$$y(t) = \int_{-\infty}^{\infty} s(\tau) s_i^*(t + \tau) \, d\tau \quad (1\text{–}93a)$$

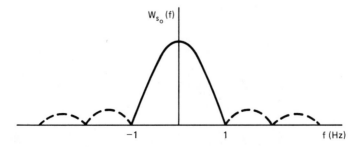

Figure 1–56 Energy density of output, Example 1–11.

by following the procedure of Section 1.10. If $s(t)$ is a real function, $s_i(t) = s_i^*(t)$. This gives us

$$y(t) = \int_{-\infty}^{\infty} s_i(\tau) s_i(t + \tau) \, dt \qquad (1\text{–}93b)$$

which is identical in form to the autocorrelation integral, Equation 1–78b [i.e., $y(t) = R_{s_i}(\tau)$]. If $s_i(t)$ is a member of an orthonormal set of signals, $\{s(t)\}$, the output of the filter matched to $s_i(t)$ will be greater than zero if the received signal is $s_i(t)$, and equal to zero if the received signal is not $s_i(t)$.

Matched filters are arranged in a group, one for each signal of the transmitted signal set. The filter with the largest output is assumed to indicate which signal of the set was received. A practical application of this is in a radar system, where each filter in the group is matched to the transmitted signal plus a Doppler frequency shift, ω_{d_i}, in the carrier. A representation of the Doppler filter bank is given in Figure 1–57, where $r(t)$ is the received signal (including noise), ω_c is the carrier radian frequency, and ω_{d_i} is the ith Doppler frequency.

Because each filter represents an autocorrelation with a member of the set $\{s_i\}$, its output, $y_i(t)$, is the autocorrelation function, $R_{s_i}(\tau)$. As the energy content of the signal is given by $R_{s_i}(0)$, the filter that is most closely matched to the frequency of $r(t)$, $\omega_c + \omega_{c_{ig}}$, will indicate the total energy content of the signal, the other filter outputs indicating zero (i.e., the filter with the largest output indicates the frequency of the received signal). In this manner, not only is target detection indicated by the presence of an output at a given filter, but its velocity can be determined by the value of the Doppler frequency,

$$\dot{R}_0 = \frac{v_0 c}{2 f_c} \qquad (1\text{–}94)$$

where $\dot{R}_0$ is the range rate (velocity), v_0 is the Doppler frequency, f_c is the transmitted frequency, and c is the velocity of the transmitted signal. This technique is used in the orbiter space shuttle Ku-band radar system for target detection and parameter estimation.

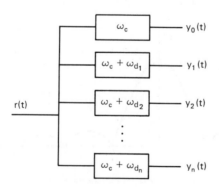

Figure 1–57 Doppler filter bank.

1.14 DIFFERENTIAL EQUATION REPRESENTATION OF A LINEAR SYSTEM

As the last topic on linear systems, we will consider the representation of a linear system as a sum of derivatives of its input and output signals. The purpose of this is to serve as an introduction to discrete systems in a later chapter where we represent the discrete system in terms of a difference equation relating the discrete input and output of the system. The general form of the I/O relations of an LTIC system is given in Figure 1–58 and Equation 1–95.

$$\sum_{j=0}^{N} a_j \frac{d^j x(t)}{dt^j} = \sum_{j=0}^{M} b_j \frac{d^j y(t)}{dt^j} \qquad (1\text{--}95)$$

where a_j and b_j are constants, and $x(t)$ and $y(t)$ represent the I/O voltage or current signals of the system.

We will not provide a proof of Equation 1–95; rather, an example will be given that demonstrates the validity of the representation above.

Example 1–12

For the circuit shown in Figure 1–59, express the loop voltages in terms of their derivatives.

Solution Using Kirchhoff's voltage law (KVL), we have

$$e_{in}(t) = v_L(t) + v_r(t) + v_C(t) \qquad (1\text{--}96a)$$

This can be expressed as

$$e_{in}(t) = L \frac{di}{dt} + iR + \frac{1}{C} \int_{-\infty}^{t} i\, dt \qquad (1\text{--}96b)$$

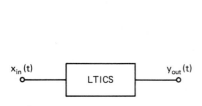

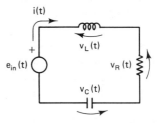

Figure 1–58 General form of I/O for a linear system.

Figure 1–59 Linear circuit, RLC

based on the definition of inductor and capacitor voltages in terms of the current through them. Taking derivatives of both sides of Equation 1-96b, we have

$$\frac{de_{in}}{dt} = L\frac{d^2i}{dt^2} + R\frac{di}{dt} + \frac{1}{C}i \qquad (1\text{-}96c)$$

If we assume that the output signal is the current, $i(t)$, Equation 1-96c is in the form of Equation 1-95, where $a_0 = 0$, $a_1 = 1$, and $a_i = 0$, $i > 1$; and $b_0 = 1/C$, $b_1 = R$, $b_2 = L$, and $b_j = 0$, $j > 2$.

As the circuit of Figure 1-59 was shown to have its I/O relations in the form of Equations 1-95, we will assume that all circuits have these equations as valid I/O representations.

1.15 FOURIER AND LAPLACE TRANSFER FUNCTIONS

Given $h(t)$, we can find the transfer function of a system in the frequency or complex frequency representations by performing the Fourier or the Laplace transform, respectively. The desired transfer function can also be obtained by choosing the system input as either exp $(j\omega t)$ or exp (st). These functions have the characteristic that if they are used as the input signal to a system, the output is the same as the input, but multiplied by a constant. This is shown in Figure 1-60 and Equation 1-97.

$$x(t) = e^{j\omega t} \Longrightarrow y(t) = He^{j\omega t} \qquad (1\text{-}97a)$$

$$x(t) = e^{st} \Longrightarrow y(t) = He^{st} \qquad (1\text{-}97b)$$

The interpretation of H as a constant in Equation 1-97 means that H is not a function of t: $H \neq H(t)$; however, H can depend on other variables, such as ω or s. Equation 1-97 can be expressed as

$$x(t) = e^{j\omega t} \Longrightarrow y(t) = H(\omega)e^{j\omega t} \qquad (1\text{-}98a)$$

$$x(t) = e^{st} \Longrightarrow y(t) = H(s)e^{st} \qquad (1\text{-}98b)$$

From this, we see that the transfer function of any system can be obtained simply by applying either of the "special" signals as an input. This is shown in the following example.

Figure 1-60 Representation of system I/O.

Example 1–13

Using the circuit of Figure 1–59 and its I/O equation (1–96c), find the system transfer function in the ω and s domains.

Solution

1. **ω domain.** Using $e_{in}(t) = \exp(j\omega t)$ in Equation 1–96c, we obtain

$$\frac{d}{dt}\exp(j\omega t) = L\frac{d^2 H \exp(j\omega t)}{dt^2}$$

$$+ R\frac{dH \exp(j\omega t)}{dt} + \frac{1}{C}H \exp(j\omega t)$$

where, by Ohm's law, $i(t)$ is proportional to $e_{in}(t)$, that is, $i(t) = He_{in}(t) = H\exp(j\omega t)$. Taking all derivatives, we have

$$j\omega \exp(j\omega t) = LH(j\omega)^2 \exp(j\omega t)$$

$$+ RH(j\omega)\exp(j\omega t) + \frac{H}{C}\exp(j\omega t)$$

Factoring $\exp(j\omega t)$, we are left with

$$j\omega = LH(j\omega)^2 + RH(j\omega) + \frac{H}{C}$$

We then have

$$H = \frac{1}{j\omega L + R + 1/j\omega C} = H(\omega)$$

where H is a function of ω. Therefore, we can interpret H as the current-voltage transfer function of the system $H = H(\omega)$.

2. **s-domain.** Repeating the steps above with $e_{in}(t) = \exp(st)$, we obtain

$$H = \frac{1}{sL + R + 1/s\omega} = H(s)$$

where $H = H(s)$, the s-domain current-voltage transfer function.

This example has shown that the system transfer function can be obtained from its differential equation by applying either of the "special" signals as input. We will make use of this technique when we wish to obtain the transfer function of a discrete system.

1.16 SUMMARY

This chapter has introduced linear systems from a nonrigorous point of view so that when they are discussed, the reader will have some background. For additional, in-depth material, the reader is referred to the References at the end of this chapter.

PROBLEMS

1. For the following functions, state the domain, the range, and the rule.

   (a) $y = y(t) = \begin{cases} \dfrac{3t^2}{2} - 4, & -15 \leq t \leq +30 \\ 0, & \text{otherwise} \end{cases}$

   (b) $z = z(t) = \begin{cases} \dfrac{1}{t^2} + t - 2, & 1 \leq t \leq 5 \\ 10, & \text{otherwise} \end{cases}$

   (c) $V = V(f) = \begin{cases} 10 \sin [2\pi f(1 \text{ ms})], & -50 \text{ Hz} \leq f \leq +100 \text{ Hz} \\ 20 \cos [2\pi f(1 \text{ ms})], & 200 \text{ Hz} \leq f \leq 400 \text{ Hz} \\ 0, & \text{otherwise} \end{cases}$

   (d) $H = H(\omega) = \begin{cases} 3 \exp(-j\omega \, 1 \text{ ms}), & 0 \leq \omega \\ 0, & \text{otherwise} \end{cases}$

2. Given $v(t) = 3 \cos(\omega t - 30°)$ V, form the discrete domain ($t = nT$) by dividing the interval between -2 and 2 ms into 0.2 ms intervals. Find $v(t = nT)$ for your values where $\omega = 2\pi \, 1000$ rad/sec.

3. Is the following function given discrete? Explain.

   $$C_n = \begin{cases} \dfrac{3}{nf_0}, & n = \text{integer}, \; n \neq 0 \\ 0, & \text{otherwise} \end{cases}$$

4. Evaluate the following implicit functions.
   (a) $V(\omega) = 10 \cos(\omega t - 45°)$ for $t = 1$ ms and $1 \text{ kHz} < f < 1.5 \text{ kHz}$.
   (b) $H(\Omega) = 3 \cos 0.2\pi\Omega - 4 \sin 0.7\pi\Omega$ for $2\pi\Omega = \omega T$, $T = 0.002$ s and $f = 1500$ Hz.

5. Plot $y = y(z) = 1.5z^2 + 4$. Then, using $z = x/2$, plot $y = y'(x)$ on the same graph. What is $y'(x)$?

6. (a) Does $f(t) = 3 \sin[2\pi(1 \text{ kHz})t]$ satisfy Equation 1–2? Explain.
   (b) Does $f(t) = t \exp(-t) u(t)$ satisfy Equation 1–2? Explain.

7. A given system produces $y(t) = 0.5[x(t)^2] + 2$. Demonstrate that it is not linear.

8. Show that the rectifier equation $y(t) = |x(t)|$ is not linear.

9. The equation of a straight line is $y = y(x) = mx + b$. If $b \neq 0$, does this represent a linear system? Explain. What if $b = 0$?

10. Which of the following represents a linear system?

    (a) $y(t) = c|x(t)|$
    (b) $y(t) = 3 \exp[x(t)]$
    (c) $y(t) = \displaystyle\int_{-\infty}^{t} x(u) \, du$
    (d) $y(t) = ax(t) + b\dfrac{dx(t)}{dt}$

Chap. 1 Problems

11. $x_1 = 5 + 3t$ and produces $y_1 = 4t + 1$. $x_2 = 1.5 - 7t$ and produces $y_2 = 5t - 3$. What is the output if the input is $x_1 + x_2$ and the system is linear?

12. Using the I/O relations of Problem 11, what is the output of the system if the input is $7x_1$? $-9x_2$?

13. What is $s(t - 0.5 \text{ ms})$ if

$$s(t) = \begin{cases} 3 \cos [2\pi(1 \text{ kHz})t] & 0 \leq t \\ 0, & \text{otherwise} \end{cases}$$

14. Repeat Problem 13 and find $s(t + 0.9 \text{ ms})$.

15. Does a causal system allow batteries, charged capacitors, or inductors with initial currents for $t < 0$? Explain.

16. Evaluate the following integrals.

(a) $\int_2^7 \delta(t - 3) \, dt$

(b) $\int_{-\infty}^{\infty} (3t^2 - 7)\delta(t + 2) \, dt$

(c) $\int_{-9}^{-1} 21 t \delta(t + 5) \, dt$

(d) $\int_{-\infty}^{\infty} (\cos 6280t)\delta(t - 0.3 \text{ ms}) dt$

(e) $\int_{-10}^{+15} \exp(j\omega t)\delta(t - 5) \, dt$

(f) $\int_{+15}^{+25} (t - 6)\delta(t - 14) \, dt$

17. Using Laplace transforms, find $h(t)$ for the circuit shown.

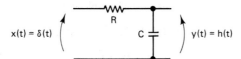

Figure P1–17

18. The impulse response of an LTIC system is as shown. An input signal, $x(t)$, is applied to the system.

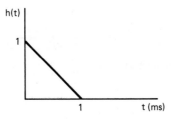

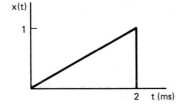

Figure P1–18

(a) Approximate $x(t)$ by a sum of time-shifted δ functions weighted by $x(kt)$. Choose $\tau = n(0.2 \text{ ms})$, $n = 0, 1, \ldots, 19$.
(b) Apply the sum from part (a) to the system and determine the output, $y(kt)$ (see Equation 1–23).
(c) What happens if we choose $\tau = n(0.02 \text{ ms})$, $n = 0, 1, \ldots, 190$? Consider the input and output form.
(d) What happens if we choose $\tau = n(0.002 \text{ ms})$, $n = 0, 1, \ldots, 1900$?
(e) Comment on the form of the input and output as we increase the number of points.

19. Find the convolution of $s(t)$ and $h(t)$.

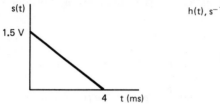 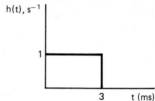

Figure P1–19

20. Find the convolution of $h(t) = \exp(-t)u(t)$ and $s(t)$.

$$s(t) = \begin{cases} 2t, & 0 \le t \le 2 \text{ s} \\ 0, & \text{otherwise} \end{cases}$$

21. Given $h(t) = (1/RC) \exp(-t/RC)$, if $RC = 0.1$ ms, what is the output if $s(t)$ is the input: where $T_1 = 0.2$ μs, $T_2 = 0.2$ ms, and $T_3 = 200.2$ μs?

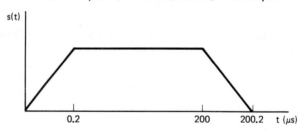

Figure P1–21

22. Find the convolution of:

(a)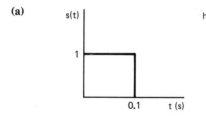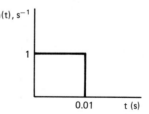

Figure P1–22a

(b) $s(t) = \begin{cases} 1 \text{ V}, & 0 \le t \le 0.1 \text{ s} \\ 0, & \text{otherwise} \end{cases}$

$h(t) = \begin{cases} \dfrac{1}{s}, & 0 \le t \le 10 \text{ ms} \\ 0, & \text{otherwise} \end{cases}$

23. Express $10 \exp[-j2\pi(500 \text{ Hz})t]$ as a complex number.
24. Express $7 \sin[2\pi(300 \text{ Hz})t]$ in exponential form.
25. A periodic function is expressed as

$$f(t) = \sum_{n=1}^{\infty} b_n \sin 2\pi nft + \frac{20}{2}$$

where $b_n = -20 \text{ Vs}/n\pi$.
 (a) Sketch the magnitude of the coefficients as a function of frequency.
 (b) Express the series in exponential form.
 (c) Sketch $|C_n|$ for $-6 \le n \le 6$.
 (d) Compare values of $|C_n|$ with $|b_n|$.

26. A periodic function is given by

$$f(t) = \sum_{n=-\infty}^{\infty} \frac{-8 \text{ V}}{(4n^2 - 1)\pi} \exp(j2\pi nf_0 t)$$

 (a) Expand the series. Write the terms for $-6 \le n \le 6$.
 (b) Sketch $|C_n|$ versus f, $-6 \le n \le 6$.
 (c) Form the sine/cosine series.
 (d) Sketch $|a_n|$ and $|b_n|$ versus f.
 (e) Compare the values of $|a_n|$ and $|b_n|$ with $|C_n|$.

27. Find the Fourier series of an 8-V (peak-to-peak) 1-kHz triangle wave.
28. Find the Fourier series of a 4-V (peak-to-peak) 500-kHz square wave. Plot the magnitude and phase of the frequency components.
29. Using the results of Problem 28, take the first five terms of the series expansion and find their sum at intervals of $t = n(0.2 \text{ }\mu\text{s})$, $0 \le n \le 10$. Plot the result versus t.
30. Referring to Problem 28, find the exponential form of the series. Plot the magnitude and phase of the components and compare with the results of Problem 28.
31. Convert the following sequence to exponential form.

$$f(t) = a_1 \cos \omega_0 t + a_2 \cos 2\omega_0 t + a_3 \cos 3\omega_0 t \\ + b_3 \sin 3\omega_0 t + b_2 \sin 2\omega t + b_1 \sin \omega_0 t$$

32. Convert the following sequence to sine/consine form.

$$f(t) = C_{-3} \exp(j3\omega_0 t) + \cdots + C_3 \exp(-j3\omega_0 t)$$

33. Find the Fourier transform of $s(t)$. Sketch the magnitude and phase of $S(f)$.

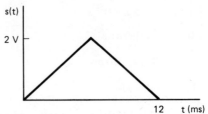

Figure P1-33

34. Repeat Problem 33 for $f(t)$.

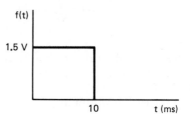

Figure P1-34

35. Repeat Problem 34 for $f(t - 15 \text{ ms})$. Compare results.
36. Find the Fourier transform of $h(t) = \exp(-t)u(t)$. Sketch the magnitude and phase of $H(f)$.
37. Find $Y(f)$ for Problems 33 and 36 using convolution in the frequency domain.
38. Find the cross-correlation function of $s_1(t)$ and $s_2(t)$.

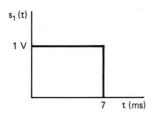 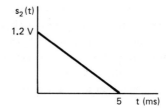

Figure P1-38

39. Find the autocorrelation of $s_1(t)$. Plot the result as a function of τ. (See Problem 38.)
40. Refer to Problem 39.
  (a) What is $W_s(f)$?
  (b) What is the energy content of the signal?
  (c) What is $S(f)$?
  (d) Does $s(t)$ satisfy Equations 1-2 and 1-79?
41. The signal whose FT is shown is fed to an ideal LP filter with a cutoff frequency of 750 Hz. What is the energy density of the input and output signals?

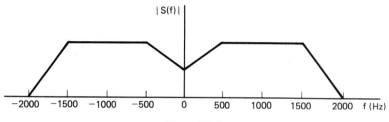

Figure P1-41

42. A set of orthonormal signals, $\{s_i(t)\}$, have an equal probability of being transmitted:

$$s_0 = \sqrt{12}\ \phi_1(t)$$
$$s_1 = -\sqrt{3}\ \phi_1(t) + \sqrt{8}\ \phi_2(t)$$
$$s_2 = +\sqrt{3}\ \phi_1(t) - \sqrt{2}\ \phi_2(t)$$
$$s_3 = -\sqrt{3}\ \phi_1(t) - \sqrt{8}\ \phi_2(t)$$

where $\phi_1(t)$:

and $\phi_2(t)$:

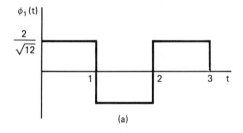

(a)

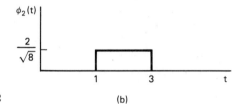

(b)

Figure P1-42

(a) Set up the bank of matched filters.
(b) What are their impulse responses?
(c) If s_1 is transmitted, what is the output of the filters?

43. For the RC circuit shown, what is the differential equation?

Figure P1-43

44. Find $H(\omega)$ and $H(s)$ for the system of Problem 43 by choosing:
   (a) $e_{in}(t) = \exp(j\omega t)$
   (b) $e_{in}(t) = \exp(st)$
45. Replace R with L in Problem 43 and repeat Problem 44.

REFERENCES

AHMED, N., and NATARAJAN, T. *Discrete-Time Signals and Systems.* Reston, Va.: Reston, 1983.

LATHI, B. P. *Signals, Systems and Communications.* New York: Wiley, 1965.

WOZENCRAFT, J. M., and JACOBS, I. W. *Principles of Communication Engineering.* New York: Wiley, 1965.

AN OVERVIEW OF A DIGITAL SIGNAL PROCESSING SYSTEM

2

The rapid pace at which digital signal processing (DSP) is beginning to appear in many items of military, space, and commercial equipment can be attributed to the advantages over and improvements to units that employ analog signal processing. The recent introduction of audio and video equipment that uses the techniques of DSP is an example. By having the sound or image signal recorded and stored in digital, rather than analog, format, the quality of the reproduction is greatly improved. More important, this quality will be maintained regardless of the number of plays, as the record condition will not affect the stored information: record/tape noise is analog, while the playback system is digital.

Another advantage of DSP is its use of digital components. These units require and provide binary signal levels, causing them to be immune to external analog noise and relatively stable. Drifting of their characteristics with time and temperature will have a minimal effect on their performance, resulting in output repeatability from component to component. The latter characteristic almost guarantees that the performance of each unit will be identical to every other one, providing a high level of confidence in their output values over a period of time. In addition, the accuracy of the output is fixed and not subject to component drift.

58 Chap. 2   An Overview of a Digital Signal Processing System

As the DSP operation will be based on software (i.e., an instructional set), it is relatively simple to modify and test any changes in a design. As an example, any changes in the characteristics of a digital filter can be brought about by a few changes in a processor program. In addition, the parameters of the filters can have values usually unattainable using analog components, such as op-amps, resistors, and capacitors, and can be modified periodically an in the adaptive filter.

To indicate how extensive the use of DSP has become, the orbiter space shuttle employs it in its radar system when it is establishing contact with the TDRS satellite. Also, several manufacturers of oscilloscopes have used DSP as the basis of their systems and have increased the scope's capability by allowing it to serve also as a spectrum analyzer.

This chapter presents an overview of a digital signal processing system. It will be presented in block form and the function of each block will be discussed. In the chapters that follow, characteristics of the blocks comprising the system will be looked at in more detail. In addition, an attempt will be made to define signal processing and to see how this is accomplished from both the analog and digital viewpoints. These approaches to signal processing will be compared as to their characteristics and the problems they each pose. It will be shown that a digital signal processing system requires both analog and digital circuits to process and obtain a signal in a desired format.

2.1 DEFINITION OF SIGNAL PROCESSING

We will begin by defining signal processing as any operation that changes the characteristics of an electrical signal. These characteristics include the amplitude, shape, phase, and frequency content of the signal. These changes are brought about by passing the signal through a system that contains components that can alter one or more of its characteristics.

Many signals of interest are ones whose amplitude is continuously changing, its value being dependent on the time at which the signal is observed. These are termed *analog signals* and are usually represented by lowercase letters [e.g., $v(t)$]. This notation indicates a voltage whose amplitude might vary with time. The example that readily comes to mind is the standard sine-wave variation of voltage. This is given in Equation 2–1.

$$v(t) = V_P \sin 2\pi ft \qquad (2\text{–}1)$$

where V_p is the maximum value of $v(t)$ and f is its frequency.

A more complex signal (i.e., one composed of many sine waves) is an audio signal. This contains frequency components that can range from dc to 20 kHz. Additional examples of analog signals are those generated by TV and AM/FM stations. These contain frequency components in the high kHz to MHz ranges. Also of interest are the returns from radar and sonar transmissions, together with those used in seismic studies. All of these complex signals

Sec. 2.1   Definition of Signal Processing

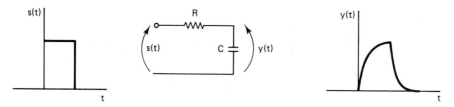

Figure 2–1 Input/output of an LP filter: *RC* circuit.

contain some form of information that we will attempt to extract by signal processing.

It is assumed that the effect of processing the signal will be to put it in a form more suitable to our needs or that its information content will be more easily extracted. In some cases, processing will result in the generation of a signal (i.e., the processor serves as a waveform generator).

As a simple example of analog signal processing, we have only to think of the *RC* circuit and its effect on a single pulse. If we connect the *RC* circuit as shown in Figure 2–1, it will show low-pass filter characteristics to an input signal: The high-frequency components of the signal will be attenuated and shifted in phase much more than will the low-frequency ones. In Figure 2–1, a signal, $s(t)$, is applied to the *RC* circuit. Its output, $y(t)$, is also shown, where $y(t)$ is obtained by convolving $s(t)$ and the transfer function of the *RC* circuit, $h(t)$. The effect of removing the high-frequency components of the input signal was to increase the rise and fall times of the signal at the output. In this example of signal processing, the characteristics of $s(t)$ that changed were its shape (waveform), amplitude, the amplitude of its frequency components, and their relative phase.

While Figure 2–1 shows the effect of processing on the first two characteristics, being a representation in the time domain, a representation in the frequency domain is required to see the effect of processing on the latter two characteristics. In Figure 2–2, the frequency and phase response of the *RC* circuit of Figure 2–1 are presented as a function of frequency. This represents the transfer function of the *RC* circuit from the frequency-domain point of view, where $|H(2\pi f)|$

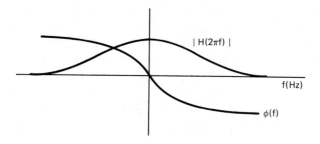

Figure 2–2 Phase and amplitude response of an *RC* circuit.

indicates the amount of attenuation that each frequency component of the input signal will receive as it passes through the system and $\theta(f)$ is the number of degrees that each component is shifted, both as a function of frequency. $H(f)$ can be obtained by a Fourier transform of $h(t)$ or from the circuit differential equation (see Section 1.15). Figure 2–2 clearly shows that the high-frequency components of the pulse signal of Figure 2–1 are attenuated much more than those of the lower-frequency components. In addition, the higher-frequency components will receive more of a phase shift.

The actual changes in the frequency content of $s(t)$ as it passes through the system can be determined by finding its FT and multiplying it by $H(f)$ as in Equation 2–2.

$$Y(f) = S(f)H(f) \qquad (2-2)$$

where $Y(f)$ is the frequency content and relative phase of $y(t)$.

By combining the time- and frequency-domain representations, a complete picture of the processing effect of the *RC* circuit can be obtained. Each one will provide a given amount of information: In the time domain, the convolution of $s(t)$ and $h(t)$ will provide $y(t)$; in the frequency domain, $S(f)$ and $H(f)$ will provide $Y(f)$, where $Y(f) = S(f)H(f)$.

2.2 ANALOG SIGNAL PROCESSING

To process an analog signal, devices are required that will respond to the continuous variation of the input signal's instantaneous amplitude. Among these devices currently used are linear or analog ones such as the transistor and operational amplifier (op amp). These are used to form analog circuits, such as the amplifier, filter, and waveform generator, among others.

Although analog devices and circuits are in daily use in all forms of electrical systems, they have some characteristics that result in the final signal we receive not being the best possible. Among these characteristics is the drifting of device parameters with time, due to temperature, and power supply variations. An example of this is the change in the transistor beta (β). Also, there is a good chance that if one transistor is replaced with another of the same part number, their betas will be different. This effect is termed the *batch effect*. Another example is the drift of the offset voltage and current of an op amp with temperature and power supply variation. Add to this the problems involved in matching components, their tolerance requirements, the cost of high accuracy, and the effect of noise on the circuit and we readily see how processing a signal using analog devices can introduce some error and provide an output signal of less-than-optimum quality.

2.3 DIGITAL SIGNAL PROCESSING

The above-mentioned drawbacks of analog systems can be reduced, if not eliminated, by performing as much signal processing as possible with devices whose properties are constant with time, temperature, and power supply voltages and are not affected by part replacement, variation of external component values, and external noise. Devices with these properties are available if we look at the group of components with digital characteristics. These electrical systems and components respond to signals that are in a digital (binary) format and provide digital signals at their output. These devices are relatively immune to time and temperature variations and noise, resulting in an input/output (I/O) relation that is constant. Elimination of I/O variations due to time, temperature, part replacement, and noise together with the growth of digital technology are some of the major reasons why digital signal processing has grown so rapidly over the past few years.

Digital signal processing uses devices whose characteristics are not affected by time and temperature to generate or alter analog signals. This results in a system stability that is usually unattainable with analog devices. The importance of this lies in the fact that it enables the result of a measurement taken at one time to be related to a measurement taken at another time or place and with different equipment. No correction is required to account for parameter drift or unit-to-unit differences. The need for this in a system used to monitor a patient in a hospital is evident. It would not be beneficial to a patient to have the system output drift, leading the nurse or doctor who is monitoring it to believe that the patient's condition was changing when it was not. On a lesser scale, a digital format for an audio or television system would result in the elimination of most noise. Its use in tape systems would see the degradation of sound or image quality, due to the effect of time and wear on the tape, greatly reduced if not eliminated.

Although the foregoing discussion has focused upon the hardware used in DSP, it should be mentioned that much of the advantage of DSP arises from its use of software. Many of the circuits we will examine, such as digital filters, have their parameters expressed as part of a sequence of instructions to a processor integrated circuit (IC). Not only does this provide identical results from device to device, but also the ability to readily change the circuit parameters to meet design changes.

With this in mind, we will see below that the system used to process the signal will be mostly digital in form, with the major processing effort accomplished with digital units; however, the input portion of the system will require analog circuits and devices to put the analog signal in a form that will allow its conversion to digital format. Analog circuits and devices will also be required at the output to put the processed signal in a form that allows us readily to extract information from it.

2.4 DESCRIPTION OF A DIGITAL PROCESSING SYSTEM

The units that comprise a digital processing system are introduced below. Figure 2–3 presents a block diagram of a typical digital processing system, where $s(t)$ is the analog input signal to be processed by the system and $r(t)$ is the analog processed output. The system is divided into five sections. The input and output sections perform the required analog processing of the signal. The input section processes the signal so that it is in a state that allows it to be readily converted to a digital format; the output section processes the signal so that it can be used. The digital processor is the main portion of the system and is where the major processing occurs. The analog-to-digital converter (ADC) and the digital-to-analog converter (DAC) serve as the interface between the digital and analog portions of the system.

The sections that comprise a digital processing system were introduced above and will be discussed in more detail in later chapters. The sections that follow present an overview of the contents of each section.

Figure 2–3 Block diagram of a digital signal processor.

2.4.1 Input Section

The analog signal is processed in this section; that is, it is put in a form that allows it to be readily converted to digital format. Figure 2–4 shows the block diagram of a typical input section. The source of the signal is usually the output of a transducer. This is a device that converts energy (e.g., thermal, radiation, or kinetic) into electrical energy. Examples of transducers are thermocouples, photodetectors, microphones, phonograph cartridges, tape heads, flow meters, pressure-sensitive devices, strain gauges, and accelerometers.

In some situations there may be no need for a transducer when the system under investigation is able to generate its own voltage. This is the case when an electroencephalogram (EEG) or electrocardiogram (EKG) is performed to investigate brain or heart activity. Another example is the return signal from a radar transmission. Here, the reflected electromagnetic energy is received by an antenna, amplified, and down-converted in frequency. It can then serve as the signal to be processed to extract the information relating to the target's range and velocity.

The remaining units of the input section will include filters, an amplifier and a sample-and-hold (SH) device. The filters are used to remove high-frequency noise and to band-limit the signal. Band limiting is a required part of DSP to prevent the appearance of extraneous components in the output. This effect, called aliasing, will be discussed in more detail later. In addition

Figure 2–4 Block diagram of an input section.

to the LP filter, it may also be advantageous to include a 60-Hz notch filter that can remove the power-frequency component, which is a large part of external noise.

The amplifier is required to bring the signal up to the voltage range that is required by the input of the analog-to-digital conversion (ADC) unit; the SH device provides the input to the ADC and will be required if the input signal must remain relatively constant during the conversion of the analog signal to digital format. The term track-and-hold (TH) is replacing the term sample-and-hold in the current literature and will be used in this text.

2.4.2 Analog-to-Digital Converter

The output of the track-and-hold circuit serves as the input to the ADC. The major concerns are the time it takes to perform the conversion, its accuracy, and the timing of the conversion to correlate with other device activities (e.g., the processor). The output of the ADC is an n-bit binary number that is related to the value of the analog signal at its input. The larger n is, the higher the accuracy, that is, the closer the agreement between the analog value and its binary equivalent. Lack of accuracy between the n-bit number and its analog value will give rise to processor noise. This noise is different from analog noise and arises from variations in the sampling process: sampling rate error; signal variation during sampling; quantization effect due to the conversion of a continuous signal into a discrete signal; and the effect of sampling at too low a rate, referred to as aliasing. All of these will be discussed later. The ADC input signal is limited to a range of either 0 to $+10$ V (or 0 to $+20$ V) if unipolar, and -5 to $+5$ V (or -10 to $+10$ V) if bipolar, where it is the function of the preceding amplifier to provide a signal in this range. Once converted to digital form, the signal can be processed using digital techniques.

2.4.3 Processor

The actual processor used will be determined by what is to be done to the input signal, its frequency content, and how fast it is required to perform the processing. For our purposes, we will require the processor to either filter the signal (alter its spectral content), determine its frequency content (spectral analysis), or compare it with another signal (correlation). In either case, this will involve transform techniques which require the determination of the sums of many products. The rapidity with which these sums are required will determine the actual integrated-circuit (IC) device. These are currently divided into three groups of IC devices: (1) the microprocessor chips (e.g., Intel's 80xx,

Motorola's 68xxx, and Zilog's Z-80xxx; (2) processor chips such as Intel's 2920; and (3) FFT chips, manufactured by NEC, TRW, AMI, and TI.

With the current growth rate of IC fabrication technology into the area of very large scale integration (VLSI) along with advances in processing techniques, it is relatively easy to predict that digital processors will soon be fabricated that can accept and process the high-frequency content of audio, radio, and television systems, ushering in a set of commercial units that will provide a level of quality in sound and image that will far exceed what is available today and, most important, at a relatively moderate cost.

In addition to the processor ICs, the many personal and industrial computers available today can also be used to process the signal in nonreal time. Software packages that allow the user to design filters, perform convolution and correlation operations on signals, and obtain amplitude versus frequency data can either be written easily or are readily found in the literature. Regardless of the processor used, the result of the processing operation is an output signal in digital form. This output can be stored in a memory or sent directly to the DAC.

2.4.4 Digital-to-Analog Converter

The digital signal from the processor is applied to the input of a DAC and generates a related, quantized analog value at its output. This signal, due to the discrete nature of the digital-to-analog conversion, is continuous but not smooth, containing many jumps in its value. This is represented in Figure 2–5.

The output of the DAC is a continuous signal whose output varies with time (i.e., an analog signal); however, as each n-bit number from the processor will generate a particular voltage, the discrete change in the input will cause a jump in the DAC output. These jumps in the output voltage give rise to many high-frequency components that are unwanted (i.e., they are noise). Prior to applying this signal to the output section, where it is put in usable form, the high-frequency jumps must be removed.

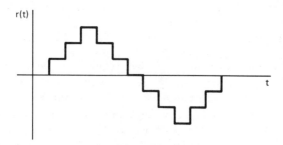

Figure 2–5   DAC output signal.

Sec. 2.5 Summary

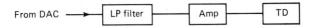

Figure 2-6 Block diagram of an output section.

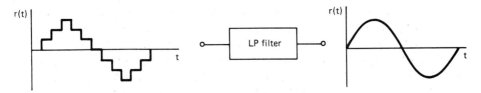

Figure 2-7 Effect of an LP filter on DAC output.

Figure 2-8 Block diagram of an output transducer.

2.4.5 Output Section

As the output of the DAC cannot be used in its present form, it is the function of the output section to perform the final processing prior to usage. Figure 2-6 shows a block diagram of a typical output section. To smooth out the waveform (eliminate the high-frequency components due to the jumps), a low-pass filter usually follows the DAC. This presents a sharp cutoff characteristic to the DAC output signal and allows only the desired frequency components to pass, resulting in a smoothly varying signal at its output. The output of the LP analog filter represents a (digitally) filtered form of the original analog input signal or provides its frequency content. In either case, it has been converted to a form that we can use; however, it may require some further analog processing prior to its application. Figure 2-7 shows the input and output of the LP filter.

The output signal of the DAC/LP filter section may require amplification so that it can be in usable form. After any analog processing has occurred, the signal is applied to the output transducer. This will be a device that receives an electrical signal and converts it to another form of energy (sound or light). Examples that readily come to mind are the loudspeaker or cathode-ray tube (CRT). In this form, it can readily be used. Figure 2-8 shows the block diagram of an output transducer.

2.5 SUMMARY

The paragraphs above give a brief description of what is provided by each block in the digital processing system. The chapters that follow will discuss each block in more detail together with the basic equations and terminology

PROBLEMS

1. Determine the output of the circuit shown if $s(t)$ is a 1-V, 1-ms pulse.

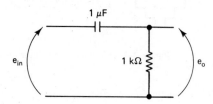

Figure P2–1

2. Find the frequencies that constitute a "brain wave."
3. Find the FT of $s(t)$, Figure 2–1, if the signal extends from 0 to 10 ms.
4. Find $H(f)$ for the circuit of Figure 2–1 if $RC = 1.6$ ms.
5. Using the results of Problems 3 and 4, find $Y(f)$. Compare with the results of Problem 3 to see the frequency domain effect of the filter on the signal.

ANALOG SIGNAL PROCESSING

— 3 —

The signal to be processed cannot be applied directly to an ADC, as (1) it contains noise, (2) it contains unwanted (high) frequency components, (3) its level (mV) is too low, and (4) it is continuously changing. Processing is required to alter the signal, putting it in a format acceptable to the ADC. As the signal is analog in form, the processing, of necessity, must be analog.

The analog processing might include the following: (1) filtering, (2) amplification, and (3) sampling. The output of the sampling circuit is a voltage that is within the input range required by the type of ADC used, and "constant" long enough to permit a conversion to digital form to take place. The circuits that perform the processing are discussed in the sections that follow.

3.1 ANALOG FILTERS

The use of analog filters is dictated by several requirements that must be met by the processing system: (1) to reduce the high frequency noise content of the signal, (2) to reduce the amount of interference from power lines, and (3) to band-limit the signal in preparation for sampling. The first and third requirements can usually be met with a low-pass (LP) filter; the second one can be

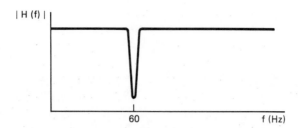

Figure 3-1 Gain magnitude versus frequency for a 60-Hz notch filter.

met with a notch (or band-reject) filter, with the center of the reject (or stop) band tuned to the power frequency, 60 Hz. The frequency response of the notch filter is shown in Figure 3-1. The width of the stopband should be relatively narrow (high Q) and is determined by the choice of filter circuit components. This will be discussed below.

A sharply tuned (i.e., narrow bandwidth) notch filter with a high level of rejection at the center frequency is almost a necessity when performing EEG or EKG measurements. Strong fields may exist in the vicinity of power lines or fluorescent lighting, resulting in a patient "picking up" the 60-Hz signal. The signal generated by brain activity has frequency components up to 30 Hz which must be separated from the noise. If large enough, the power signal can mask the desired signal and possibly cause the amplifiers of the system to saturate, further degrading the signal of interest. The effect of the notch filter on a signal spectrum is shown in Figure 3-2. The notch filter has generally reduced the unwanted signal component at 60 Hz while allowing the desired signal to pass through unattenuated.

Many of the filters used today are active ones, using op amps, with no inductors. This results in a simpler design technique and a reduction of filter size. Use of op amps requires circuitry to compensate for offset voltages and currents, together with their drift. In addition, depending on the tolerance of resistors used in the circuit, it will require adjustments so as to provide the LP characteristics at the desired frequencies. It is just for these reasons that digital filters are so attractive; however, at this point in the system we have no alternative but to use analog filters.

Some advantages to using active over passive filters are ease of design, the ability to cascade several stages without affecting the design of a single stage, and the ability to achieve gain rather than attenuation. Unlike the filter

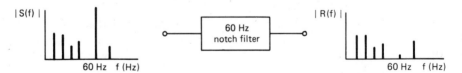

Figure 3-2 Effect of a 60-Hz notch filter on a 60-Hz component.

Sec. 3.1   Analog Filters   69

fabricated from passive components (R's, L's, and C's), the active filter can provide gain to the input signal. In some designs, gains of 10 or more can be obtained, reducing the amount of gain required from an amplifier in the system.

In designing passive filters, it is not possible to cascade two second-order filter circuits to obtain a fourth-order circuit without loading each one and changing the passband frequency parameters. If a fourth-order filter is desired, the design must be carried out with that end in view. Any change in filter order requires a new design. With active filters, the use of the op amp with its high input and low output impedance essentially isolates one filter section from the others. This allows stages to be cascaded with negligible effects on the design frequency. In this manner, the basic second-order section is designed and constructed, with several of them cascaded to obtain the desired $2n$-order filter. The cutoff frequency is essentially that of the second-order filter, but the roll-off of gain will be that of a $2n$-order filter. This will be discussed in Section 3.2.5, and Appendix A.

Figure 3–3 shows the block diagram of a filter with its input and output relationship together with its transfer function. The I/O relation is presented in the time and frequency domains as well as the complex frequency (s) domain. The output is given by the convolution of the input signal $s(t)$ and the filter transfer function, $h(t)$. This is expressed as

$$r(t) = s(t) * h(t) \tag{3-1}$$

which is shorthand notation for convolution. Equation 1–26 is presented here for reference.

$$r(t) = s(t) * h(t) = \int_{-\infty}^{\infty} s(\tau) h(t - \tau)\, d\tau \tag{1-26}$$

The Fourier and Laplace transforms of Equation 1–26 yield

$$R(f) = S(f)H(f) \tag{3-2}$$

$$R(s) = S(s)H(s) \tag{3-3}$$

that is, the product of the transforms of the input signal and the filter transfer function yield the transform of the filter output signal. We will refer to this I/O relation when we discuss digital filtering, as this will serve as the starting point in our derivations of the digital filter characteristics.

The filter transfer function, $h(t)$, is called the *impulse response* of the

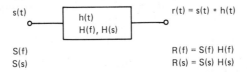

Figure 3–3 Time and frequency domain representation of the filter I/O relationship.

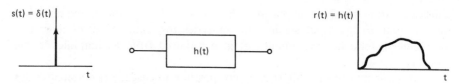

Figure 3-4 $h(t)$ as the response of a filter to an impulse.

system. This states that an impulse (i.e., a delta function) applied to the filter will result in an output signal $r(t) = h(t)$. This is shown in Figure 3-4. The filter cannot respond to an input signal before it is applied. A filter with this characteristic is termed *causal*. In addition, the filter will also be assumed to be linear and time invariant (LTI).

The derivation of the convolution intergral (the filter I/O relationship in the time domain) was presented in Chapter 1 together with the derivation of the FT relation, Equation 3-2, and the definitions of other filter and linear system terms. It should be mentioned that the Laplace transform is used for a causal filter only and that the Fourier transform can be obtained from $H(s)$ by letting $\sigma = 0$ for $s = \sigma + j\omega$. The system transfer function, $h(t)$, can be determined from either $R(f)/S(f)$ or $R(s)/S(s)$ by an IFT or ILPT.

3.2 FILTER CHARACTERISTICS

In choosing a filter, some considerations are: (1) the passband characteristics, (2) the cutoff frequency, f_c, (3) the stopband characteristics, (4) the stopband frequency, f_s, and (5) the roll-off of gain between the pass and stopbands. These are discussed in the following section.

3.2.1 Passband Characteristics

The range of frequencies allowed to pass through the filter with maximum gain or minimum attenuation is called the *passband*. The gain versus frequency relationship of the filter will be determined by the type chosen. We will consider two of them.

Butterworth filter. The passband characteristics of a (two-pole) Butterworth filter are shown in Figure 3-5. The gain drops slowly in the passband and is sufficiently high in the stopband to allow many unwanted components of the input to "pass through" the filter. These characteristics make this filter far from ideal. In addition, the phase shift versus frequency response is nonlinear, introducing further distortion in the envelope of the signal.

Chebychev filter. Figure 3-6 presents the passband characteristics of a four pole Chebychev filter. The ripple in the passband will provide unequal

Sec. 3.2   Filter Characteristics

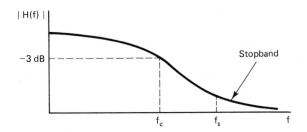

Figure 3-5 Gain versus frequency response of a Butterworth filter.

attenuation to the desired frequency components of the input signal; however, the variation in the passband (ripple width) can be minimized by proper choice of external components. As with the Butterworth filter, the phase is nonlinear.

3.2.2 Cutoff Frequency

The value of the frequency at which the voltage gain of the filter is down by 0.707 (-3 dB) from its maximum value in the passband is termed the *cutoff frequency*, f_c. Components of the input signal with frequencies greater than this value will have their power reduced by 50% or more in passing through the filter; those frequencies less than f_c are relatively unattenuated.

3.2.3 Stopband Characteristics

The stopband will have the same characteristics as the passband: smoothly decreasing for the Butterworth; ripples for the Chebychev. The stopband is delineated by the stopband frequency, f_s, and the amplitude of the frequency components in this region must be brought as close to zero as possible.

3.2.4 Stopband Frequency

This locates the range of frequencies that are to be attenuated. It is usually specified by requiring the filter gain to be reduced by a stated amount from its value in the passband for frequencies greater than f_s.

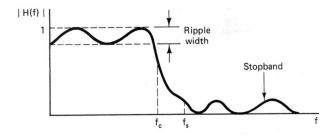

Figure 3-6 Gain versus frequency response of a Chebychev filter.

3.2.5 Gain Roll-off

The Butterworth and Chebychev filters have their gain roll-off at a given rate past the cutoff frequency. Ideally, the roll-off should be a vertical drop; practically, it is not, but should be made as steep as possible to ensure that frequency components of the input signal greater than the cutoff value do not appear at the output. This implies that the cutoff and stopband frequencies are relatively close. This is shown in Figure 3-7. The closer f_c is to f_s, the steeper the roll-off of filter gain. The rate of roll-off is determined by the order of the filter (i.e., the number of poles or active components the filter possesses). A single filter section with two capacitors will provide two poles in the complex frequency plane. This will cause the gain to drop at a 40 dB/decade rate. Each pole added to the filter provides an additional 20 dB/decade.

To obtain the desired roll-off, several stages are cascaded, each providing one or more poles. The increase in roll-off with the number of poles (order) is shown in Figure 3-8 for a Butterworth filter of order 2 and 10. The roll-off is at the rate of 20 dB/decade for each pole or active component.

As will be shown, the frequency content of a signal prior to sampling must be limited to a predetermined range (i.e., the frequencies above a given value must be eliminated). This is referred to as *band limiting,* and it must be performed to prevent the introduction of unwanted frequency components (noise) into the desired signal. This requires a steep roll-off of gain to limit

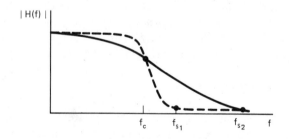

Figure 3-7 Gain response showing the effect of stopband location on rolloff.

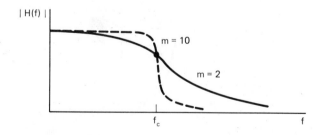

Figure 3-8 Effect of the order number on rolloff.

Sec. 3.2 Filter Characteristics

sharply the band of frequencies contained in the signal prior to sampling and reduce greatly the amplitude of those components in the stopband.

3.2.6 Filter Circuits

Several circuit connections using op amps are available to form the various filter types mentioned above. These circuits allow one to construct the Butterworth or Chebychev filters with either the low-pass (LP), bandpass or reject, or high-pass (HP) characteristics, by the proper choice of the type and value of the external components. The basic circuits are shown below, with design examples available in the References at the end of this chapter.

Voltage-controlled voltage source. This circuit uses the op amp as a voltage amplifier with high input and low output impedance. The basic voltage-controlled voltage source (VCVS) configuration is shown in Figure 3–9. Depending on the choice of values for the admittances, Y_i, a filter with any passband/stopband configuration and characteristics, including gain, can be fabricated. To construct a LP filter, Y_1 through Y_4 are conductances and Y_5 and Y_6 are capacitive susceptances. Figure 3–10 shows the basic circuit of a 2-pole LP VCVS filter.

The actual choice of component values will determine the passband characteristics (e.g., Butterworth or Chebychev), the cutoff frequency, and the gain. The desired roll-off is obtained by cascading several of these filter circuits. The high input and low output impedances of the op amp serve to reduce the loading of one stage upon another.

A VCVS configuration can also be used to design the 60-Hz notch filter referred to above. The design procedure requires the notch frequency and bandwidth (or Q) of the response. A design for a notch filter at 60 Hz is

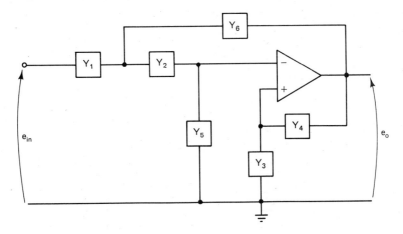

Figure 3–9 VCVS active filter configuration.

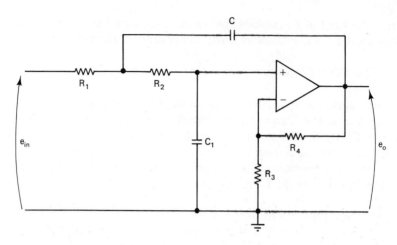

Figure 3-10 LP VCVS filter configuration.

shown in Figure 3-11. Due to component tolerance and op-amp drift, adjustment of the circuit is required to obtain the design characteristics.

The design procedure has been simplified to a table-look up approach, once the filter characteristics are specified. The actual procedure has been described in several textbooks together with providing the tables required to perform the design. One of these is listed in the References at the end of this chapter.

Infinite-gain multiple-feedback filter (IGMF). A different filter configuration is presented in Figure 3-12. The choice of external component values will determine the passband characteristics and cutoff frequency. Its design is also simple, requiring tables to determine the component values.

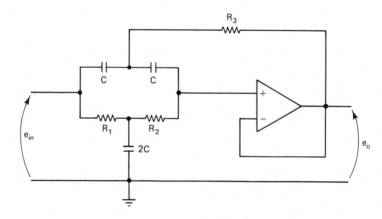

Figure 3-11 VCVS configuration for a notch filter.

Sec. 3.2   Filter Characteristics   75

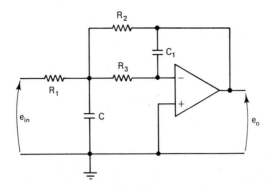

Figure 3-12 LP IGMF filter configuration.

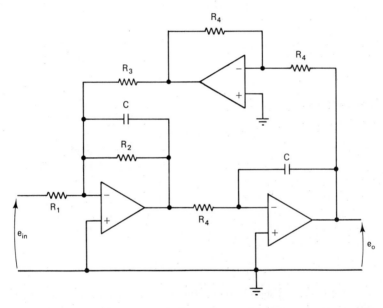

Figure 3-13 LP biquad filter configuration.

Biquad filter.   Another circuit choice to realize the desired transfer function is the biquad filter. Although using more components, it does offer features different from the previous ones. Figure 3-13 shows the LP biquad configuration. As with the other circuits, design tables are available providing the values of components that yield the desired response.

A comparison of the three filter circuits discussed above is provided.

VCVS:

1. Noninverted gain
2. Few external components

3. Adjustment of characteristics possible
4. Low output impedance
5. Low spread of component values
6. High gain

IGMF:

1. Inverted gain
2. Fewest external components
3. Good stability
4. Low output impedance

Biquad:

1. Good tuning capabilities
2. High stability

3.3 AMPLIFIERS

The signal of interest, whether the output of a transducer, the difference of potential due to brain or heart activity, or the transmitted signal from a communication system, is usually a low-level one, typical values being in the millivolt region for transducers and in the microvolt region for brain waves. In addition, the signal is corrupted with noise, due to either the nature of the signal source or the pickup of electromagnetic interference (EMI) as the signal travels from its source to the point where it is to be processed. This is shown in Equation 3–4.

$$r(t) = s(t) + n(t) \qquad (3\text{–}4)$$

where $s(t)$ is the desired signal, $n(t)$ is the noise signal, and $r(t)$ is the combined signal plus noise.

We will see in Chapter 4 that the ADC will accept a voltage that lies between 0 and up to $+20$ V if unipolar, or between -10 and $+10$ V if bipolar. Based on the comments above, it is clear that amplification of the signal is required. At the same time that the amplifier is increasing the amplitude of the desired signal, it should reduce the relative value of the noise signal, and not add any additional noise of its own.

3.3.1 Ideal Amplifier

The ideal amplifier will provide the desired gain with flat frequency response over the range of interest. The input impedance, Z_{in}, of the amplifier should be high, preventing it from drawing excessive current from the signal source and loading it (i.e., causing its terminal voltage to drop). The output impedance, Z_o, should be as close to 0 Ω as possible so as to provide an output voltage that is independent of the current drawn from the amplifier. It should be stable, showing little or no drift of its parameters with time, due to aging and temperature or power supply variation, allowing it to be part of a system that is on for long periods of time. It should add no noise to the signal and, most important, should have as high a noise rejection capability as possible.

3.3.2 Differential Amplifier

As a single-input amplifier will amplify noise with the same amount of gain as it does the desired signal and add noise of its own, it will not be considered here on the grounds of poor noise rejection capability.

The circuit that does show a high level of noise rejection capability is the differential amplifier (diff amp) shown in Figure 3–14. This circuit uses an op amp with the desired signal connected across the (−) and the (+) terminals. This is termed the *differential mode* and the signal receives the differential gain, A_d. In this mode of connection, the unwanted (noise) signal, e_c, is applied between both terminals and ground. This is the *common mode* connection, which causes the signal to receive the common mode gain, A_c. The ratio of A_d to A_c is defined as the *common mode rejection ratio* (CMRR) and is given by Equation 3–5.

$$\text{CMRR (dB)} = 20 \log \frac{A_d}{A_c} \qquad (3\text{–}5)$$

Ideally, this should be as large as possible, resulting in the desired signal being amplified much more than the noise signal. In this manner, the signal-to-noise ratio (SNR) of the signal is increased.

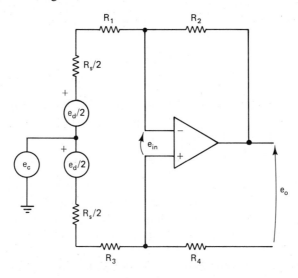

Figure 3–14   Differential amplifier circuit.

Using superposition and the gain equations for an inverting and noninverting op-amp configuration, the differential output of the circuit is obtained.

$$e_o = -\frac{e_d}{2}\frac{R_2}{(R_s/2) + R_1} - \frac{e_d}{2}\frac{R_4}{(R_s/2) + R_3 + R_4}\frac{R_1 + R_2 + R_s/2}{R_1 + R_s/2} \quad (3\text{-}6)$$

where R_s represents the source resistance.
The differential gain is given by Equation 3–7.

$$A_d = \frac{e_o}{e_d} = -\frac{1}{2}\frac{R_2}{(R_s/2) + R_1} - \frac{1}{2}\frac{R_4}{(R_s/2) + R_3 + R_4}\frac{R_1 + R_2 + R_s/2}{R_1 + R_s/2} \quad (3\text{-}7)$$

Using similar techniques, the common mode output and gain are determined.

$$e_o = e_c\frac{-R_2}{(R_s/2) + R_1} + e_c\frac{R_4}{(R_s/2) + R_3 + R_4}\frac{R_1 + R_2 + R_s/2}{R_1 + R_s/2} \quad (3\text{-}8)$$

$$A_c = \frac{e_o}{e_c} = \frac{-R_2}{(R_s/2) + R_1} + \frac{R_4}{(R_s/2) + R_3 + R_4}\frac{R_1 + R_2 + R_s/2}{R_1 + R_s/2} \quad (3\text{-}9)$$

By proper choice of resistors, the desired amount of differential gain can be obtained and the common mode gain can be set close to zero. At this point, the ability to match resistors becomes important. A mismatch in the resistors might provide a slight variation in the differential mode gain; however, that same mismatch can cause a large difference in A_c and, hence, the CMRR. This circuit requires much less than 1% tolerance resistors to obtain values of CMRR close to the theoretical limit. A mismatch of 1% in the resistor values can cause the CMRR to be reduced from its maximum value.

Another problem with this circuit is the fact that the input impedance is limited to R_1. As this is usually set low to obtain some gain, there is a possibility of loading the signal source. In addition, if the signal source resistance is not balanced, it could further reduce the CMRR.

3.3.3 Instrumentation Amplifier

The circuit above can be improved to provide a high CMRR and input impedance. This circuit can be fabricated using resistors and 3 op amps. It is also the basis for commercial instrumentation amplifiers (IAs) and is available from several manufacturers in packaged form.

Discrete IA. Figure 3–15 shows the circuit of an IA fabricated from discrete components. As the differential voltage, e_d, is applied to the (+) terminals of the input op amps, the input impedance of the circuit becomes that of the op amp. Depending on the devices used, this can be as high as several megohms.

The gain of the circuit can be determined by superposition and KVL.

Sec. 3.3 Amplifiers

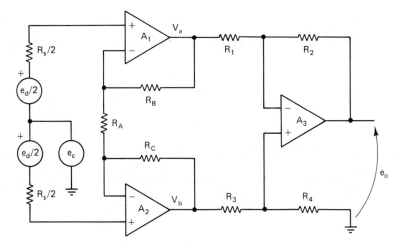

Figure 3–15 Discrete instrumentation amplifier.

If we assume ideal op amps, the equations for V_a and V_b of Figure 3–15 are obtained.

$$V_a = \left(1 + \frac{R_B}{R_A}\right)\frac{e_d}{2} - \frac{R_B}{R_A}\frac{-e_d}{(2)} + e_{ca} \qquad (3\text{–}10\text{a})$$

$$V_b = \left(1 + \frac{R_C}{R_A}\right)\frac{-e_d}{2} - \frac{R_C}{R_A}\frac{e_d}{(2)} + e_{cb} \qquad (3\text{–}10\text{b})$$

If $e_d = 0$ V, the common mode signal is given a gain of unity: $V_a = V_b = e_c$. For nonideal and nontracking op amps, the common mode signal will not cancel at the output: $e_{ca} \neq e_{cb}$. This produces a nonzero input to A_3 which will reduce the CMRR from its ideal value.

The input to the second stage is $V_a - V_b$, which is given by Equation 3–11.

$$V_a - V_b = \frac{R_A + R_B + R_C}{R_A} e_d \qquad (3\text{–}11)$$

This indicates that any mismatch between R_B and R_C affects only the gain, not the CMRR. If $R_2 = R_4$ and $R_1 = R_3$, we obtain the differential output voltage and gain.

$$e_o = \frac{R_2}{R_1}\frac{R_A + R_B + R_C}{R_A} e_d \qquad (3\text{–}12\text{a})$$

$$A_d = \frac{R_2}{R_1}\frac{R_A + R_B + R_C}{R_A} = \frac{e_o}{e_d} \qquad (3\text{–}12\text{b})$$

Integrated Circuit
Precision Instrumentation Amplifier
AD521

FEATURES
Programmable Gains from 0.1 to 1000
Differential Inputs
High CMRR: 110dB min
Low Drift: $2\mu V/°C$ max (L)
Complete Input Protection, Power ON and Power OFF
Functionally Complete with the Addition of Two Resistors
Internally Compensated
Gain Bandwidth Product: 40MHz
Output Current Limited: 25mA
Very Low Noise: $0.5\mu V$ p-p, 0.1Hz to 10Hz, RTI @ G = 1000
Extremely Low Cost

AD521 FUNCTIONAL BLOCK DIAGRAM

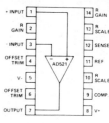

TO-116

PRODUCT DESCRIPTION

The AD521 is a second generation, low cost, monolithic IC instrumentation amplifier developed by Analog Devices. As a true instrumentation amplifier, the AD521 is a gain block with differential inputs and an accurately programmable input/output gain relationship.

The AD521 IC instrumentation amplifier should not be confused with an operational amplifier, although several manufacturers (including Analog Devices) offer op amps which can be used as building blocks in variable gain instrumentation amplifier circuits. Op amps are general-purpose components which, when used with precision-matched external resistors, can perform the instrumentation amplifier function.

An instrumentation amplifier is a precision differential voltage gain device optimized for operation in a real world environment, and is intended to be used wherever acquisition of a useful signal is difficult. It is characterized by high input impedance, balanced differential inputs, low bias currents and high CMR.

As a complete instrumentation amplifier, the AD521 requires only two resistors to set its gain to any value between 0.1 and 1000. The ratio matching of these resistors does not affect the high CMRR (up to 120dB) or the high input impedance ($3 \times 10^9 \Omega$) of the AD521. Furthermore, unlike most operational amplifier-based instrumentation amplifiers, the inputs are protected against overvoltages up to ±15 volts beyond the supplies.

The AD521 IC instrumentation amplifier is available in four different versions of accuracy and operating temperature range. The economical "J" grade, the low drift "K" grade, and the lower drift, higher linearity "L" grade are specified from 0 to $+70°C$. The "S" grade guarantees performance to specification over the full MIL-temperature range: $-55°C$ to $+125°C$ and is available screened to MIL-STD-883, Class B. All versions are packaged in a hermetic 14-pin DIP.

PRODUCT HIGHLIGHTS

1. The AD521 is a true instrumentation amplifier in integrated circuit form, offering the user performance comparable to many modular instrumentation amplifiers at a fraction of the cost.

2. The AD521 has low guaranteed input offset voltage drift ($2\mu V/°C$ for L grade) and low noise for precision, high gain applications.

3. The AD521 is functionally complete with the addition of two resistors. Gain can be preset from 0.1 to more than 1000.

4. The AD521 is fully protected for input levels up to 15V beyond the supply voltages and 30V differential at the inputs.

5. Internally compensated for all gains, the AD521 also offers the user the provision for limiting bandwidth.

6. Offset nulling can be achieved with an optional trim pot.

7. The AD521 offers superior dynamic performance with a gain-bandwidth product of 40MHz, full peak response of 100kHz (independent of gain) and a settling time of $5\mu s$ to 0.1% of a 10V step.

Figure 3–16 AD521 data sheets. (From *Data-Acquisition Databook*, 1982, Vol. 1 Courtesy of Analog Devices, Inc., Two Technology Way, Norwood, MA.)

Sec. 3.3 Amplifiers

SPECIFICATIONS (typical @ $V_S = \pm15V$, $R_L = 2k\Omega$ and $T_A = 25°C$ unless otherwise specified)

MODEL	AD521JD	AD521KD	AD521LD	AD521SD (AD521SD/883B)
GAIN				
Range (For Specified Operation, Note 1)	1 to 1000	*	*	*
Equation	$G = R_S/R_G$ V/V	*	*	*
Error from Equation	$(\pm 0.25 - 0.004G)\%$	*	*	*
Nonlinearity (Note 2)				
$1 \leq G \leq 1000$	0.2% max	*	0.1% max	*
Gain Temperature Coefficient	$\pm(3 \pm 0.05G)$ ppm/°C	*	*	$\pm(15 \pm 0.4G)$ ppm/°C
OUTPUT CHARACTERISTICS				
Rated Output	±10V, ±10mA min	*	*	*
Output at Maximum Operating Temperature	±10V @ 5mA min	*	*	*
Impedance	0.1Ω	*	*	*
DYNAMIC RESPONSE				
Small Signal Bandwidth (±3dB)				
G = 1	>2MHz	*	*	*
G = 10	300kHz	*	*	*
G = 100	200kHz	*	*	*
G = 1000	40kHz	*	*	*
Small Signal, ±1.0% Flatness				
G = 1	75kHz	*	*	*
G = 10	26kHz	*	*	*
G = 100	24kHz	*	*	*
G = 1000	6kHz	*	*	*
Full Peak Response (Note 3)	100kHz	*	*	*
Slew Rate, $1 \leq G \leq 1000$	10V/μs	*	*	*
Settling Time (any 10V step to within 10mV of Final Value)				
G = 1	7μs	*	*	*
G = 10	5μs	*	*	*
G = 100	10μs	*	*	*
G = 1000	35μs	*	*	*
Differential Overload Recovery (±30V Input to within 10mV of Final Value) (Note 4)				
G = 1000	50μs	*	*	*
Common Mode Step Recovery (30V Input to within 10mV of Final Value) (Note 5)				
G = 1000	10μs	*	*	*
VOLTAGE OFFSET (may be nulled)				
Input Offset Voltage (V_{OS_I})	3mV max (2mV typ)	1.5mV max (0.5mV typ)	1.0mV max (0.5mV typ)	**
vs. Temperature	15μV/°C max (7μV/°C typ)	5μV/°C max (1.5μV/°C typ)	2μV/°C max	**
vs. Supply	3μV/%	*	*	**
Output Offset Voltage (V_{OS_O})	400mV max (200mV typ)	200mV max (30mV typ)	100mV max	**
vs. Temperature	400μV/°C max (150μV/°C typ)	150μV/°C max (50μV/°C typ)	75μV/°C max	**
vs. Supply (Note 6)	$0.005 V_{OS_O}/\%$	*	*	**
INPUT CURRENTS				
Input Bias Current (either input)	80nA max	40nA max	**	**
vs. Temperature	1nA/°C max	500pA/°C max	**	**
vs. Supply	2%/V	*	**	**
Input Offset Current	20nA max	10nA max	**	**
vs. Temperature	250pA/°C max	125pA/°C max	**	**
INPUT				
Differential Input Impedance (Note 7)	$3 \times 10^9 \Omega \| 1.8pF$	*	*	*
Common Mode Input Impedance (Note 8)	$6 \times 10^{10} \Omega \| 3.0pF$	*	*	*
Input Voltage Range for Specified Performance (with respect to ground)	±10V	*	*	*
Maximum Voltage without Damage to Unit, Power ON or OFF Differential Mode (Note 9)	30V	*	*	*
Voltage at either input (Note 9)	$V_S \pm 15V$	*	*	*
Common Mode Rejection Ratio, DC to 60Hz with 1kΩ source unbalance				
G = 1	70dB min (74dB typ)	74dB min (80dB typ)	**	**
G = 10	90dB min (94dB typ)	94dB min (100dB typ)	**	**
G = 100	100dB min (104dB typ)	104dB min (114dB typ)	**	**
G = 1000	100dB min (110dB typ)	110dB min (120dB typ)	**	**
NOISE				
Voltage RTO (p-p) @ 0.1Hz to 10Hz (Note 10)	$\sqrt{(0.5G)^2 + (225)^2}$ μV	*	*	*
RMS RTO, 10Hz to 10kHz	$\sqrt{(1.2G)^2 + (50)^2}$ μV	*	*	*
Input Current, rms, 10Hz to 10kHz	15pA (rms)	*	*	*
REFERENCE TERMINAL				
Bias Current	3μA	*	*	*
Input Resistance	10MΩ	*	*	*
Voltage Range	±10V	*	*	*
Gain to Output	1	*	*	*
POWER SUPPLY				
Operating Voltage Range	±5V to ±18V	*	*	*
Quiescent Supply Current	5mA max	*	*	*
TEMPERATURE RANGE				
Specified Performance	0 to +70°C	*	*	-55°C to +125°C
Operating	-25°C to +85°C	*	*	-55°C to +125°C
Storage	-65°C to +150°C	*	*	*
PACKAGE OPTION:[1] TO-116 Style (D14A)	AD521JD	AD521KD	AD521LD	AD521SD

[1] See Section 20 for package outline information.

*Specifications same as AD521JD.
**Specifications same as AD521KD.
Specifications subject to change without notice.

Figure 3–16 (cont.)

Applying the AD521

NOTES:

1. Gains below 1 and above 1000 are realized by simply adjusting the gain setting resistors. For best results, voltage at either input should be restricted to ±10V for gains equal to or less than 1.

2. Nonlinearity is defined as the ratio of the deviation from the "best straight line" through a full scale output range of ±9 volts. With a combination of high gain and ±10 volt output swing, distortion may increase to as much as 0.3%.

3. Full Peak Response is the frequency below which a typical amplifier will produce full output swing.

4. Differential Overload Recovery is the time it takes the amplifier to recover from a pulsed 30V differential input with 15V of common mode voltage, to within 10mV of final value. The test input is a 30V, 10μs pulse at a 1kHz rate. (When a differential signal of greater than 11V is applied between the inputs, transistor clamps are activated which drop the excess input voltage across internal input resistors. If a continuous overload is maintained, power dissipated in these resistors causes temperature gradients and a corresponding change in offset voltage, as well as added thermal time constant, but will not damage the device.)

5. Common Mode Step Recovery is the time it takes the amplifier to recover from a 30V common mode input with zero volts of differential signal to within 10mV of final value. The test input is 30V, 10μs pulse at a 1kHz rate. (When a common mode signal greater than $V_S - 0.5V$ is applied to the inputs, transistor clamps are activated which drop the excessive input voltage across internal input resistors. Power dissipated in these resistors causes temperature gradients and a corresponding change in offset voltage, as well as an added thermal time constant, but will not damage the device.)

6. Output Offset Voltage versus Power Supply Change is a constant 0.005 times the unnulled output offset per percent change in either power supply. If the output offset is nulled, the output offset change versus supply change is substantially reduced.

7. Differential Input Impedance is the impedance between the two inputs.

8. Common Mode Input Impedance is the impedance from *either* input to the power supplies.

9. Maximum Input Voltage (differential or at either input) is 30V when using ±15V supplies. A more general specification is that neither input may exceed either supply (even when $V_S = 0$) by more than 15V and that the difference between the two inputs must not exceed 30V. (See also Notes 4 and 5.)

10. 0.1Hz to 10Hz Peak-to-Peak Voltage Noise is defined as the maximum peak-to-peak voltage noise observed during 2 of 3 separate 10 second periods with the test circuit of Figure 10.

DESIGN PRINCIPLE

Figure 1 is a simplified schematic of the AD521. A differential input voltage, V_{IN}, appears across R_G causing an imbalance in the currents through Q_1 and Q_2, $\Delta I = V_{IN}/R_G$. That imbalance is forced to flow in R_S because the collector currents of Q_3 and Q_4 are constrained to be equal by their biasing (current mirror). These conditions can only be satisfied if the differential voltage across R_S (and hence the output voltage of the AD521) is equal to $\Delta I \times R_S$. The feedback amplifier, A_{FB}, performs that function. Therefore, $V_{OUT} = \dfrac{V_{IN}}{R_G} \times R_S$ or $\dfrac{V_{OUT}}{V_{IN}} = \dfrac{R_S}{R_G}$.

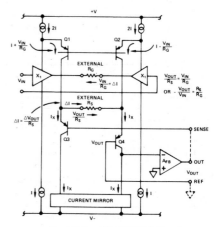

Figure 1. Simplified AD521 Schematic

Figure 3–16 (*cont.*)

Sec. 3.3 Amplifiers

APPLICATION NOTES FOR THE AD521
These notes ensure the AD521 will achieve the high level of performance necessary for many diversified IA applications.

1. Gains below 1 and above 1000 are realized by adjusting the gain setting resistors as shown in Figure 2 (the resistor, R_S between pins 10 and 13 should remain 100kΩ ±15%, see application note 3). For best results, the input voltage should be restricted to ±10V especially for gain equal to or less than 1.

2. Provide a return path to ground for input bias currents. The AD521 is an instrumentation amplifier, not an isolation amplifier. When using a thermocouple or other "floating" source, this return path may be provided directly to ground or indirectly through a resistor to ground from pins 1 and/ or 3, as shown in Figure 3. If the return path is not provided, bias currents will cause the output to saturate. The value of the resistor may be determined by dividing the maximum allowable common mode voltage for the application by the bias current of the instrumentation amplifier.

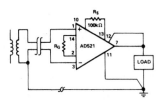

a). Transformer Coupled, Direct Return

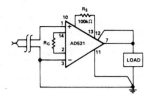

b). Thermocouple, Direct Return

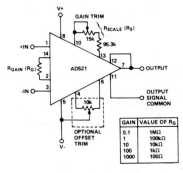

Figure 2. Operating Connections for AD521

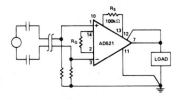

c). AC Coupled, Indirect Return

Figure 3. Ground Returns for "Floating" Transducers

3. The resistors between pins 10 and 13, (R_{SCALE}) must equal 100kΩ ±15% (Figure 2). If R_{SCALE} is too low (below 85kΩ) the output swing of the AD521 is reduced. At values below 80kΩ and above 120kΩ the stability of the AD521 may be impaired.

4. Do not exceed the allowable input signal range. The linearity of the AD521 decreases if the inputs are driven within 5 volts of the supply rails, particularly when the device is used at a gain less than 1. To avoid this possibility, attenuate the input signal through a resistive divider network and use the AD521 as a buffer, as shown in Figure 4. The resistor R/2 matches the impedance seen by both AD521 inputs so that the voltage offset caused by bias currents will be minimized.

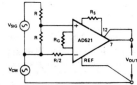

Figure 4. Operating Conditions for $V_{IN} \approx V_S = 10V$

Figure 3–16 (*cont.*)

5. Use the compensation pin (pin 9) and the applicable compensation circuit when the amplifier is required to drive a capacitive load. It is worth mentioning that coaxial cables can "invisibly" provide such capacitance since many popular coaxial cables display capacitance in the vicinity of 30pF per foot.

This compensation (bandwidth control) feature permits the user to fit the response of the AD521 to the particular application as illustrated by Figure 5. In cases of extremely high load capacitance the compensation circuit may be changed as follows:

1. Reduce 680Ω to 24Ω
2. Reduce 330Ω to 7.5Ω
3. Increase 1000pF to 0.1μF
4. Set C_X to 1000pF if no compensation was originally used. Otherwise, do not alter the original value.

This allows stable operation for load capacitances up to 3000pF, but limits the slew rate to approximately 0.16V/μs.

6. Signals having frequency components above the Instrumentation Amplifier's output amplifier closed-loop bandwidth will be transmitted from V- to the output with little or no attenuation. Therefore, it is advisable to decouple the V- supply line to the output common or to pin 11.[1]

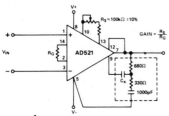

$$C_X = \frac{1}{100\pi f_t} \text{ when } f_t \text{ is the desired bandwidth.}$$

(f_t in kHz, C_X in μF)

Figure 5. Optional Compensation Circuit

INPUT OFFSET AND OUTPUT OFFSET

When specifying offsets and other errors in an operational amplifier, it is often convenient to refer these errors to the inputs. This enables the user to calculate the maximum error he would see at the output with any gain or circuit configuration. An op amp with 1mV of input offset voltage, for example, would produce 1V of offset at the output in a gain of 1000 configuration.

In the case of an instrumentation amplifier, where the gain is controlled in the amplifier, it is more convenient to separate errors into two categories. Those errors which simply add to the output signal and are unaffected by the gain can be classified as output errors. Those which act as if they are associated with the input signal, such that their effect at the output is proportional to the gain, can be classified as input errors.

As an illustration, a typical AD521 might have a +30mV output offset and a -0.7mV input offset. In a unity gain configuration, the *total* output offset would be +29.3mV or the sum of the two. At a gain of 100, the output offset would be -40mV or: 30mV + 100(-0.7mV) = -40mV.

By separating these errors, one can evaluate the total error independent of the gain settings used, similar to the situation with the input offset specifications on an op amp. In a given gain configuration, both errors can be combined to give a total error *referred to the input* (R.T.I.) or *output* (R.T.O.) by the following formula:

Total Error R.T.I. = input error + (output error/gain)

Total Error R.T.O. = (Gain x input error) + output error

The offset trim adjustment (pins 4 and 6, Figure 2) is associated primarily with the output offset. At any gain it can be used to introduce an output offset equal and opposite to the input offset voltage multiplied by the gain. As a result, the total output offset can be reduced to zero.

As shown in Figure 6, the gain range on the AD521 can be extended considerably by adding an attenuator in the sense terminal feedback path (as well as adjusting the ratio, R_S/R_G). Since the sense terminal is the inverting input to the output amplifier, the additional gain to the output is controlled by R_1 and R_2. This gain factor is $1 + R_2/R_1$.

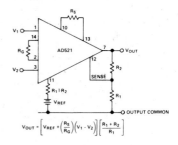

Figure 6. Circuit for utilizing some of the unique features of the AD521. Note that gain changes introduced by changing R1 and R2 will have a minimum effect on output offset if the offset is carefully nulled at the highest gain setting.

---

[1] For further details, refer to "An I.C. User's Guide to Decoupling, Grounding, and Making Things Go Right for a Change," by A. Paul Brokaw. This application note is available from Analog Devices without charge upon request.

Figure 3-16 (cont.)

Where offset errors are critical, a resistor equal to the parallel combination of R_1 and R_2 should be placed between pin 11 and V_{REF}. This minimizes the offset errors resulting from the input current flowing in R_1 and R_2 at the sense terminal. Note that gain changes introduced by changing the R_1/R_2 attenuator will have a minimum effect on output offset if the offset is carefully nulled at the highest gain setting.

When a predetermined output offset is desired, V_{REF} can be placed in series with pin 11. This offset is then multiplied by the gain factor $1 + R_2/R_1$ as shown in the equation of Figure 6.

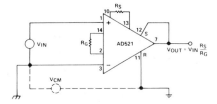

Figure 7. Ground loop elimination. The reference input, Pin 11, allows remote referencing of ground potential. Differences in ground potentials are attenuated by the high CMRR of the AD521.

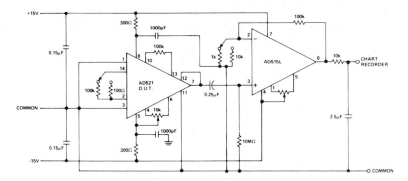

Figure 8. Test circuit for measuring peak to peak noise in the bandwidth 0.1Hz to 10Hz. Typical measurements are found by reading the maximum peak to peak voltage noise of the device under test (D.U.T.) for 3 observation periods of 10 seconds each.

Figure 3–16 (*cont.*)

As the first stage can provide most of the gain, the second-stage resistors need not be too large and will be easier to match. As the common mode gain is relatively constant, the CMRR will increase with gain.

Commercial IAs. Due to its excellent amplifier characteristics, several manufacturors have fabricated the circuit above in a single package. The gain is set by an external resistor and can be varied between 1 and 1000, with some units providing gain as high as 10,000. Input impedances as high as 10 MΩ are available together with CMRRs of 110 dB. Typical examples are the Burr-Brown 36xx series of IAs and the AD521, 522, and 524 from Analog Devices, among many. Data sheets of the AD521 are presented in Figure 3–16.

3.4 TRACK-AND-HOLD CIRCUITS

To convert the analog signal to digital form, the ADC requires a finite amount of time. As the signal at the ADC input must remain constant during the conversion, a circuit is required to (1) acquire the analog signal level when

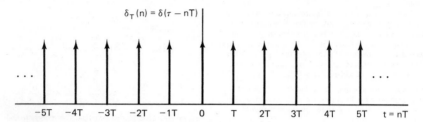

Figure 3–17 Train of impulse signals.

the conversion is to take place, and (2) maintain that level until the conversion is completed. This is accomplished by the track-and-hold (TH) circuit. The ideal tracking process occurs instantaneously and at a precise rate. This allows us to consider it as the equivalent of multiplying the analog signal to be sampled, $s(t)$, by a train of impulse functions.

3.4.1 Ideal Sampling

The ideal hold command acts like a train of impulses, $\delta(t - nT)$, where n is an integer and T is the sampling rate. This is shown in Figure 3–17. Unlike the practical pulse train, to be discussed below, the FT of the impulse train is itself a discrete sequence of components due to the periodicy of the input. In addition, the amplitude of all the frequency components is the same, regardless of frequency, due to the zero pulse width. This is shown in Figure 3–18. The sequence of pulses in the time domain is represented by Equation 3–13a.

$$\sum_{n=\infty}^{\infty} \delta(t - nT) \equiv \delta_T(t) \qquad (3\text{–}13a)$$

while the sequence of pulses in the frequency domain is represented by Equation 3–13b.

$$\frac{1}{T} \sum_{n=-\infty}^{\infty} \delta(f - kf_{Sa}) \equiv \delta_{fSa}(f) \qquad (3\text{–}13b)$$

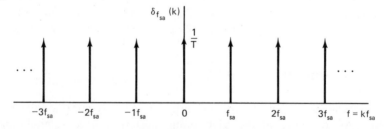

Figure 3–18 FT of $\sigma_T(t)$.

Sec. 3.4 Track-and-Hold Circuits

f_{sa} is the spacing between the frequency domain pulses and the reciprocal of T, the spacing between the sampling (time-domain) pulses.

Sampled output. If $s(t)$ is the analog signal to be sampled, the output is the product of $s(t)$ and $\delta_T(n)$

$$r(t) = s(t)\delta_T(n) = s(t) \sum_{n=-\infty}^{\infty} \delta(t - nT)$$
$$= \sum_{n=-\infty}^{\infty} s(t) \delta(t - nT) \qquad (3\text{-}14)$$

As the terms of the sum are zero except for $t = nT$, we have (see Equation 1-20)

$$r(t) = \sum_{n=-\infty}^{\infty} s(nT) \delta(t - nT) \qquad (3\text{-}15)$$

This implies that the output of the ideal sampling circuit is a sequence of numbers whose values are the values of $s(t)$ evaluated at $t = nT$.

$$s(nT) = r(t = nT) = r(nT) \qquad \text{for } t = nT \qquad (3\text{-}16)$$

The question that arises is: How does the information contained in the sampled signal compare with that of the original? The answer is that the sampled signal contains the same information as the original signal with some additional frequency content (i.e., noise). If the sampling is performed at a high-enough rate, the unwanted frequencies can be removed by filtering. It is this fact that allows us to use DSP: There is no loss of information when a signal is properly sampled.

Sampling rate. To see the effect of ideal sampling on a signal, the FT of a sampled signal (Equation 3-14) is obtained.

$$FT\{r(t)\} = FT\left\{ \sum_{n=-\infty}^{\infty} s(t)\delta(t - nT) \right\}$$
$$= \sum_{n-\infty}^{\infty} FT\{s(t)\delta(t - nT)\} = R(f) \qquad (3\text{-}17)$$

From Table 1-1, property 8, the FT of a product is the convolution of the FTs of each term in the product:

$$FT\{s(t)\delta(t - nT)\} = S(f) * \delta_{f_{sa}}(k) = R(f) \qquad (3\text{-}18a)$$

where

$$S(f) * \delta_{f_{sa}}(k) = \int_{-\infty}^{\infty} \frac{1}{T} \sum_{k=-\infty}^{\infty} \delta(\beta - kf_{sa}) S(f - \beta) \, d\beta \qquad (3\text{-}18b)$$

where we have made use of Equations 3-13b. In effect, this convolution will slide the spectrum of $s(t)$ past each of the frequency components of Figure

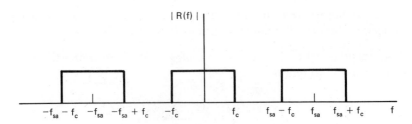

Figure 3–19 Frequency spectrum of TH output.

3–18. Using the properties of the delta function discussed in Chapter 1, we have the very important result that

$$R(f) = S(f) * \delta_{f_{sa}}(k) = \frac{1}{T} \sum_{k=-\infty}^{\infty} S(f - kf_{sa}) \qquad (3\text{-}18c)$$

The result of ideal sampling is the replication of the spectrum of $s(t)$ for each value of nf_{sa} (multiples of the sampling frequency) with constant amplitude. This is shown in Figure 3–19 where $s(t)$ is assumed to be band-limited to f_c and f_{sa} is greater than $2f_c$. This is an example of *Nyquist's theorem*, which states that for effective sampling, the sampling rate must be greater than twice the highest-frequency component of the input signal.

Example 3–1

A signal, $s(t)$, has an input spectrum with a maximum frequency of 15 kHz. What is the minimum sampling frequency?

Solution Using Nyquist's theorem, the sampling frequency should be 30 kHz or higher.

As can be seen, the spectrum associated with adjacent values of nf_{sa} will overlap if f_{sa} is less than $2f_c$. This is termed *aliasing*. Its effect is to change the relative amplitude of $s(t)$'s frequency content at the output of the TM. This is shown in Figure 3–20 which should be compared with Figure 3–19. To reduce the effect of aliasing, sampling must be at the Nyquist rate or, preferably, greater. To provide a "guardband" between adjacent spectrum segments,

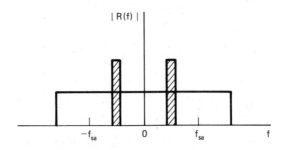

Figure 3–20 Effect of sampling at too low a frequency.

the practical sampling rate is usually three to four times the highest-frequency component of the input signal. This will reduce the roll-off requirements of a LP filter used to remove the unwanted frequency components located at nf_{sa}.

From the discrete-time domain, the effect of ideal sampling is to generate a sequence of discrete numbers at the sampler output $\{r(nT) = s(t = nT)\}$. At this point, we can see that any system used to process the output of the sampler must also operate in the discrete-time domain, even though the sampler output is a continuous range of voltages.

3.4.2 Basic TH Circuit

Figure 3–21 shows the basic circuit of a TH unit. The FET acts as a switch that is turned off upon receipt of the hold command, effectively isolating the capacitor from the input signal, $s(t)$. Due to the high OFF impedance of the FET and the input impedance of the unity-gain buffer amplifier, the capacitor will hold the voltage across it at a relatively constant value. The voltage will then appear at the output of the buffer amplifier and serve as the constant signal to the ADC.

When the FET is turned ON by the track command, its low on resistance from source to drain allows the capacitor to charge or discharge rapidly to the level of $s(t)$, and follow its variations from there to the next hold command. This is shown in Figure 3–22. During the tracking period, the buffer output (ADC input) is following $s(t)$. The ADC is deactivated during this time, and the input has no effect on its output. The ADC will require a "start conversion" signal to begin. To increase the impedance seen by $s(t)$, an input buffer stage is added. This is shown in Figure 3–23.

As with the circuit of Figure 3–20, the capacitor voltage will follow the level of $s(t)$ until a hold command opens the switch. In most TH devices, the hold command is a TTL-compatible signal as is the "data ready" signal. Data sheets for the AD346 SH device are presented in Figure 3–24.

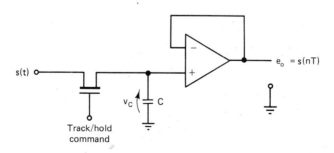

Figure 3–21 Track-and-hold circuit.

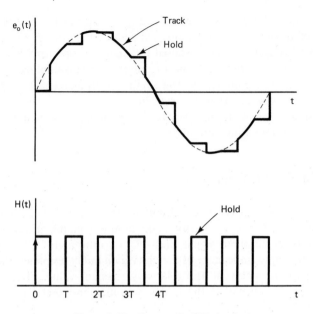

Figure 3-22 Output of a TH circuit.

3.4.3 TH Terminology

Like all electronic devices, there is a list of terms that describes the device characteristics. These provide an indication of how much the results obtained with the practical circuit will differ from the ideal, causing noise to be introduced in the system output.

Aperture time. The interval between the application of the hold command and the switch opening is termed the *aperture time*. As the signal is usually changing during this interval, the actual value of $s(t)$ sampled is not

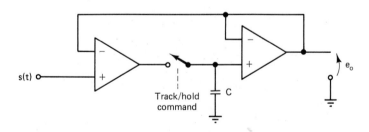

Figure 3-23 Buffer input TH circuit.

Sec. 3.4   Track-and-Hold Circuits

High Speed
Sample and Hold Amplifier
AD346

FEATURES
Fast 2.0μs Acquisition Time to ±0.01%
Low Droop Rate: 0.5mV/ms
Low Offset
Low Glitch: <40mV
Aperture Jitter: 400ps
Military Temperature Range: −55°C to +125°C
Internal Hold Capacitor

AD346 FUNCTIONAL BLOCK DIAGRAM

PRODUCT DESCRIPTION
The AD346 is a high speed (2μs to 0.01%), adjustment free sample and hold amplifier designed for high throughput rate data acquisition applications. The fast acquisition time (2μs to 0.01%) and low aperture jitter (400ps) make it suitable for use with fast A/D converters to digitize signals up to 97kHz.

The AD346 is complete with an internal hold capacitor and it incorporates a compensation network which minimizes the sample to hold charge offset. The AD346 is also laser trimmed to eliminate the need for external trimming potentiometers.

Typical applications for the AD346 include sampled data systems, D/A deglitchers, peak hold functions, strobed measurement systems and simultaneous sampling converter systems.

The device is available in two versions: the "J" specified for operation over the 0 to +70°C commercial temperature range and the "S" specified over the full military temperature range, −55°C to +125°C with processing to MIL-STD-883, Level B screening available.

PRODUCT HIGHLIGHTS
1. The AD346 is an improved second source for other sample and holds of the same pin configuration.
2. The AD346 provides separate analog and digital grounds, thus improving the device's immunity to ground and switching transients.
3. The droop rate is only 0.5mV/ms so that it may be used in slower high accuracy systems without the loss of accuracy.
4. The fast acquisition time and low aperture make it suitable for very high speed data acquisition systems.

PIN CONFIGURATIONS

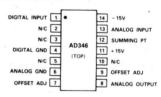

PIN OUT

Figure 3–24 AD346 SH data sheets. (From *Data-Acquisition Databook*, 1982, Vol. 1. Courtesy of Analog Devices, Inc., Two Technology Way, Norwood, MA.)

SPECIFICATIONS (typical @ +25°C, $V_S = \pm 15V$ unless otherwise noted)

Model	AD346JD	AD346SD	Units
ANALOG INPUT			
Voltage Range	±10.0	*	Volts
Input Impedance	3.0	*	kΩ
DIGITAL INPUT			
"0" Input Threshold Voltage (Hold)	+0.8 max	*	Volts
"1" Input Threshold Voltage (Sample)	2.0 min	*	Volts
"0" Input Current	50.0	*	μA
"1" Input Current	1.0	*	μA
TRANSFER CHARACTERISTICS			
Gain	−1.0	*	
Gain Error	±0.02 max (±0.01 typ)	*	% FSR
Gain Error, $T_{min} - T_{max}$	±0.05 max (±0.03 typ)	*	% FSR
Offset Voltage	±3 max (±1 typ)	*	mV
Offset Voltage, $T_{min} - T_{max}$	±20 max (±6 typ)	*	mV
Pedestal	±4 max (±2 typ)	*	mV
Pedestal, $T_{min} - T_{max}$	±20 max (±8 typ)	±20 max (±10 typ)	mV
Droop Rate	0.5 max (0.1 typ)	*	mV/μs
Droop Rate, $T_{min} - T_{max}$	60 max (20 typ)	700 max (200 typ)	mV/μs
DYNAMIC CHARACTERISTICS			
Full Power Bandwidth			
$V_{OUT} = +10V, -3dB$	1.4	*	MHz
Output Slew Rate	50	*	V/μs
Acquisition Time			
To ±0.01% 10V Step	2.0 max (1.0 typ)	*	μs
To ±0.01% 20V Step	2.5 max (1.6 typ)	*	μs
Aperture Delay	60 max (30 typ)	*	ns
Settling Time			
Sample Mode (10V Step)	2.0 max (1.0 typ)	*	μs
Sample to Hold	500 max (150 typ)	*	ns
Feedthrough (Hold Mode)			
at 1kHz	0.02 max (0.005 typ)	*	% FSR
Transient Peak Amplitude			
Sample/Hold/Sample	40	*	mV
ANALOG OUTPUT			
Output Voltage Swing[1]	±10.0 min	*	Volts
Output Current	3.0	*	mA
POWER REQUIREMENTS			
Operating Voltage Range	±12 to ±18	*	Volts
Supply Current			
+V	18 max (9 typ)	*	mA
−V	−10 max (−3 typ)	*	mA
Power Supply Rejection Ratio	100	*	μV/V
Power Consumption	500 max (200 typ)	*	mW
PACKAGE OPTION[2]			
14 Pin DIP	HY14C	*	

[1] Maximum output swing is 4V less than $+V_S$.
[2] See Section 20 for package outline information.
*Specifications same as AD346JD.
Specifications subject to change without notice.

Figure 3–24 (*cont.*)

Sec. 3.4   Track-and-Hold Circuits   93

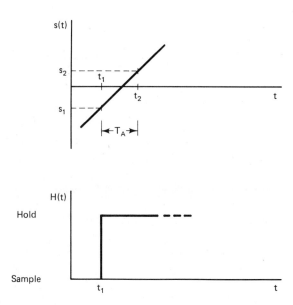

Figure 3–25 Effect of aperture time on TH output.

the value when the hold command was issued. This is shown in Figure 3–25. t_1 is the time when the hold command is received by the TH (corresponding to nT seconds), and t_2 is the time at which the switch is fully opened. As a result of the delay, s_2 is the value of $s(t)$ that will be fed to the ADC, not s_1. This will introduce some error (noise) into the final output if $s(t)$ is rapidly changing at this time. This can be reduced by advancing the hold command by an amount equal to the aperture time, allowing s_1 to be sent to the ADC. Depending on the unit used, the aperture time can be between 25 and 150 ns.

Aperture jitter. The time at which the switch actually closes is not fixed, but has a distribution about a given time. This causes a sample-to-sample variation in the aperture time. As the variations are random, it appears as a noise at the system output. Figure 3–26 shows the effect of *aperture jitter*.

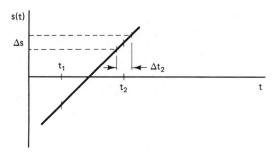

Figure 3–26 Effect of aperture jitter.

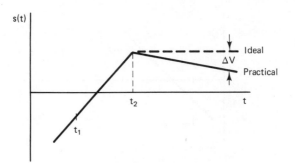

Figure 3-27 Effect of droop on TH output.

Droop. Once in the hold mode, the voltage of the practical capacitor will change due to leakage currents. These currents are caused by the FET leakage (OFF) resistance and by the (very low) bias current of the buffer amplifier. The effect of *droop* is shown in Figure 3-27. Droop is usually given in units of mV/μs.

Feedthrough. If $s(t)$ is changing, stray capacitance in the system connected to the main capacitor can couple the input signal to it, causing its voltage to change while in the hold mode. *Feedthrough* is measured as a percent of the input signal and shown in Figure 3-28.

Acquisition time. When the conversion of the analog signal to digital form is completed, the TH circuit must acquire and follow the signal in order to be ready for the next hold command. The interval between issuance of the track command and the settling of the output to the present value of $s(t)$ is called the *acquisition time*. This interval consists of the hold-and-track delay, the slew rate limitations of the buffer op amp (i.e., how fast the output voltage can change), and the settling time. This is presented in Figure 3-29. t_3 is the time the track command is issued, and t_4 is the time at which the output equals $s(t)$.

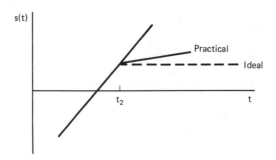

Figure 3-28 Effect of feedthrough on TH output.

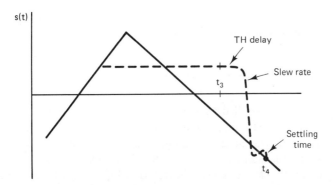

Figure 3–29 TH acquisition time.

3.4.4 Sampling Noise

If a signal is sampled at a rate well above the Nyquist rate, it can usually be recovered from its samples. As stated above, ideal sampling is equivalent to multiplying the analog signal by a train of impulses. Any deviation from the ideal serves to introduce extraneous frequency components (i.e., noise) in the output. Section 3.4.2 presented the characteristics of the sampling circuit that resulted in a deviation from ideal sampling. In this section we will look at variations in the sampling signal that introduce noise.

Aliasing. It was demonstrated that the sampling frequency must be at least twice the highest-frequency component of the analog signal to enable the signal to be recovered with a minimum of distortion. If this rule is not followed, the frequency content of the sampled signal will be altered, introducing erroneous information into the recovered signal. The introduction of noise into the output due to sampling at less than required rates is termed *aliasing*. Due to the practical nature of filters (finite gain roll-off) and signals with components that reach into the high-frequency ranges, aliasing noise will be present in most outputs; however, practical sampling rates of at least three to four times the highest significant frequency component will reduce this noise.

Sampling rate variation. Ideal sampling requires the sampling rate to be stable. Variations in this rate will result in the introduction of noise. The noise due to sample rate instability can be reduced by the use of a crystal oscillator to generate the hold command.

Nonzero pulse width. As the hold time must be long enough to allow the ADC to perform the conversion to digital, the sampling pulse width, of necessity, is not an impulse (i.e., it does not have zero width). This will introduce noise in two ways: (1) the signal at the TH ouput can change due to droop or feedthrough, causing the digital output to be in error if the change is large enough; and (2) variations in the pulse width from sample to sample act as noise and will appear in the output.

3.5 PRACTICAL SAMPLING THEORY

If an analog signal is applied to a track-and-hold (TH) circuit, the output is a sequence of pulses. The amplitude of the pulses is equal to the value of the analog signal when the hold command is applied. This is shown in Figure 3–30. The hold command is issued every T seconds or at a frequency $f_{sa} = (1/T)$. We would like to know the relationship between the analog signal, $f(t)$, and the sampled signal at the output of the TH, $f(mT)$].

3.5.1 Track-and-Hold Output Characteristics

The output of the TH is a periodic sequence of pulses with varying amplitude whose frequency is that of the hold command. As the ADC, which receives the output, requires a minimum time to perform the conversion, the pulse will have a minimum width. During the conversion interval, the amplitude of the pulse may vary due to the variation of the input signal, introducing an error in the output.

The effect of the TH circuit is to provide a signal at its output that is the product of the input signal, $s(t)$, and a train of pulses with unit height, a width equal to or greater than the minimum conversion time, and a rate equal

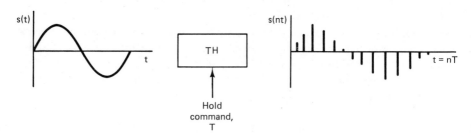

Figure 3–30 Representation of TH circuit effect on an input signal.

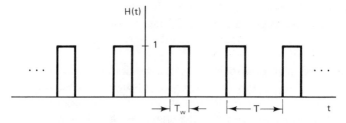

Figure 3–31 Representation of the pulse train used to sample $s(t)$.

Sec. 3.5 Practical Sampling Theory

to that of the hold command, which must be sufficiently high. Figure 3–31 shows the representation of the train of pulses.

As the output is the product of two signals, we can state that sampling acts as a multiplication process in which two signals are multiplied to produce frequency shifts. This is shown in Example 3–2.

Example 3–2

Two sine waves are applied to a multiplier circuit. What is the output if $v_1(t) = 3 \sin [2\pi(1000 \text{ Hz})t]$ and $v_2(t) = 4 \sin [2\pi(4000 \text{ Hz})t]$?

Solution Figure 3–32 shows the representation of a multiplicative modulation process.

$$v(t) = v_1(t)v_2(t) = 3 \sin [2\pi(1000 \text{ Hz})t] \cdot 4 \sin [2\pi(4000 \text{ Hz})t]$$

Using trigonometric identities, we obtain

$$\sin A \sin B = \frac{1}{2}[\cos (A - B) - \cos (A + B)] \qquad (3\text{-}19a)$$

$$V(t) = \frac{3 \cdot 4}{2} \{\cos [2\pi t(1 \text{ kHz} - 4 \text{ kHz})] - \cos [2\pi t(1 \text{ kHz} + 4 \text{ kHz})]\} \qquad (3\text{-}19b)$$

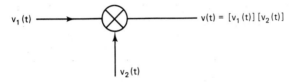

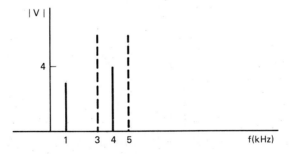

Figure 3–32 Multiplication of two time varying signals.

The output is a complex wave consisting only of components whose frequencies are the sum and difference of the input frequencies. This is shown in Figure 3–33, where the dashed lines are the sum and difference frequencies of the input signals. If either or both inputs were complex (i.e., composed of more than one sine wave), then each frequency component of $v_1(t)$ would combine to yield the sum–difference terms with each frequency component of $v_2(t)$.

Figure 3–33 Output components of a multiplier circuit.

To see what relationship exists between the input signal and the sampled output, we must determine the frequency components of the sample pulses of Figure 3–31.

3.5.2 Frequency Content of a Pulse Train

As the pulse train in Figure 3–31 is periodic, it can be represented by a Fourier series.

$$H(t) = \sum_{n=-\infty}^{\infty} C_n \exp(j2\pi n f_{sa} t) \qquad (3\text{-}20)$$

where $nf_{sa} = n/T$. The C_n are obtained using Equation 1–50d and $H(t)$.

$$C_m = \frac{1}{T} \int_{-T/2}^{T/2} H(t) \exp(-j2\pi m f_{sa} t) \, dt \qquad (3\text{-}21)$$

Since $H(t) = 1$ during the hold command, zero otherwise, the resulting integral is identical to that of Equation 1–56; that is, the amplitude of C_m will vary as the sampling function for each value of m.

$$C_m = \frac{1}{T} \int_{-T_w/2}^{+T_w/2} 1 \exp(-j2\pi m f_{sa} t) \, dt \qquad (3\text{-}22a)$$

$$= \frac{-1}{Tj2\pi m f_{sa}} \exp(-j2\pi m f_{sa} t) \Big|_{-T_w/2}^{+T_w/2} \qquad (3\text{-}22b)$$

$$= \frac{1}{T\pi m f_{sa}} \sin(\pi m f_{sa} T_w) \qquad (3\text{-}22c)$$

$$= \frac{T_w}{T} \frac{\sin(\pi m f_{sa} T_w)}{\pi m f_{sa} T_w} = \frac{T_w}{T} \operatorname{Sa}(\pi m f_{sa} T_w) \qquad (3\text{-}22d)$$

where T_w is the sampling period width. The value of $|C_m|$ is shown in Figure 3–34. The spacing between frequency components is equal to the sampling frequency, f_{sa}.

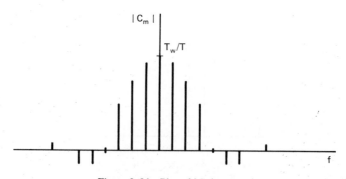

Figure 3–34 Plot of $|C_m|$ versus f.

Example 3-3

What are the values of the first three components of a sampling pulse, 1 V in amplitude, with $f_{sa} = 2$ kHz, and $T_w = 50$ μs?

Solution Using Equation 3–22d, we have

$$m = 0: \quad C_0 = \frac{T_w}{T} = \frac{50 \ \mu s}{500 \ \mu s} = 0.1$$

$$m = 1: \quad C_1 = \frac{T_w}{T_1} \frac{\sin\left[\pi(1)(2 \text{ kHz})(50 \ \mu s)\right]}{\pi(1)(2 \text{ kHz})(50 \ \mu s)}$$

$$= \frac{50 \ \mu s}{500 \ \mu s} \frac{\sin 0.314}{0.314} = 0.098$$

$$m = 2: \quad C_2 = \frac{T_w}{T} \frac{\sin\left[\pi(2)(2 \text{ kHz})(50 \ \mu s)\right]}{\pi(2)(2 \text{ kHz})(50 \ \mu s)}$$

$$= \frac{50 \ \mu s}{500 \ \mu s} \frac{\sin 0.628}{0.628} = 0.094$$

From Figure 3–34 we see that the pulse train contains an infinite number of (discrete) frequencies whose amplitudes follow the sampling function curve. This implies that sampling an analog signal with a finite-width pulse train will generate many frequencies at the output as each frequency component of the sampling signal will produce a sum–difference pair for each component of the input.

To see the possible effect that practical sampling might have on an input signal, $s(t)$, let us consider the outcome of multiplying a single-frequency sine wave with the train of pulses shown in Figure 3–31. This is represented in Figure 3–35. As each component of $H(t)$ will produce a sum–difference pair, we will obtain components with frequency $nf_{sa} - f$ and $nf_{sa} + f$. This is shown in Figure 3–36, where we have assumed that f_{sa} is sufficiently larger than f. The lines on each side of the sampling component value are the sum–difference pairs; the component at nf_{sa} is eliminated as a result of the multiplication. Note that each component of the output is multiplied by a corresponding value of C_n.

If we pass the sampled signal through an ideal LP filter, whose cutoff lies between f and $f_{sa} - f$, we can obtain the original signal, as shown in Figure 3–37. The original signal can be obtained from its sampled values; however, the ideal LP filter is not realizable as it is not causal (i.e., there is

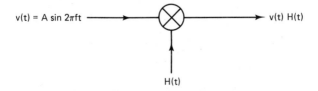

Figure 3–35 Multiplication of the signal with a train of pulses.

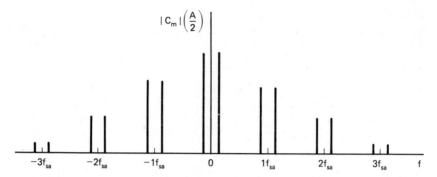

Figure 3-36 Output of the TH circuit with sine wave input and pulse train as hold command.

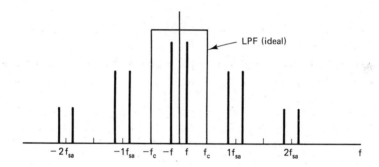

Figure 3-37 Effect of ideal LPF on the output of SH.

an output prior to an input). This can be seen by taking the IFT of the ideal LP filter. We see from Figure 1-39 that the IFT of the LPF's passband is the sampling function in the time domain. This is shown in Figure 3-38. The shaded portion represents the signal output prior to the appearance of an input (i.e., the filter is noncausal).

Use of a practical LP filter with its gradual roll-off will allow other frequency components to appear in the output, introducing some erroneous information. This is presented in Figure 3-39. This can be improved at the cost

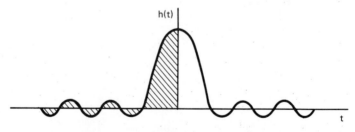

Figure 3-38 Ideal LPF impulse response.

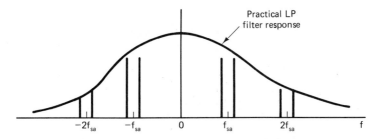

Figure 3–39 Effect of practical LPF on the output of the SH circuit.

of introducing a higher-order filter (steeper roll-off), but this may not completely eliminate the high-frequency components introduced by sampling.

3.5.3 Sampling Rate

If the sampled frequency signal, $s(t)$, is complex, containing many frequency components, each one will be reproduced as a sum–difference pair about the sampled frequency component. If we consider an arbitrary signal, its FT might have a large extent on either side of 0 Hz. To develop the concept of sampling rate, we will assume that the signal to be sampled has been passed through a LP filter with a steep roll-off (i.e., the signal is band-limited). This causes the stopband components to have a low amplitude. We can then idealize the situation and state that the signal has no components with significant amplitudes above a given frequency f_c. This is shown in Figure 3–40, where $S(f) = \text{FT}\{s(t)\}$.

When the signal is sampled, all of its components will generate sum–difference pairs about the components of the sampling signal, $H(t)$, as shown in Figure 3–41. As with ideal sampling, we form the sampled output, $r(t)$, using a practical pulse

$$r(t) = H(t)s(t) = \sum_{n=-\infty}^{\infty} H_p(t - nT)s(t) \qquad (3\text{–}23)$$

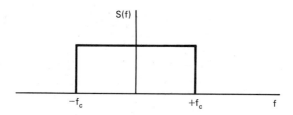

Figure 3–40 Frequency content of ideal band limited signal $s(t)$.

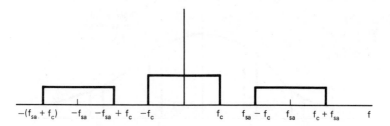

Figure 3-41 Output of the TH circuit with band-limited input.

where $H_p(t - nT)$ is the practical sampling signal of Figure 3-31 with FS given by Equation 3-20. Taking the FT of $r(t)$, we obtain

$$\text{FT}\{r(t)\} = \text{FT}\left\{\sum_{n=-\infty}^{\infty} s(t)H_p(t - nT)\right\}$$

$$= \sum_{n=-\infty}^{\infty} \text{FT}\{s(t)H_p(t - nT)\} \qquad (3\text{--}24)$$

$$= S(f) * H_p(f) = R(f)$$

As we slide $S(f)$ past each of the components of $H_p(f)$, we will obtain the original spectrum replicated about each integer multiple of the sampling frequency as for the case of ideal sampling; however, the amplitude of each replica is multiplied by the corresponding value of C_m. From Equations 3-20, 3-23, and 3-24, we obtain

$$R(f) = \text{FT}\{r(t)\} = \int_{-\infty}^{\infty} r(t) \exp(-j2\pi ft)\, dt$$

$$= \int_{-\infty}^{\infty} s(t) \left\{\sum_{n=-\infty}^{\infty} C_n \exp(j2\pi f_{sa}nt)\right\} \exp(-j2\pi ft)\, dt \qquad (3\text{--}25)$$

$$= \sum_{n=-\infty}^{\infty} C_n \int_{-\infty}^{\infty} s(t) \exp(-j2\pi(f - nf_{sa})t)\, dt$$

$$= \sum_{n=-\infty}^{\infty} C_n\, S(f - nf_{sa})$$

The frequency spectrum of $s(t)$ is replicated about each value of nf_{sa}.

In Figure 3-41, we have assumed f_{sa} to be "sufficiently large." We immediately can see that a practical LP filter will require a very steep gain roll-off to separate the components about $f = 0$ (the baseband) from those about f_{sa}. More important, if f_{sa} is not "sufficiently large," there is a possibility that the frequency components from one band will overlap those on either side of it, causing aliasing. This is shown in Figure 3-42. The overlap has caused the amplitude of those components in the overlap region to be larger than they

Sec. 3.5   Practical Sampling Theory   103

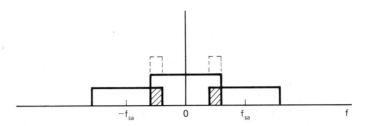

Figure 3-42   Effect of too low a sampling rate.

should, introducing distortion in the output (i.e., noise). In addition, a practical filter (or an ideal one for that matter) cannot separate the bands. This is presented in Figure 3-43. As can be seen, many unwanted (noise) components will pass through the filter, providing false information to the user.

From the discussion above, we see that there must be a reasonable amount of separation between the frequency-component bands of the output signal. This is accomplished by two steps: (1) the sampled signal is band-limited (i.e., components above a given frequency have zero amplitude); and (2) from Figure 3-41, we see that no overlap occurs if the sampling frequency is at least twice the maximum frequency component of the sampled signal as given by Nyquist's sampling rate theorem: To reduce the noise content of a sampled signal introduced by aliasing, the sampling rate must be at least twice the maximum frequency contained in the signal. Figure 3-44 shows how the higher sampling frequency will reduce the noise content by allowing less of the unwanted frequency components (shaded area) to pass through.

In some instances, a signal can be sampled at a much higher rate than required to reduce aliasing. Then, by using a portion of the samples (e.g., every second or third one), the original signal information is retained with a reduction in noise due to aliasing. This is termed *decimation*.

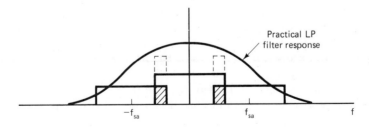

Figure 3-43   Inability of the LP filter to separate overlappping bands.

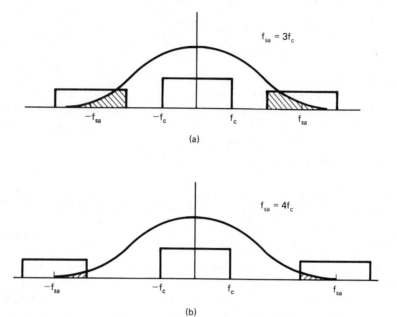

Figure 3-44 (a) Effect of the filter on sampled output: $f_{sa} = 3f_c$; (b) effect of the filter on sampled output: $f_{sa} = 4f_c$.

3.6 SUMMARY

This chapter has considered the analog processing required to alter a signal so that it can be converted to digital format. It has obtained the very important result that if a signal $s(t)$ is to be recovered from its samples with a minimum of distortion and noise, it must be sampled at a rate which is at least twice the rate of the highest-frequency component of $s(t)$. In addition, except for band limiting, the reader will see that most filtering can be accomplished digitally.

PROBLEMS

1. Using the circuit of Figure 3–14, with $R_s = 50\ \Omega$, $R_1 = R_3 = 1\ k\Omega$, and $R_2 = R_4 = 50\ k\Omega$, find the CMRR (dB).
2. Referring to Problem 1, if $R_4 = 50\ k\Omega$, but $R_2 = 49,300\Omega$, what is the CMRR (dB)?
3. Using the circuit of Figure 3–15, with $R_s = 50\ \Omega$, $R_A = 1\ k\Omega = R_B = R_C$, $R_1 = R_3 = 1\ k\Omega$, and $R_2 = R_4 = 2.5\ k\Omega$, find the differential mode gain and the CMRR (dB).
4. $v(t) = 10 \sin (2\pi[1\ \text{kHz}]t)$ is ideally sampled at 4.5 kHz.
   (a) Find the sample points for two cycles.
   (b) Sketch $|V(f)|$.

(c) Sketch the frequency content of $v(nT)$.
(d) If $v(nT)$ is passed through an ideal LPF, $f_c = 1.5$ kHz, sketch the output spectrum.
5. Repeat Problem 4 for practical sampling with a 1V pulse at 4.5 kHz and a duty cycle of 15%.
6. For $S(f)$ as shown (Figure P3–6), sketch the spectrum of the sampled signal for:
   (a) Ideal sampling at 10 kHz, 6 kHz, and 5 kHz.
   (b) Practical sampling at 10 kHz with a pulse whose duty cycle is 10%.

Figure P3–6

7. Repeat Problem 6(a) for a sampling frequency of (a) 2 kHz, (b) 3 kHz, and (c) 8 kHz.
8. If the practical pulse used to sample a signal is a 10-kHz triangle wave (Figure P3–8) with a duty cycle of 15%, what is the spectrum of the sampled output? Compare your results with Problem 5(b).

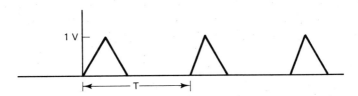

Figure P3–8

REFERENCE

JOHNSON, D. E., and HILBURN, J. L. *Rapid Practical Designs of Active Filters.* New York: Wiley, 1975.

ANALOG-TO-DIGITAL CONVERTERS

4

The output of the amplifier or track-and-hold unit is still analog in form and cannot serve as input to the digital processor section. To convert this analog signal to digital at the specified sampling intervals requires a unit to serve as an interface between the two sections of the system. This unit is the analog-to-digital converter (ADC). It accepts an analog signal at its input and provides an n-bit binary signal at its output. In addition to the above-mentioned I/O signals, the ADC usually requires a signal to commence the conversion, and it will generate a signal to indicate that the conversion is completed, termed the end-of-conversion (EOC) signal.

The important characteristics of an ADC are the speed and accuracy with which it can perform a conversion: the former will place a limit on the number of conversions that will occur per second, limiting the upper value of the sampling frequency and, in turn, the upper-frequency component of the analog signal being converted to digital; the latter will affect the overall fidelity of the process, introducing extraneous components into the output signal of the processor system. We will see that the accuracy of conversion will be increased as the number of bits in the digital output is increased, but at the cost of extending the conversion time.

In this chapter we look at the characteristics of the conversion process and the ADC circuits presently in use. In addition, it includes a section on the data acquisition devices presently available. These units contain most of the input circuitry required to convert an analog signal to digital form.

4.1 CHARACTERISTICS OF CONVERSION

Regardless of the actual circuitry or device used to obtain the conversion, there are specific relations that exist between the I/O values. Knowledge of these relations will allow the user to determine how close to the analog input value the digital output will be. We will assume that the input value at the beginning of conversion is the one we wish to change to digital form and that the input is constant for a length of time sufficient to perform the conversion. In this section we consider the conversion event and the terminology associated with it.

4.1.1 Accuracy

The primary concern of a conversion is accuracy, that is, the agreement between the I/O values. If the decimal equivalent of the n-bit output number is determined, we want to know how close it is to the value of voltage at the input.

Immediately, we run into a problem. The smallest change in the digital output is equal to the weight of the least significant bit (LSB) while the input voltage is continuous. This implies that a given single digital output must correspond to a range of input values. This effect, called *quantization*, introduces an error (noise) into the output signal. Quantization has taken a range of the TH output voltage and made it relate to a single binary output: Not all values are available at the ADC output. To determine the effect of quantization, we must know how the input signal is divided. If the output is n bits, the input range is divided into 2^n units. This is given by Equation 4–1.

$$\text{input range division} = \frac{\text{FSR (volts)}}{2^n \text{ (divisions)}} \quad (4\text{–}1)$$

where FSR is the full-scale range of the input. This is shown in Example 4–1.

Example 4–1

If the FSR is 0 to 10 V and n is 8, what is the range of each division?

Solution

$$\text{input range division} = \frac{\text{FSR}}{2^n}$$

$$= \frac{10 \text{ V}}{2^8 \text{ div}} \quad (4\text{–}1)$$

$$= 39.062 \frac{\text{mV}}{\text{div}}$$

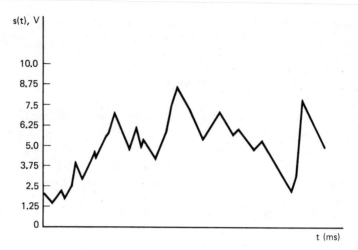

Figure 4–1 Division of FSR by 2^n units: unipolar.

Starting with the value of 0 V, the input is divided into 256 (=2^8) sections, each corresponding to a variation in the input of 39.062 mV. To see this graphically, let us assume an output 3 bits in length and an input range of 10 V. Each division of the input corresponds to 1.25 V. This is shown in Figure 4–1.

As the input is always positive (i.e., 0 to 10 V), it is termed *unipolar*. An input that is positive and negative (e.g., −5 to +5 V) is termed *bipolar*. This is shown in Figure 4–2. Each division of the input still consists of 1.25 V.

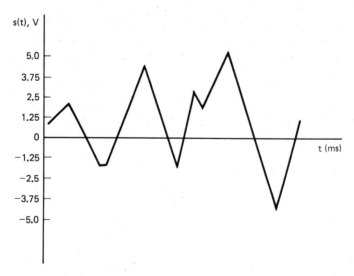

Figure 4–2 Division of FSR by 2^n units: bipolar.

Sec. 4.1   Characteristics of Conversion

Each value of the input range given on the ordinate of Figure 4–1 corresponds to a fraction of the FSR. 7.5 V corresponds to $\frac{3}{4}$ FSR; 3.75 V corresponds to $\frac{3}{8}$ FSR. With this in mind, the output values that correspond to these decimal fractional input values can be binary fractional values. If the output ranges from 000 to 111, each binary value will be considered as equal to the numerator of the input decimal fraction, the denominator being equal to 2^3 or 8 (1000 in binary). The relation between the input value given in Figure 4–1 and the binary output values is given in Table 4–1.

Even though the FSR is 10 V, there will be no 3-bit output for this value of input; however, some ADCs do have an overrange output, which is nothing more than an additional bit beyond the most significant bit (MSB) of the output. It will indicate that the input has exceeded the maximum voltage measurable by the output number.

Each binary output corresponds to a particular analog input; for example, 011 corresponds to 3.75 V for a unipolar input signal. If the input is bipolar, there is a different correlation given by Table 4–2.

The most negative voltage of the input range is assigned 000, 0 V at the input is assigned 100, and 3.75 V is assigned 111. This correlation of I/O values is termed "positive time offset binary." Again, the upper limit has no corresponding output value unless there is an over-range terminal: +5 V corresponds to 1000. Figure 4–3 shows the ideal graphical relation between the input as a fraction of FSR and the binary output.

We must now ask about input signal values that do not correspond to the "ideal" fractions of the FSR (e.g., 6.913 V). This is handled by defining a range of voltages equivalent to 1 LSB (1.25 V for the example above) about each fractional input value. This is shown in Figure 4–4. Any value of input voltage whose fractional value of FSR lies within $\frac{1}{2}$ LSB of the ideal fractions of the FSR (specified with 2^n as the denominator) is assigned the binary output

TABLE 4–1 I/O Relations for a 3-Bit ADC: Unipolar

Input		Output
Voltage (V)	Fraction FSR	Binary
0	0	000
1.25	$\frac{1}{8}$	001
2.5	$\frac{2}{8}$	010
3.75	$\frac{3}{8}$	011
5.0	$\frac{4}{8}$	100
6.25	$\frac{5}{8}$	101
7.5	$\frac{6}{8}$	110
8.75	$\frac{7}{8}$	111
10.0	$\frac{8}{8}$	—

TABLE 4-2 I/O Relations for a 3-Bit ADC: Bipolar

Input		Output
Voltage (V)	FSR	Binary
−5	$-\frac{4}{8}$	000
−3.75	$-\frac{3}{8}$	001
−2.5	$-\frac{2}{8}$	010
−1.25	$-\frac{1}{8}$	011
0	−0	100
1.25	$\frac{1}{8}$	101
2.5	$\frac{2}{8}$	110
3.75	$\frac{3}{8}$	111
5.0	$\frac{4}{8}$	—

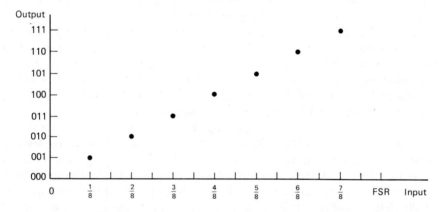

Figure 4-3 Output/input relations for a three bit ADC.

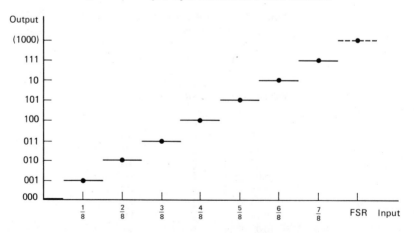

Figure 4-4 Output value of the input range.

Sec. 4.1 Characteristics of Conversion   111

corresponding to the fraction. For the I/O relations given in Figure 4–4, a voltage of 6.913 V corresponds to a fractional value of 0.6913 for an FSR of 10 V. As this is within 0.0625 V ($\pm\frac{1}{2}$ LSB in fractional form) of 0.75 (6/8), the value of 6.913 V is assigned the binary number 110. Table 4–3 shows the range of input voltages that correspond to a given binary output.

Each range of input voltages corresponds to 1 LSB, and an input voltage must be within that range to generate its assigned binary output. This approach to AD conversion causes a continuous range of (analog) input values to be assigned a specific binary output (i.e., they have been quantized). This indicates that the error in the output, which is the difference between the actual input and the binary output value, can be as large as $\frac{1}{2}$ LSB. We can see why the accuracy of the conversion is improved with n: The higher n is, the smaller the range corresponding to $\frac{1}{2}$ LSB. For $n = 8$, FSR = 10 V, $\frac{1}{2}$ LSB will correspond to 19.53 mV. For $n = 12$ and 16, $\frac{1}{2}$ LSB will correspond to 1.22 mV and 76.3 μV, respectively.

A measure of the effect of increasing n is the signal-to-quantization noise ratio, SQNR. This is given in decibels and defined in Equation 4–2.

$$\text{SQNR (dB)} = 20 \log \frac{\text{FSR(V)}}{\text{LSB(V)}} \quad (4\text{–}2)$$

For the 3-bit output discussed above, we find that the SQNR = 18 dB. The general relationship for a value of n bits is given by Equation 4–3.

$$\text{SQNR (dB)} = 6(n) \text{ dB} \quad (4\text{–}3)$$

The 12- and 16-bit outputs will have SQNR values of 72 dB and 96 dB, respectively. Although these values yield higher accuracies, they might cause problems when the ADC is connected to a processor that does not accept more than 8 bits at a time. Additional logic circuitry and time might be required to enter the 12- or 16-bit number into the processor. The cost in time and circuitry must be weighed against the advantages of high accuracy; however, the appearance of 16- and 32-bit processor ICs has reduced that problem.

TABLE 4–3 Input Range for a Given Binary Output

Input Voltage (V)	Output Binary
0.000–0.625	000
0.625–1.875	001
1.875–3.125	010
3.125–4.375	011
4.375–5.625	100
5.625–6.875	101
6.875–8.125	110
8.125–9.375	111

4.1.2 Additional I/O Errors

Figure 4–4 showed the ideal transfer curve for 3-bit digital output. In the practical device or circuit, this curve may not be obtained, causing the output to deviate from its expected value.

Offset errors. As the input signal of a unipolar ADC increases from zero volts, a value is reached which causes the output to increase from 00 . . . 000 to 00 . . . 001. For an ideal ADC, this value corresponds to $\frac{1}{2}$ LSB. For the practical ADC, the transition might occur at another level of voltage. This occurrence, termed *offset error*, is shown in Figure 4–5. Each binary output now corresponds to a range of input values that is still equivalent to an LSB, but shifted upward: where 011 ideally corresponds to a range of inputs 3.125 to 4.375 V (see Table 4–3), it might now correspond to a range of 3.6 to 4.85 V.

Gain error. We have assumed that the ratio of output binary value to input analog value is unity: whenever the input is at a fractional value of FSR, the binary output is at the corresponding fractional value. That this is not the situation is indicated by the amount of gain error of the ADC.

The result of gain error is that the output binary number is not related to the corresponding fractional FSR input. This is shown in Figure 4–6. The line labeled A is drawn through the transfer points of an ideal ADC, while that labeled B passes through the practical values. We see that binary 110 can be generated by a voltage range which differs from the ideal one of 6.875 to 8.125, resulting in an incorrect conversion.

Nonlinearity. If we plot the midpoints of the analog input ranges against the corresponding binary codes, an ideal ADC will have the points lie on a straight line going through the origin. This was shown in Figure 4–3.

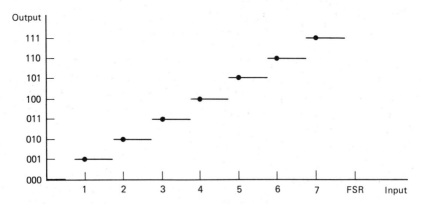

Figure 4–5 Effect of offset error on conversion.

Sec. 4.1   Characteristics of Conversion

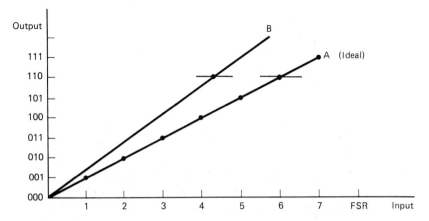

Figure 4-6   Practical versus ideal transfer curve.

In the practical ADC, if the points deviate from a straight line, the ADC is nonlinear, as shown in Figure 4-7. The result of nonlinearity is that the range of input voltages corresponding to different codes will vary from range to range: a code may be associated with a relatively large range, while another may be related to a small one. Either case will result in an erroneous conversion of the analog input to digital form.

Differential linearity.   This is defined as the difference between the actual and ideal ranges that correspond to a given binary code. This ideal range is equal to the weight of 1 LSB. If the differential linearity is more than the equivalent of 1 LSB, the possibility exists that a specific code may be missed. This is shown in Figure 4-8. The code due to $\frac{5}{8}$ FSR input signal, 101, is missing and cannot appear at the output. If the ADC is free of this

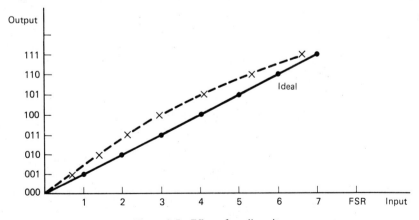

Figure 4-7   Effect of nonlinearity.

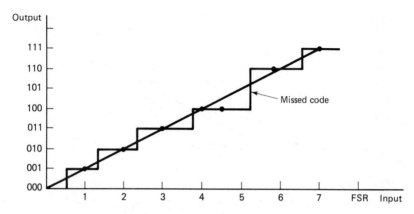

Figure 4-8 Effect of differential nonlinearity.

characteristic, the data sheet will specify that there are "no missing codes," implying that the differential linearity is less than 1 LSB.

4.1.3 Conversion Time

The amount of time it takes the ADC to convert an analog signal to digital form is termed the *conversion time*. This is provided in the data sheets and does not include the time required by external components (e.g., TH) that are part of the conversion process.

The reciprocal of conversion time is the conversion rate and indicates how many words (conversions) the ADC can provide per second. This is shown in the following example.

Example 4-2

An audio signal is to be sampled at 80 kHz. What is the minimum conversion in time an ADC must possess?

Solution

$$\text{conversion time} = \frac{1}{\text{conversion rate}}$$

$$= \frac{1}{80 \text{ kHz}}$$

$$= 12.5 \text{ } \mu\text{s}$$

The ADC chosen must have a maximum conversion time which is usually much less than this value to ensure that the conversion rate is obtained. It will require a control pulse to start the conversion process. When the conversion is completed, a signal is generated indicating that fact. At that time, the control pulse ends the conversion and readies the circuit for the next conversion.

4.2 ADC TYPES

The majority of ADC devices in use today are either of the successive approximations type or the dual-slope integrator type. Additional ones use the flash (or parallel) conversion technique when very high conversion speeds are required. The basic operation of the first two of these devices is considered below, followed by a discussion of several IC devices that are available today.

4.2.1 Successive Approximations ADC

Conversion takes place by having the input compared with an internally generated signal and setting an internal register, bit by bit, until all n bits of the output are determined. The bits are set starting with the most significant bit (MSB) down to the least significant bit (LSB). As the analog signal must remain constant during the conversion, this type of ADC will require a TH circuit. Variation of the TH output during the conversion will result in an error; that is, noise will be introduced in the final output. Figure 4–9 shows a block diagram of a successive approximations ADC.

The circuitry of the ADC includes a comparator, an n-bit register, and a digital-to-analog (DAC) converter. The input to the DAC is the current register output. This is converted to an analog signal and compared with the input signal: If the input is greater than the DAC output, the current register position will receive a logic one; if not, it will receive a zero. In this manner, each bit position of the register will be assigned a one or a zero, depending on the relative size of the input signal. When all bits have been assigned a value, the binary output will be equivalent to the analog input signal. As all

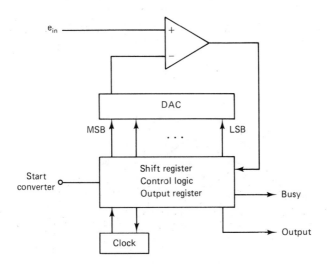

Figure 4–9 Successive approximation ADC.

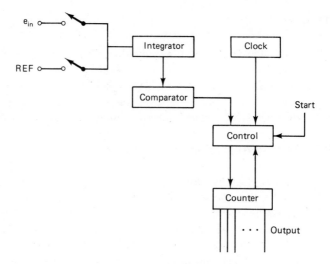

Figure 4-10 Dual slope converter.

bits must be assigned a value, the time for a conversion is fixed by the clock rate and the number of bits in the output.

4.2.2 Dual-Slope Integrator

In this ADC, the analog input signal is fed to an integrator circuit for a fixed number of clock pulses. At the end of the integration period, the integrator input is switched to a reference voltage of opposite polarity to the input. Whereas the input signal caused the integrator output to rise to a given level, the voltage of opposite polarity causes it to fall (hence the name "dual slope") to zero. Figure 4-10 shows the block diagram of a dual-slope integrator and Figure 4-11 shows the integrator output as a function of time.

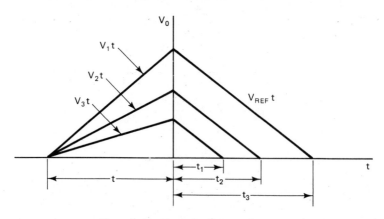

Figure 4-11 Dual slope integrater output.

The input signal generates a rising integrator output for a fixed count of the clock, reaching a value proportional to the input signal: The larger e_{in} is, the larger the integrator output. When the reference voltage is applied to the integrator at the end of the first segment, its output drops toward zero. The point at which the integrator output reaches zero is determined by a zero-crossing detector in the form of a comparator. During the time that the integrator output was dropping, the clock signal was fed to a counter. When the integrator output reaches zero, the count is terminated and its value at that time will be the binary form of the input signal.

As the slope for the second portion of the conversion interval is constant, the time it takes to reach zero is determined solely by the initial value of the integrator output at the end of the first portion. This was determined by the value of the input signal during the conversion time.

Due to the fact that the input signal is fed to an integrator, low-level noise signals will provide a random component that integrates to zero. In addition, any power signal noise (60 Hz) can also be removed from the output by integration if the clock frequency is chosen as a multiple of the power period. This characteristic of the dual-slope integrator makes it the preferred choice in a noise-filled environment.

The use of an integrator as part of the ADC conversion circuit eliminates the need for a TH circuit at the ADC input. As the conversion time is relatively short, the input signal will change slightly during this interval. The action of the integrator will convert the average of the input signal, over the length of the conversion time, to binary. Due to the short time interval, the average value will be a good approximation to the signal level at the midpoint of the conversion.

4.3 IC ADCs

Although the ADC circuits discussed above can be put together using discrete components, they are available in IC package form from manufacturers such as Analog Devices, Burr-Brown, and Datel-Intersil. In the sections that follow, a commercial ADC of each type is discussed and its data sheets are presented.

4.3.1 AD 571 10-Bit Successive Approximation ADC

The AD 571 is a general-purpose ADC and will convert an analog signal, in either the range 0 to +10 V, or −5 to +5 V, into a 10-bit binary output signal. The relation between the unipolar input and output values is given by Equation 4-4

$$\frac{B_1 2^9 + B_2 2^8 + \cdots + B_{10} 2^0}{2^{10}} = \frac{E_{in}}{FSR} \qquad (4\text{-}4)$$

where $B_1B_2B_3\cdots B_{10}$ is the binary output, E_in is the input signal, $0 < E_\text{in} < 10$ V and, FSR is the full-scale range of the input (e.g., 10 V).

With the analog signal applied to the input, the CONVERT command is brought low to start the conversion. When each bit has been set, the AD 571 sends out a DATA READY signal (active low). This also serves to enable the tri-state buffers, allowing the output pins to assume their binary values. As the successive approximation technique requires a relatively constant voltage at its input, a TH circuit may be required. The ADC and the TH can both be controlled by the same signal; that is, the TH goes into a hold mode just before the ADC begins its conversion, which is determined by the sampling rate. Figure 4–12 is the specification sheet for the AD 571.

Among the many terms in the specifications, the conversion time is given as 25 ms (typical) and the relative accuracy is given as ±1 LSB (maximum). The former term is defined (by Analog Devices) as the time required for a complete conversion by an ADC. The relative accuracy is defined as the deviation of the analog value at any code (relative to the full analog range of the device transfer characteristic) from its theoretical value (relative to the same range) after the FSR has been calibrated. These values are used to determine the maximum-frequency component of the input signal and how well the output represents the value of the analog input.

Among the additional information provided by the manufacturers is the circuit description, control and timing, TH amplifier connection, and connection to a microprocessor.

4.3.2 AD 7550 13-Bit Monolithic A/D Converter

A device that performs the A/D conversion using the dual-slope technique is the AD 7550. This unit uses CMOS technology for low power dissipation and accomplishes the conversion using an additional two slopes of integration prior to the actual dual-slope conversion (i.e., it is a quad-slope converter). Figure 4–13 provides the specifications for the AD 7550 together with a block diagram of the internal circuitry. As the device has a 13-bit output, relative accuracy of 1 LSB now corresponds to a value of $FSR/2^{13}$ or 1.22 mV. The conversion time is 40 ms (typical) with a 1-MHz clock, which is longer than that of the AD571.

The quad-slope conversion begins with a reset operation (phase 0) that brings the integrator output to zero. This initiates the first dual-slope interval (phases 1 and 2). Its purpose is to reduce the errors caused by leakage, op-amp offset voltage, and bias currents, together with their drifting due to temperature, resulting in a more accurate and stable conversion.

At the end of phases 1 and 2, the actual conversion process (as described earlier) begins. This process comprises phases 3 and 4. Figure 4–14 shows the quad-slope timing diagram.

Sec. 4.3   IC ADCs

Integrated Circuit 10-Bit
Analog to Digital Converter
AD571*

FEATURES
Complete A/D Converter with Reference and Clock
Fast Successive Approximation Conversion — 25µs
No Missing Codes Over Temperature
  0 to +70°C — AD571K
  −55°C to +125°C — AD571S
Digital Multiplexing — 3 State Outputs
18-Pin Ceramic DIP
Low Cost Monolithic Construction

AD571 FUNCTIONAL BLOCK DIAGRAM

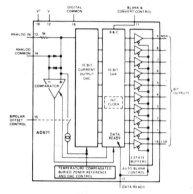

18-PIN DIP

PRODUCT DESCRIPTION
The AD571 is a 10-bit successive approximation A/D converter consisting of a DAC, voltage reference, clock, comparator, successive approximation register and output buffers — all fabricated on a single chip. No external components are required to perform a full accuracy 10-bit conversion in 25µs.

The AD571 incorporates the most advanced integrated circuit design and processing technology available today. It is the first complete converter to employ I²L (integrated injection logic) processing in the fabrication of the SAR function. Laser trimming of the high stability SiCr thin film resistor ladder network at the wafer stage (LWT) insures high accuracy, which is maintained with a temperature compensated, subsurface Zener reference.

Operating on supplies of +5V to +15V and −15V, the AD571 will accept analog inputs of 0 to +10V, unipolar or ±5V bipolar, externally selectable. As the BLANK and CONVERT input is driven low, the three state outputs will be open and a conversion will commence. Upon completion of the conversion, the DATA READY line will go low and the data will appear at the output. Pulling the BLANK and CONVERT input high blanks the outputs and readies the device for the next conversion. The AD571 executes a true 10-bit conversion with no missing codes in approximately 25µs.

The AD571 is available in two versions for the 0 to +70°C temperature range, the AD571J and K. The AD571S guarantees 10-bit accuracy and no missing codes from −55°C to +125°C.

PRODUCT HIGHLIGHTS
1. The AD571 is a complete 10-bit A/D converter. No external components are required to perform a conversion. Full scale calibration accuracy of ±0.3% is achieved without external trims.

2. The AD571 is a single chip device employing the most advanced IC processing techniques. Thus, the user has at his disposal a truly precision component with the reliability and low cost inherent in monolithic construction.

3. The AD571 accepts either unipolar (0 to +10V) or bipolar (−5V to +5V) analog inputs by simply grounding or opening a single pin.

4. The device offers true 10-bit accuracy and exhibits no missing codes over its entire operating temperature range.

5. Operation is guaranteed with −15V and +5V to +15V supplies. The device will also operate with a −12V supply.

6. The AD571S is also available with processing to MIL-STD-883, Class B. The single chip construction and functional completeness make the AD571 especially attractive for high reliability applications.

*Covered by Patent Nos. 3,940,760; 4,213,806; 4,136,349.

Figure 4–12 AD571 ADC data sheets. (From *Data-Acquisition Databook*, 1982, Vol. 1. Courtesy of Analog Devices, Inc., Two Technology Way, Norwood, MA.)

SPECIFICATIONS

(typical @ +25°C with V+ = +5V, V- = -15V, all voltages measured with respect to digital common, unless otherwise indicated)

MODEL	AD571J	AD571K	AD571SD/AD571SD-883[1]
RESOLUTION	10 Bits	*	*
RELATIVE ACCURACY @ 25°C[2]	±1LSB max	±1/2LSB max	±1LSB max
T_{min} to T_{max}	±1LSB max	±1/2LSB max	±1LSB max
FULL SCALE CALIBRATION[3] (With 15Ω Resistor In Series With Analog Input	±2LSB (typ)	*	*
UNIPOLAR OFFSET (max)	±1LSB	±1/2LSB	*
BIPOLAR OFFSET (max)	±1LSB	±1/2LSB	*
DIFFERENTIAL NONLINEARITY (Resolution for Which no Missing Codes are Guaranteed)			
+25°C	10 Bits	*	*
T_{min} to T_{max}	9 Bits	10 Bits	10 Bits
TEMPERATURE RANGE	0 to +70°C	*	−55°C to +125°C
TEMPERATURE COEFFICIENTS Guaranteed max Change T_{min} to T_{max}			
Unipolar Offset	±2LSB (44ppm/°C)	±1LSB (22ppm/°C)	±2LSB (20ppm/°C)
Bipolar Offset	±2LSB (44ppm/°C)	±1LSB (22ppm/°C)	±2LSB (20ppm/°C)
Full Scale Calibration (With 15Ω Fixed Resistor or 50Ω Trimmer)	±4LSB (88ppm/°C)	±2LSB (44ppm/°C)	±5LSB (50ppm/°C)
POWER SUPPLY REJECTION Max Change In Full Scale Calibration			
CMOS Positive Supply (K only) +13.5≤V+≤+16.0V	N.A.	±1LSB max	N.A.
TTL Positive Supply +4.5V≤V+≤+5.5V	±2LSB max	±1LSB max	*
Negative Supply −16.0V≤V-≤−13.5V	±2LSB max	±1LSB max	*
ANALOG INPUT RESISTANCE	3kΩ min	*	*
	5kΩ typ	*	*
	7kΩ max	*	*
ANALOG INPUT RANGES (Analog Input to Analog Common)			
Unipolar	0 to +10V	*	*
Bipolar	−5V to +5V	*	*
OUTPUT CODING			
Unipolar	Positive True Binary	*	*
Bipolar	Positive True Offset Binary	*	*
LOGIC OUTPUT[4] Bit Outputs and $\overline{\text{Data Ready}}$			
Output Sink Current (V_{OUT} = 0.4V max, T_{min} to T_{max})	3.2mA min (2TTL Loads)	* *	* *
Output Source Current (Bit Outputs)[5] (V_{OUT} = 2.4V min, T_{min} to T_{max})	0.5mA min	*	*
Output Leakage When Blanked	±40μA max	*	*
LOGIC INPUT Blank and Convert Input			
0≤V_{in}≤V+	±40μA max	*	*
Blank − Logic "1"	2.0V min	*	*
Convert − Logic "0"	0.8V max	*	*
CONVERSION TIME	15μs min	*	*
	25μs typ	*	*
	40μs max	*	*

Figure 4–12 (cont.)

Sec. 4.3 IC ADCs

MODEL	AD571J	AD571K	AD571SD/AD571SD-883[1]
POWER SUPPLY			
Absolute Maximum			
V+	+7V	+16.5V	*
V−	−16.5V	*	*
Specified Operating − Rated Performance			
V+	+5V	+5V to +15V	*
V−	−15V	*	*
Operating Range			
V+	+4.5V to +5.5V	+4.5V to +16.5V	*
V−	−12.0V to −16.5V	*	*
Operating Current			
Blank Mode			
V+ = +5V	2mA typ (10mA max)	*	*
V+ = +15V	5mA typ (10mA max)	*	*
V− = −15V	9mA typ (15mA max)	*	*
Convert Mode			
V+ = +5V	5mA	*	*
V+ = +15V	10mA	*	*
V− = −15V	10mA	*	*
PACKAGE OPTIONS[6]			
D18A	AD571JD	AD571KD	AD571SD
N18A[7]	AD571JN	AD571KN	

*Specifications same as AD571J
**Specifications same as AD571K
Specifications subject to change without notice.

NOTES
[1] The AD517S is available processed and screened to the requirements of MIL-STD-883, Class B. When ordering, specify the AD571SD/883B.
[2] Relative accuracy is defined as the deviation of the code transition points from the ideal transfer point on a straight line from the zero to the full scale of the device.
[3] Full scale calibration is guaranteed trimmable to zero with an external 50Ω potentiometer in place of the 15Ω fixed resistor. Full scale is defined as 10 volts minus 1LSB, or 9.990 volts.
[4] Logic Input and Output Thresholds and Levels are a function of V+. They are guaranteed TTL compatible at V+ = +5V, CMOS compatible at V+ = 15V for the AD571K.
[5] The data output lines have active pull-ups to source 0.5mA. The DATA READY line is open collector with a nominal 6kΩ internal pull-up resistor.
[6] See Section 20 for package outline information.
[7] To be available June, 1982.

```
                ABSOLUTE MAXIMUM RATINGS
                V+ to Digital Common   AD571J, S . . . . . . . . . . . 0 to +7V
                                       AD571K . . . . . . . . . . 0 to +16.5V
                V− to Digital Common . . . . . . . . . . . . . . . . . . 0 to −16.5V
                Analog Common to Digital Common. . . . . . . . . . . . . . ±1V
                Analog Input to Analog Common. . . . . . . . . . . . . . . ±15V
                Control Inputs . . . . . . . . . . . . . . . . . . . . . . . 0 to V+
                Digital Outputs (Blank Mode). . . . . . . . . . . . . . . . 0 to V+
                Power Dissipation. . . . . . . . . . . . . . . . . . . . . . . 800mW
```

Figure 4–12 (*cont.*)

CIRCUIT DESCRIPTION

The AD571 is a complete 10-bit A/D converter which requires no external components to provide the complete successive-approximation analog-to-digital conversion function. A block diagram of the AD571 is shown in Figure 1. Upon receipt of the $\overline{\text{CONVERT}}$ command, the internal 10-bit current output DAC is sequenced by the I^2L successive-approximation register (SAR) from its most-significant bit (MSB) to least-significant bit (LSB) to provide an output current which accurately balances the input signal current through the 5kΩ input resistor. The comparator determines whether the addition of each successively-weighted bit current causes the DAC current sum to be greater or less than the input current; if the sum is less the bit is left on, if more, the bit is turned off. After testing all the bits, the SAR contains a 10-bit binary code which accurately represents the input signal to within ±½LSB (0.05%).

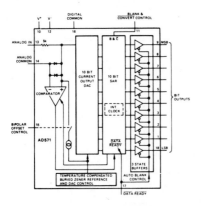

Figure 1. AD571 Functional Block Diagram

Upon completion of the sequence, the SAR sends out a $\overline{\text{DATA READY}}$ signal (active low), which also brings the three-state buffers out of their "open" state, making the bit output lines become active high or low, depending on the code in the SAR. When the BLANK and $\overline{\text{CONVERT}}$ line is brought high, the output buffers again go "open", and the SAR is prepared for another conversion cycle. Details of the timing are given further.

The temperature compensated buried Zener reference provides the primary voltage reference to the DAC and guarantees excellent stability with both time and temperature. The bipolar offset input controls a switch which allows the positive bipolar offset current (exactly equal to the value of the MSB less ½LSB) to be injected into the summing (+) node of the comparator to offset the DAC output. Thus the nominal 0 to +10V unipolar input range becomes a –5V to +5V range. The 5kΩ thin film input resistor is trimmed so that with a full scale input signal, an input current will be generated which exactly matches the DAC output with all bits on. (The input resistor is trimmed slightly low to facilitate user trimming, as discussed on the next page.)

POWER SUPPLY SELECTION

The AD571 is designed for optimum performance using a +5V and –15V supply, for which the AD571J and AD571S are specified. AD571K will also operate with up to a +15V supply, which allows direct interface to CMOS logic. The input logic threshold is a function of V+ as shown in Figure 2. The supply current drawn by the device is a function of both V+ and the operating mode (BLANK or CONVERT). These supply current variations are shown in Figure 3. The supply currents change only moderately over temperature as shown in Figure 7.

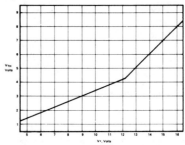

Figure 2. Logic Threshold (AD571K Only)

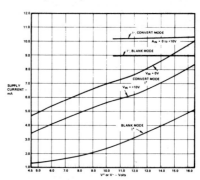

Figure 3. Supply Currents vs. Supply Levels and Operating Modes

Figure 4–12 (cont.)

Sec. 4.3 IC ADCs

Applying the AD571

CONNECTING THE AD571 FOR STANDARD OPERATION

The AD571 contains all the active components required to perform a complete A/D conversion. Thus, for most situations, all that is necessary is connection of the power supply (+5 and −15), the analog input, and the conversion start pulse. But, there are some features and special connections which should be considered for achieving optimum performance. The functional pin-out is shown in Figure 4.

Figure 4. AD571 Pin Connections

FULL SCALE CALIBRATION

The 5kΩ thin film input resistor is laser trimmed to produce a current which matches the full scale current of the internal DAC−plus about 0.3%−when a full scale analog input voltage of 9.990 volts (10 volts − 1LSB) is applied at the input. The input resistor is trimmed in this way so that if a fine trimming potentiometer is inserted in series with the input signal, the input current at the full scale input voltage can be trimmed down to match the DAC full scale current as precisely as desired. However, for many applications the nominal 9.99 volt full scale can be achieved to sufficient accuracy by simply inserting a 15Ω resistor in series with the analog input to pin 13. Typical full scale calibration error will then be about ±2LSB or ±0.2%. If the more precise calibration is desired, a 50Ω trimmer should be used instead. Set the analog input at 9.990 volts, and set the trimmer so that the output code is just at the transition between 1111111110 and 1111111111. Each LSB will then have a weight of 9.766mV. If a nominal full scale of 10.24 volts is desired (which makes the LSB have weight of exactly 10.00mV), a 100Ω resistor in series with a 100Ω trimmer (or a 200Ω trimmer with good resolution) should be used. Of course, larger full scale ranges can be arranged by using a larger input resistor, but linearity and full scale temperature coefficient may be compromised if the external resistor becomes a sizeable percentage of 5kΩ.

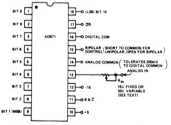

Figure 5. Standard AD571 Connections

BIPOLAR OPERATION

The standard unipolar 0 to +10V range is obtained by shorting the bipolar offset control pin to digital common. If the pin is left open, the bipolar offset current will be switched into the comparator summing node, giving a −5V to +5V range with an offset binary output code. (−5.00 volts in will give a 10-bit code of 0000000000; an input of 0.00 volts results in an output code of 1000000000 and 4.99 volts at the input yeilds the 1111111111 code). The bipolar offset control input is not directly TTL compatible, but a TTL interface for logic control can be constructed as shown in Figure 6.

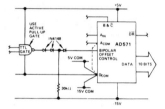

Figure 6. Bipolar Offset Controlled by Logic Gate

Gate Output = 1   Unipolar 0 - 10V Input Range
Gate Output = 0   Bipolar ±5V Input Range

COMMON MODE RANGE

The AD571 provides separate Analog and Digital Common connections. The circuit will operate properly with as much as ±200mV of common mode range between the two commons. This permits more flexible control of system common bussing and digital and analog returns.

In normal operation the Analog Common terminal may generate transient currents of up to 2mA during a conversion. In addition, a static current of about 2mA will flow into Analog Common in the unipolar mode after a conversion is complete. An additional 1mA will flow in during a blank interval with zero analog input. The Analog Common current will be modulated by the variations in input signal.

The absolute maximum voltage rating between the two commons is ±1 volt. We recommend the connection of a parallel pair of back-to-back protection diodes between the commons if they are not connected locally.

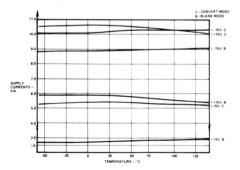

Figure 7. AD571 Power Supply Current vs. Temperature

Figure 4−12 (cont.)

ZERO OFFSET

The apparent zero point of the AD571 can be adjusted by inserting an offset voltage between the Analog Common of the device and the actual signal return or signal common. Figure 8 illustrates two methods of providing this offset. Figure 8A shows how the converter zero may be offset by up to ±3 bits to correct the device initial offset and/or input signal offsets. As shown, the circuit gives approximately symmetrical adjustment in unipolar mode. In bipolar mode R2 should be omitted to obtain a symmetrical range.

high accuracy is required, a 5Ω potentiometer (connected as a rheostat) can be used as R1. Additional negative offset range may be obtained by using larger values of R1. Of course, if the zero transition point is changed, the full scale transition point will also move. Thus, if an offset of ½LSB is introduced, full scale trimming as described on previous page should be done with an analog input of 9.985 volts.

NOTE: During a conversion transient currents from the Analog Common terminal will disturb the offset voltage. Capacitive decoupling should not be used around the offset network. These transients will settle as appropriate during a conversion. Capacitive decoupling will "pump up" and fail to settle resulting in conversion errors. Power supply decoupling which returns to analog signal common, should go to the signal input side of the resistive offset network.

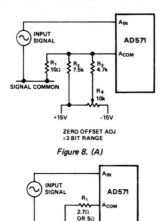

Figure 8. (A)

Figure 8. (B)

Figure 9 shows the nominal transfer curve near zero for an AD571 in unipolar mode. The code transitions are at the edges of the nominal bit weights. In some applications it will be preferable to offset the code transitions so that they fall between the nominal bit weights, as shown in the offset characteristics. This offset can easily be accomplished as shown in Figure 8B. At balance (after a conversion) approximately 2mA flows into the Analog Common terminal. A 2.7Ω resistor in series with this terminal will result in approximately the desired ½ bit offset of the transfer characteristics. The nominal 2mA Analog Common current is not closely controlled in manufacture. If

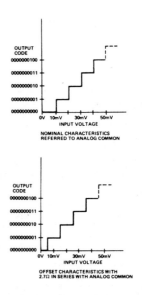

Figure 9. AD571 Transfer Curve – Unipolar Operation (Approximate Bit Weights Shown for Illustration, Nominal Bit Weights ~9.766mV)

Figure 4–12 *(cont.)*

Control and Timing of the AD571

CONTROL AND TIMING OF THE AD571

There are several important timing and control features on the AD571 which must be understood precisely to allow optimal interfacing to microprocessor or other types of control systems. All of these features are shown in the timing diagram in Figure 10.

The normal stand-by situation is shown at the left end of the drawing. The BLANK and $\overline{\text{CONVERT}}$ (B & $\overline{\text{C}}$) line is held high, the output lines will be "open", and the $\overline{\text{DATA READY}}$ ($\overline{\text{DR}}$) line will be high. This mode is the lowest power state of the device (typically 150mW). When the (B & $\overline{\text{C}}$) line is brought low, the conversion cycle is initiated; but the $\overline{\text{DR}}$ and Data lines do not change state. When the conversion cycle is complete (typically 25μs), the $\overline{\text{DR}}$ line goes low, and within 500ns, the Data lines become active with the new data.

About 1.5μs after the B & $\overline{\text{C}}$ line is again brought high, the $\overline{\text{DR}}$ line will go high and the Data lines will go open. When the B & $\overline{\text{C}}$ line is again brought low, a new conversion will begin. The minimum pulse width for the B & $\overline{\text{C}}$ line to blank previous data and start a new conversion is 2μs. If the B & $\overline{\text{C}}$ line is brought high during a conversion, the conversion will stop, and the $\overline{\text{DR}}$ and Data lines will not change. If a 2μs or longer pulse is applied to the B & C line during a conversion, the converter will clear and start a new conversion cycle.

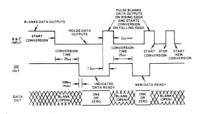

Figure 10. AD571 Timing and Control Sequence

CONTROL MODES WITH BLANK AND $\overline{\text{CONVERT}}$

The timing sequence of the AD571 discussed above allows the device to be easily operated in a variety of systems with differing control modes. The two most common control modes, the Convert Pulse Mode, and the Multiplex Mode, are illustrated here.

Convert Pulse Mode — In this mode, data is present at the output of the converter at all times except when conversion is taking place. Figure 11 illustrates the timing of this mode. The BLANK and $\overline{\text{CONVERT}}$ line is normally low and conversions are triggered by a positive pulse. A typical application for this timing mode is shown in Figure 14 in which μP bus interfacing is easily accomplished with three-state buffers.

Multiplex Mode — In this mode the outputs are blanked except when the device is selected for conversion and readout; this timing is shown in Figure 12. A typical AD571 multiplexing application is shown in Figure 15.

This operating mode allows multiple AD571 devices to drive common data lines. All BLANK and $\overline{\text{CONVERT}}$ lines are held high to keep the outputs blanked. A single AD571 is selected, its BLANK and $\overline{\text{CONVERT}}$ line is driven low and at the end of conversion, which is indicated by $\overline{\text{DATA READY}}$ going low, the conversion result will be present at the outputs. When this data has been read from the 10-bit bus, BLANK and $\overline{\text{CONVERT}}$ is restored to the blank mode to clear the data bus for other converters. When several AD571's are multiplexed in sequence, a new conversion may be started in one AD571 while data is being read from another. As long as the data is read and the first AD571 is cleared within 15μs after the start of conversion of the second AD571, no data overlap will occur.

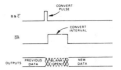

Figure 11. Convert Pulse Mode

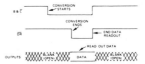

Figure 12. Multiplex Mode

SAMPLE-HOLD AMPLIFIER CONNECTION TO THE AD571

Many situations in high-speed acquisition systems or digitizing of rapidly changing signals require a sample-hold amplifier (SHA) in front of the A-D converter. The SHA can acquire and hold a signal faster than the converter can perform a conversion. A SHA can also be used to accurately define the exact point in time at which the signal is sampled. For the AD571, a SHA can also serve as a high input impedance buffer.

Figure 13 shows the AD571 connected to the AD582 monolithic SHA for high speed signal acquisition. In this configuration, the AD582 will acquire a 10 volt signal in less than 10μs with a droop rate less than 100μV/ms. The control signals are arranged so that when the control line goes low, the AD582 is put into the "hold" mode, and the AD571 will begin its conversion cycle. (The AD582 settles to final value well in advance of the

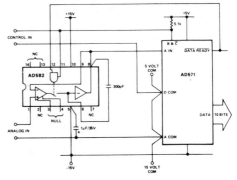

Figure 13. Sample-Hold Interface to the AD571

Figure 4–12 (*cont.*)

first comparator decision inside the AD571). The $\overline{\text{DATA READY}}$ line is fed back to the other side of the differential-input control gate so that the AD582 cannot come out of the "hold" mode during the conversion cycle. At the end of the conversion cycle, the $\overline{\text{DATA READY}}$ line goes low, automatically placing the AD582 back into the sample mode. This feature allows simple control of both the SHA and the A-D converter with a single line. Observe carefully the ground, supply, and bypass capacitor connections between the two devices. This will minimize ground noise and interference during the conversion cycle to give the most accurate measurements.

INTERFACING THE AD571 TO A MICROPROCESSOR

The AD571 can easily be arranged to be driven from standard microprocessor control lines and to present data to any standard microprocessor bus (4-, 8-, 12-or 16-bit) with a minimum of additional control components. The configuration shown in Figure 14 is designed to operate with 8-bit bus and standard 8080 control signals.

The input control circuitry shown is required to insure that the AD571 receives a sufficiently long B & $\overline{\text{C}}$ input pulse. When the converter is ready to start a new conversion, the B & $\overline{\text{C}}$ line is low, and $\overline{\text{DR}}$ is low. To command a conversion, the start address decode line goes low, followed by $\overline{\text{WR}}$. The B & $\overline{\text{C}}$ line will now go high, followed about 1.5μs later by $\overline{\text{DR}}$. This resets the external flip-flop and brings B & $\overline{\text{C}}$ back to low, which initiates the conversion cycle. At the end of the conversion cycle, the $\overline{\text{DR}}$ line goes low, the data outputs will become active with the new data and the control lines will return to the stand-by state. The new data will remain active until a new conversion is commanded. The self-pulsing nature of this circuit guarantees a sufficient convert pulse width.

This new data can now be presented to the data bus by enabling the three-state buffers when desired. A data word (8-bit or 2-bit) is loaded onto the bus when its decoded address goes low and the $\overline{\text{RD}}$ line goes low. This arrangement presents data to the bus "left-justified," with highest bits in the 8-bit word; a "right-justified" data arrangement can be set up by a simple re-wiring. Polling the converter to determine if conversion is complete can be done by addressing the gate which buffers the $\overline{\text{DR}}$ line, as shown. In this configuration, there is no need for additional buffer register storage since the data can be held indefinitely in the AD571, since the B & $\overline{\text{C}}$ line is continually held low.

BUS INTERFACING WITH A PERIPHERAL INTERFACE CIRCUIT

An improved technique for interfacing to a μP bus involves the use of special peripheral interfacing circuits (or I/O devices), such as the MC6821 Peripheral Interface Adapter (PIA). Shown in Figure 15 is a straighforward application of a PIA to multiplex up to 8 AD571 circuits. The AD571 has 3-state outputs, hence the data bit outputs can be paralleled, provided that only one converter at a time is permitted to be the active state. The $\overline{\text{DATA READY}}$ output of the AD571 is an open collector with resistor pull-up, thus several $\overline{\text{DR}}$ lines can be wire-ored to allow indication of the status of the selected device. One of the 8-bit ports of the PIA is combined with 2-bits from the other port and programmed as a 10-bit input port. The remaining 6-bits of the second port are programmed as outputs and along with the 2 control bits (which act as outputs), are used to control the 8 AD571's. When a control line is in the "1" or high state, the ADC will be automatically blanked. That is, its outputs will be in the inactive open state. If a single control line is switched low, its ADC will convert and the outputs will automatically go active when the conversion is complete. The result can be read from the two peripheral ports; when the next conversion is desired, a different control line can be switched to zero, blanking the previously active port at the same time. Subsequently, this second device can be read by the microprocessor, and so-forth. The status lines are wire-ored in 2 groups and connected to the two remaining control pins. This allows a conversion status check to be made after a convert command, if necessary. The ADC's are divided into two groups to minimize the loading effect of the internal pull-up resistors on the $\overline{\text{DATA READY}}$ buffers. See the Motorola MC6821 data sheet for more application detail.

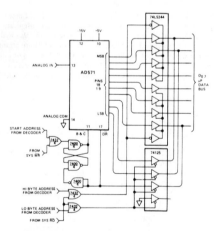

Figure 14. Interfacing AD571 to an 8-Bit Bus (8080 Control Structure)

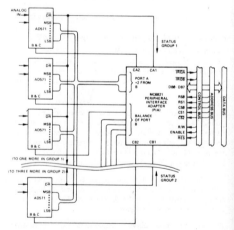

Figure 15. Multiplexing 8 AD571s Using Single PIA for μP Interface. No Other Logic Required (6800 Control Structure).

Figure 4–12 (*cont.*)

Sec. 4.3   IC ADCs   127

CMOS
13-Bit Monolithic A/D Converter
AD7550

FEATURES
Resolution: 13 Bits, 2's Complement
Relative Accuracy: ±1/2 LSB
"Quad Slope" Precision
  Gain Drift: 1ppm/°C
  Offset Drift: 1ppm/°C
Microprocessor Compatible
Ratiometric
Overrange Flag
Very Low Power Dissipation
TTL/CMOS Compatible
CMOS Monolithic Construction

AD7550 FUNCTIONAL BLOCK DIAGRAM

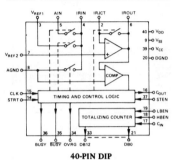

40-PIN DIP

GENERAL DESCRIPTION
The AD7550 is a 13-bit (2's complement) monolithic CMOS analog-to-digital converter on a 118 x 125 mil die packaged in a 40-pin ceramic DIP. Outstanding accuracy and stability (1ppm/°C) is obtained due to its revolutionary integrating technique, called "Quad Slope" (Analog Devices patent No. 3872466). This conversion consists of four slopes of integration as opposed to the traditional dual slope and provides much higher precision.

The AD7550 parallel output data lines have three-state logic and are microprocessor compatible through the use of two enable lines which control the lower eight LSB's (low byte enable) and the five MSB's (high byte enable). An overrange flag is also available which together with the BUSY and BUSY flags can be interrogated through the STATUS ENABLE providing easy microprocessor interface.

The AD7550 conversion time is about 40ms with a 1MHz clock. Clock can be externally controlled or internally generated by simply connecting a capacitor to the clock pin. A positive start pulse can be self-generated by having a capacitor on the start pin or can be externally applied.

For most applications, the AD7550 needs only three resistors, one capacitor, and a reference voltage since the integrating amplifier, comparator, switches and digital logic are all on the CMOS chip.

A wide range of power supply voltages (±5V to ±12V) with minuscule current requirements make the AD7550 ideal for low power and/or battery operated applications. Selection of the logic (V_{CC}) supply voltage (+5V to V_{DD}) provides direct TTL or CMOS interface on the digital input/output lines.

The AD7550 uses a high density CMOS process featuring double layer metal and silicon nitride passivation to ensure high reliability and long-term stability.

PIN CONFIGURATION

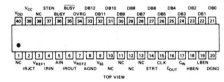

Figure 4–13 AD7550 ADC data sheets. (From *Data-Acquisition Databook*, 1982, Vol. 1. Courtesy of Analog Devices, Inc., Two Technology Way, Norwood, MA.)

SPECIFICATIONS (V_{DD} = +12V, V_{SS} = -5V, V_{CC} = +5V, V_{REF1} = +4.25V unless otherwise noted)[1]

PARAMETER	TA = +25°C	OVER SPECIFIED TEMPERATURE RANGE	TEST CONDITIONS
ACCURACY			
Resolution		13 Bits 2's Comp min	f_{CLK} = 500kHz, R_1 = 1MΩ,
Relative Accuracy	±1LSB max	±1 LSB max	C_1 = 0.01μF, IRJCT Voltage
Gain Error	±1LSB max		Adjusted to $\frac{V_{REF1}}{2}$ ±0.6%
Gain Error Drift	1ppm/°C typ		
Offset Error	±0.5LSB max		
Offset Error Drift	1ppm/°C typ		
ANALOG INPUTS			
AIN Input Resistance[2]	R1MΩ min		
V_{REF1} Input Resistance[2]	R1MΩ min		
V_{REF2} Leakage Current	10pA typ		
DIGITAL INPUTS			
CIN, LBEN, HBEN, STEN			
V_{INL}	+0.8V max	+0.8V max	V_{CC} = +5V
V_{INH}	+2.4V min	+2.4V min	
V_{INL}	+1.2V max	+1.2V max	V_{CC} = +12V
V_{INH}	+10.8V min	+10.8V min	
I_{INL}, I_{INH}	5nA typ		
START			
V_{INL}	+0.8V max	+0.8V max	V_{CC} = +5V to V_{DD}
V_{INH}	+2.4V min	+2.4V min	
I_{INL}	-1μA typ		V_{CC} = +5V to V_{DD}, BUSY = Low
I_{INH}	+150μA typ		V_{CC} = +5V to V_{DD}, BUSY = High
CLOCK			
V_{INL}	+0.8V max	+0.8V max	V_{CC} = +5V
V_{INH}	+3V min	+3V min	
V_{INL}	+1.2V max	+1.2V max	V_{CC} = +12V
V_{INH}	+10.8V min	+10.8V min	
I_{INL}	-100μA typ		V_{IN} = V_{INL}; V_{CC} = +5V to +12V
I_{INH}	+100μA typ		V_{IN} = V_{INH}; V_{CC} = +5V to +12V
DIGITAL OUTPUTS			
V_{OUTL}	+0.5V max	+0.8V max	V_{CC} = +5V, I_{SINK} = 1.6mA
V_{OUTH}	+2.4V min	+2.4V min	V_{CC} = +5V, I_{SOURCE} = 40μA
V_{OUTL}	+1.2V max	+1.2V max	V_{CC} = +12V, I_{SINK} = 1.6mA
V_{OUTH}	+10.8V min	+10.8V min	V_{CC} = +12V, I_{SOURCE} = 0.6mA
Capacitance (Floating State) (OVRG, BUSY, $\overline{BUSY}$, and DB0–DB12)	5pF typ		
I_{LKG} (Floating State) (OVRG, BUSY, BUSY, and DB0–DB12)	±5nA typ		V_{CC} = +5V to +12V, V_{OUT} = 0V and V_{CC}
DYNAMIC PERFORMANCE			
Conversion Time	90ms typ		$V_{IN(CLK)}$ = 0 to +3V, f_{CLK} = 500kHz
	40ms typ		$V_{IN(CLK)}$ = 0 to +3V, f_{CLK} = 1MHz
STEN, HBEN, LBEN Propagation Delay t_{ON}, t_{OFF}	250ns typ, 500ns max		V_{IN}(STEN, HBEN, LBEN) 0 to +3V
External STRT Pulse Duration	800ns min		$V_{IN(STRT)}$ = 0 to +3V
POWER SUPPLIES			
V_{DD} Range	+10V min, +12V max		
V_{SS} Range	-5V min, -12V max		
V_{CC} Range	+5V min, V_{DD} max		
I_{DD}	0.6mA typ, 2mA max		
I_{SS}	0.3mA typ, 2mA max		f_{CLK} = 1MHz
I_{CC}	0.06mA typ, 2mA max		

[1] Full Scale Voltage = ±V_{REF1} ÷ 2.125. For V_{REF1} = +4.25V, FS voltage is ±2.000V.
[2] The equivalent input circuit is the integrator resistor R_1 (1MΩ min, 10MΩ max) in series with a voltage source $\frac{V_{REF1}}{2}$, (see Figure 1).
Specifications subject to change without notice.

Figure 4–13 (cont.)

Sec. 4.3 IC ADCs

ORDERING INFORMATION

Model	Temperature Range	Package
AD7550BD	$-25°C$ to $+85°C$	Ceramic — (D40A)

See Section 20 for package outline information.

CAUTION:

1. The digital control inputs are zener protected; however, permanent damage may occur on unconnected units under high energy electrostatic fields. Keep unused units in conductive foam at all times. Prior to pulling the devices from the conductive foam, ground the foam to deplete any accumulated charge.
2. V_{CC} should never exceed V_{DD} by more than 0.4V, especially during power ON or OFF sequencing.

ABSOLUTE MAXIMUM RATINGS

V_{DD} to AGND . . . . 0V, +14V
V_{DD} to DGND . . . . 0V, +14V
V_{SS} to AGND . . . . 0V, −14V
V_{SS} to DGND . . . . 0V, −14V
AGND to DGND . . . . 0V, +14V
V_{CC} to DGND . . . . 0V, V_{DD}
V_{REF1} . . . . V_{SS}, V_{DD}
V_{REF2} . . . . AGND, V_{DD}
AIN . . . . V_{SS}, V_{DD}
IRIN . . . . V_{SS}, V_{DD}
IRJCT . . . . AGND, V_{DD}
IROUT . . . . V_{SS}, V_{DD}
Digital Input Voltage
   HBEN, LBEN, STEN, C_{IN} . . . . . . DGND, (DGND +27V)
   CLK, START . . . . DGND, V_{DD}
Digital Output Voltage
   DB0−DB12, OVRG, BUSY, $\overline{BUSY}$, C_{OUT} . . . DGND, V_{CC}
Power Dissipation (Package)
   Up to $+50°C$ 1000mW
   Derates above $+50°C$ by . . . . $10mW/°C$
Storage Temperature . . . . $-65°C$ to $+150°C$
Operating Temperature . . . . $-25°C$ to $+85°C$

Figure 4–13 (*cont.*)

PIN FUNCTION DESCRIPTION

PIN	MNEMONIC	DESCRIPTION
1	NC	No Connection
2	IRJCT	IntegratoR JunCTion. Summing junction (negative input) of integrating amplifier.
3	V_{REF1}	Voltage REFerence Input
4	IRIN	IntegratoR INput. External integrator input R is connected between IRJCT and IRIN.
5	AIN	Analog INput. Unknown analog input voltage to be measured. Fullscale AIN equals $V_{REF}/2.125$.
6	IROUT	IntegratoR OUTput. External integrating capacitor C_1 is connected between IROUT and IRJCT.
7	V_{REF2}	Voltage REFerence ÷ 2 Input
8	AGND	Analog GrouND
9	V_{SS}	Negative Supply (-5V to -12V)
10	NC	No Connection
11	NC	No Connection
12	NC	No Connection
13	NC	No Connection
14	STRT	STaRT Conversion. When STRT goes to a Logic "1," the AD7550's digital logic is set up and BUSY is latched "high." When STRT returns "low," conversion begins in synchronization with CLK. Reinitiating STRT during conversion causes a conversion restart. STRT can be driven from an external logic source or can be programmed for continuous conversion by connecting an external capacitor between STRT and DGND. An externally applied STRT command must be a positive pulse of at least 800 nanoseconds to ensure proper set-up of the AD7550 logic.
15	CLK	CLocK Input. The CLK can be driven from external logic, or can be programmed for internal oscillation by connecting an external capacitor between CLK and DGND.
16	C_{OUT}	Count OUT provides a number (N) of gated clock pulses given by: $$N = \left[\frac{AIN}{V_{REF1}} 2.125 + 1 \right] 4096$$
17	C_{IN}	Count IN is the input to the output counter. 2's complement binary data appears on the DB0 through DB12 output lines (if the HBEN and LBEN enable lines are "high") if C_{OUT} is connected to C_{IN}.
18	HBEN	High Byte ENable is the three-state logic enable input for the DB8-DB12 data outputs. When HBEN is low, the DB8-DB12 outputs are floating. When HBEN is "high," digital data appears on the data lines.
19	LBEN	Low Byte ENable is the three-state logic enable for DB0-DB7. When LBEN is "low," DB0-DB7 are floating. When "high," digital data appears on the data lines.
20	DGND	Digital GrouND is the ground return for all digital logic and the comparator.
21	DB0	Data Bit 0 (least significant bit)
22	DB1	
23	DB2	
24	DB3	
25	DB4	
26	DB5	
27	DB6	CODE: 2's Complement
28	DB7	
29	DB8	
30	DB9	
31	DB10	
32	DB11	
33	DB12	Data Bit 12 (most significant bit)
34	OVRG	OVerRange indicates a Logic "1" if AIN exceeds plus or minus full scale by at least 1/2 LSB. OVRG is a three-state output and floats until STEN is addressed with a Logic "1".
35	$\overline{BUSY}$	Not BUSY. $\overline{BUSY}$ indicates whether conversion is complete or in progress. $\overline{BUSY}$ is a three-state output which floats until STEN is addressed with a Logic "1." When addressed, $\overline{BUSY}$ will indicate either a "1" (conversion complete) or a "0" (conversion in progress).
36	BUSY	BUSY indicates conversion status. BUSY is a three-state output which floats until STEN is addressed with a Logic "1." When addressed, BUSY indicates a "0" (conversion complete) or a "1" (conversion in progress).
37	STEN	STatus ENable is the three-state control input for BUSY, $\overline{BUSY}$, and OVRG.
38	NC	No Connection
39	V_{CC}	Logic Supply. Digital inputs and outputs are TTL compatible if V_{CC} = +5V, CMOS compatible for V_{CC} = +10V to V_{DD}.
40	V_{DD}	Positive Supply +10V to +12V.

Figure 4-13 (*cont.*)

PRINCIPLES OF OPERATION

BASIC OPERATION

The essence of the quad slope technique is best explained through Figures 1 and 2.

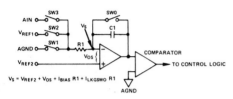

Figure 1. Quad Slope Integrator Circuit

The inputs AGND (analog ground), V_{REF1} and AIN (analog input) are applied in sequence to an integrator (Figure 1), creating four slopes (phases 1 through 4, Figure 2) at the integrator output. Voltage V_S is ideally equal to $\frac{V_{REF1}}{2}$, but if not, will create an error count "n" that will be minimized by the "quad-slope" conversion process. V_{REF1} and V_{REF2} must be positive voltages.

The equivalent integrator input voltages and their integration times are shown in Table 1.

TABLE 1
INTEGRATOR EQUIVALENT INPUT VOLTAGES AND INTEGRATION TIMES

Phase	Input Voltage	Integration Time
1	$AGND - V_S$	$t_1 = K_1 t$
2	$V_{REF1} - V_S$	$t_2 = (K_1 + n)t$
3	$AIN - V_S$	$t_3 = (2K_1 - n)t$
4	$V_{REF1} - V_S$	$t_4 = (K_3 - 2K_1 + n - 2N)t$

NOTE: Ideally $V_S = V_{REF2} = 1/2\ V_{REF1}$

where:
- t = The CLK period
- n = System error count
- K_1 = A fixed count equal to 4352 counts
- K_2 = A fixed count equal to 17408 counts ($K_2 = 4K_1$)
- K_3 = A fixed count equal to 25600 counts
- N = Digital output count corresponding to the analog input voltage, AIN

PHASE 0
After the start pulse is applied, switch SW2 is closed (all other switches open) and the integrator output is ramped to comparator zero crossing. Phase 0 can be considered the reset phase of the converter, and always has a duration $t_0 = R_1 C_1$ (integrator time constant). Upon zero crossing, counters K_1 and K_2 are started, switch SW2 is opened and SW1 is closed.

PHASE 1
Phase 1 integrates $(AGND - V_S)$ for a fixed period of time (by counter K_1) equal to $t_1 = K_1 t$. At the end of phase 1, switch SW1 is opened and SW2 is closed.

PHASE 2
The integrator input is switched to $(V_{REF1} - V_S)$ and the output ramps down until zero crossing is achieved. The integration time $t_2 = (K_1 + n)t$ includes the error count "n" due to offsets, etc. At the end of phase 2, switch SW2 is opened, SW3 is closed, and a third counter (K_3) is started.

PHASE 3
Phase 3 integrates the analog input $(AIN - V_S)$ until counter K_2 counts $4K_1 t$. At this time SW3 is opened and SW2 is closed again.

PHASE 4
Phase 4 integrates $(V_{REF1} - V_S)$ and the comparator output ramps down until zero crossing once again is achieved. Since the comparator always approaches zero crossing from the same slope, propagation delay is constant and hysteresis effect is eliminated.

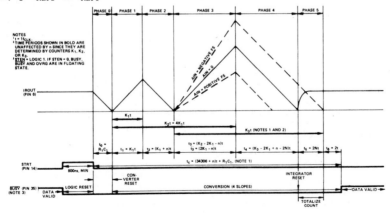

Figure 2. Quad Slope Timing Diagram

Figure 4-13 (*cont.*)

The time t_5 between the phase 4 zero crossing and the termination of counter K_3 is considered equal to 2N counts. N, the number of counts at the C_{OUT} terminal, is obtained by a divide-by-two counter stage. This reduces "jitter" effect. Barring third (and higher) order effects, it can be proven that:

$$N = \underbrace{\left(\frac{A_{IN}}{V_{REF1}} - 1\right) \cdot 2K_1 + \frac{K_3}{2}}_{\text{ideal transfer function}} + \underbrace{\left(\frac{A_{IN}}{V_{REF1}} - 1\right) \cdot \left[\frac{AGND}{V_{REF1}}(1 + 2\alpha) - \alpha^2\right] \cdot 2K_1}_{\text{error term}} \quad \text{(EQN 1)}$$

where:

$$\alpha = \frac{2V_S - V_{REF1}}{V_{REF1}}$$

The ideal case assumes:

AGND = 0V
$V_S = \frac{V_{REF1}}{2}$, therefore $\alpha = 0$

Then (EQN 1) simplifies to:

$$N = \frac{A_{IN}}{V_{REF1}} \cdot 8704 + 4096 \quad \text{(EQN 2)}$$

or

$$N = \frac{A_{IN}}{V_{FS}} \cdot 4096 + 4096 \quad \text{(EQN 3)}$$

where:

V_{FS} = full scale input voltage = $\frac{V_{REF1}}{2.125}$

The parallel output (DB0–DB12) of the AD7550 represents the number N in binary 2's complement coding when the C_{OUT} pin is connected to the C_{IN} pin (see Table II).

TABLE II
OUTPUT CODING (Bipolar 2's Complement)

Analog Input (Note 1)	N (Note 2)	Parallel Digital Output (Note 3)		
		OVRG	DB12	DB0
+Overrange	—	1	0	1111 1111 1111
+VFS (1-2⁻¹²)	8191	0	0	1111 1111 1111
+VFS (2⁻¹²)	4097	0	0	0000 0000 0001
0	4096	0	0	0000 0000 0000
−VFS (2⁻¹²)	4095	0	1	1111 1111 1111
−VFS	0	0	1	0000 0000 0000
−Overrange	—	1	1	0000 0000 0000

Notes
1 $V_{FS} = \frac{V_{REF1}}{2.125}$
2 N = number of counts at C_{OUT} pin
3 C_{OUT} strapped to C_{IN}; LBEN, HBEN and STEN = Logic 1

ERROR ANALYSIS

Equation 1 shows that only α and AGND generate error terms. Their impact can be analyzed as follows:

Case 1: AGND = 0, $\alpha \neq 0$

Error sources such as capacitor-leakage (I_L) and op amp offset (e) cause α to be different from zero.

Under this condition,

$$\alpha = \frac{2(e + I_L R_1)}{V_{REF1}}$$

where $I_L R_1$ is the equivalent error voltage generated by leakage I_L.

The evaluation of this error term is best demonstrated through the following example:

Assume:

e = 5mV, I_L = 5nA, R_1 = 1MΩ and V_{REF1} = 4.25V.

Then:

$\alpha = 4.7 \times 10^{-3}$

and:

$$N = \left[\frac{A_{IN}}{V_{REF1}} - 1\right] \times 8704 + 12800 - \underbrace{\left[\frac{A_{IN}}{V_{REF1}} - 1\right] \times 22.1 \times 10^{-6} \times 8704}_{\text{error term } N_\epsilon}$$

Therefore, the error count N_ϵ is as follows:

For AIN = −V_{FS}: N_ϵ = 0.28 counts = 0.28LSB
AIN = 0: N_ϵ = 0.19 counts = 0.19LSB
AIN = +V_{FS}: N_ϵ = 0.09 counts = 0.09LSB

The above example shows the strong reduction of the circuit errors because of the α^2 term in (EQN 1). Another consequence of this effect is that N_ϵ is always positive, regardless of the polarity of the circuit errors.

Case 2: AGND $\neq$ 0, $\alpha = 0$

When AGND is different from the signal ground, then this error will come through on a first-order basis. Indeed:

$$N = \left[\frac{A_{IN}}{V_{REF1}} - 1\right] \cdot 8704 + 12800 + \underbrace{\left[\frac{A_{IN}}{V_{REF1}} - 1\right] \cdot \frac{AGND}{V_{REF1}}}_{\text{error term } N_\epsilon}$$

The following example demonstrates the impact of AGND.

Let AGND = 1mV and V_{REF1} = 4.25V.

For AIN = −V_{FS}, then N_ϵ = 3.01 counts
AIN = 0, then N_ϵ = 2.05 counts
AIN = +V_{FS}, then N_ϵ = 1.08 counts

Therefore, ground loops should be minimized because a 330μV difference between AGND and signal ground will cause 1 count (1 LSB) of error when the analog input is at minus full scale. An optimized ground system is shown in Figure 7.

Figure 4–13 *(cont.)*

OPERATING GUIDELINES

The following steps, in conjunction with Figure 3, explain the calculations of the component values required for proper operation.

1. DETERMINATION OF V_{REF1}

When the full scale voltage requirement (VFS) has been ascertained, the reference voltage can be calculated by:

$V_{REF1} = 2.125 \, (V_{FS})$

V_{REF1} must be positive for proper operation.

2. SELECTION OF C_3 (INTERNAL CLOCK OPERATION)

For internal clock operation, connect capacitor C_3 to the clock pin as shown in Figure 3. The clock frequency versus capacitor C_3 is shown in Figure 4.

The clock frequency must be limited to 1.3MHz for proper operation.

3. SELECTION OF INTEGRATOR COMPONENTS (R_1 AND C_1)

To ensure that the integrator's output doesn't saturate to its bound (V_{DD}) during the phase (3) integration cycle, the integrator time constant ($R_1 C_1$) should be approximately equal to:

$$\pi = R_1 C_1 = \frac{V_{REF1} \, (9 \times 10^3)}{f_{CLK} \, (V_{DD} - 4V)}$$

The integrator components R_1 and C_1 can be selected by referring to Figure 5 and/or Figure 6. Figure 5 plots the time constant ($R_1 C_1$) versus clock frequency for different reference voltages. Figure 6 is a direct plot of the required C_1 versus f_{CLK} for R_1 values of 1MΩ and 10MΩ.

R_1 can be a standard 10% resistor, but must be selected between 1MΩ to 10MΩ.

The integrating capacitor "C_1" must be a low leakage, low dielectric absorption type such as teflon, polystyrene or polypropylene. To minimize noise, the outside foil of C_1 must be connected to IR_{OUT}.

4. CONVERSION TIME

As shown in Figure 2, the conversion time is independent of the analog input voltage AIN, and is given by:

$$t_{convert} = t_{STRT} + \frac{34306}{f_{CLK}} + R_1 C_1$$

where:

t_{STRT} = STRT pulse duration
$R_1 C_1$ = Integrator Time Constant
f_{CLK} = CLK Frequency

For example, if V_{EEF1} = 4.25V, R_1 = 1MΩ, C_1 = 4,000pF and CLK = 1MHz, the conversion time (not including t_{STRT}, which is normally only microseconds in duration) is approximately 40 milliseconds.

5. EXTERNAL OR AUTO STRT OPERATION

The STRT pin can be driven externally, or with the addition of C_2, made to self-start.

The size of C_2 determines the length of time from end of conversion until a new conversion is initiated. This is the "data valid" time and is given by:

$$t_{DAV} \approx (1.7 \times 10^6 \Omega) \, C_2 + 20\mu s$$

When first applying power to the AD7550, a 0V to V_{DD} positive pulse (power up restart) is required at the STRT terminal to initiate auto STRT operation.

6. INITIAL CALIBRATION

Trim R_4 (Figure 3) so that pin 2 (IRJCT) equals 1/2 V_{REF1} ±0.6%. When measuring the voltage on IRJCT, apply a Logic "1" to the STRT terminal.

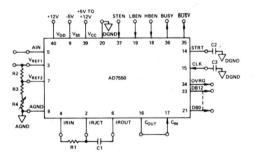

Figure 3. Operation Diagram

Figure 4–13 (cont.)

APPLICATION HINTS

When operating at f_{CLK} greater than 500kHz, the following steps are recommended to minimize errors due to noise coupling (see Figure 7).

1. Decouple AIN (pin 5), V_{REF1} (pin 3) and V_{REF2} (pin 7) through $0.01\mu F$ to signal ground.
2. Signal ground must be located as close to pin 8 (AGND) as possible.
3. Keep the lead lengths of R_1 and C_1 toward pin 2 (IRJCT) as short as possible. In addition, both components should lie over the analog ground plane. If C_1 has an outside foil, connect it to pin 6 (IROUT), not pin 2.
4. Hold the data bit enables (HBEN, LBEN) in the 0 state during conversion. This is easily accomplished by tying STEN to the 1 state and driving HBEN and LBEN with BUSY. This prevents the DB0 through DB12 outputs from coupling noise into the integrator during the phase 1–4 active integration periods.

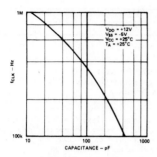

Figure 4. f_{CLK} vs. C_3

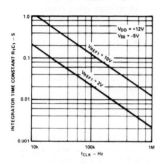

Figure 5. Integrator Time Constant $(R_1 C_1)$ vs. f_{CLK} for Different Reference Voltages

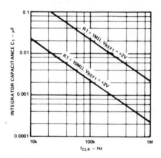

Figure 6. Integrator Capacitance (C_1) vs. f_{CLK} for Different Integrator Resistances (R_1)

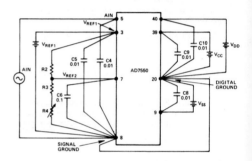

Figure 7. Ground System

Figure 4–13 (cont.)

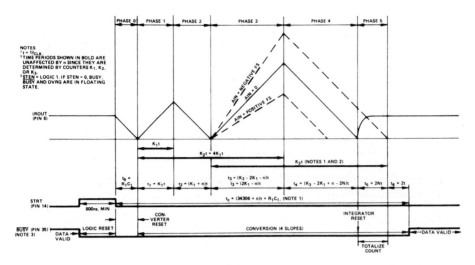

Figure 4–14 Quad slope Timing diagram. (From *Data-Acquisition Databook*, 1982, Vol. 1. Courtesy of Analog Devices, Inc., Two Technology Way, Norwood, MA.)

4.4 DATA ACQUISITION ICs

Several devices are available that provide the capability of accepting several analog signals. These are termed *data acquisition subsystems*. A typical system is the AD363, which will accept up to 16 channels of data and provide a 12-bit binary output. Figure 4–15 shows a block diagram of this device. The internal units include a multiplexer (MUX), a differential amplifier, a track-and-hold circuit, and a 12-bit successive-approximation ADC.

Subsystems such as this can be used in a system that monitors several signals, converting each one to binary. A 4-bit control signal is used to select one of the 16 channels for conversion.

Complete 16-Channel 12-Bit Integrated Circuit Data Acquisition System

AD363

FEATURES
Versatility
- Complete System in Reliable IC Form
- Small Size
- 16 Single-Ended or 8 Differential Channels with Switchable Mode Control
- Military/Aerospace Temperature Range: $-55°C$ to $+125°C$ (AD363S) MIL-STD-883B Processing Available
- Versatile Input/Output/Control Format
- Short-Cycle Capability

Performance
- True 12-Bit Operation: Nonlinearity $\leq \pm 0.012\%$
- Guaranteed No Missing Codes Over Temperature Range
- High Throughput Rate: 30kHz
- Low Power: 1.7W

Value
- Complete: No Additional Parts Required
- Reliable: Hybrid IC Construction, Hermetically Sealed. All Inputs Fully Protected
- Precision $+10.0 \pm 0.005$ Volt Reference for External Application
- Fast Precision Buffer Amplifier for External Application
- Low Cost

PRODUCT DESCRIPTION

The AD363 is a complete 16 channel, 12-bit data acquisition system in integrated circuit form. By applying large-scale linear and digital integrated circuitry, thick and thin film hybrid technology and active laser trimming, the AD363 equals or exceeds the performance and versatility of previous modular designs.

The AD363 consists of two separate functional blocks. Each in hermetically-sealed 32 pin dual-in-line packages. The analog input section contains two eight-channel multiplexers, a differential amplifier, a sample-and-hold, a channel address register and control logic. The multiplexers may be connected to the differential amplifier to perform in either an 8-channel differential or 16-channel single-ended configuration. A unique feature of the AD363 is an internal user-controllable analog switch that connects the multiplexers in either a single-ended or differential mode. This allows a single device to perform in either mode without hard-wire programming and permits a mixture of single-ended and differential sources to be interfaced to an AD363 by dynamically switching the input mode control.

AD363 FUNCTIONAL BLOCK DIAGRAM

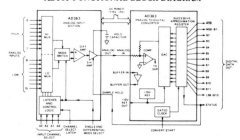

TWO-32 PIN DIPS

The Analog-to-Digital Converter Section contains a complete 12-bit successive approximation analog-to-digital converter, including internal clock, precision 10 volt reference, comparator, buffer amplifier and a proprietary-design 12-bit D/A converter. Active laser trimming of the reference and D/A converter results in maximum linearity errors of $\pm 0.012\%$ while performing a 12-bit conversion in 25 microseconds.

Analog input voltage ranges of ± 2.5, ± 5.0, ± 10, 0 to $+5$ and 0 to $+10$ volts are user-selectable. Adding flexibility and value are the precision 10 volt reference (active-trimmed to a tolerance of ± 5mV) and the internal buffer amplifier, both of which may be used for external applications. All digital signals are TTL/DTL compatible and output data is positive-true in parallel and serial form.

System throughput rate is as high as 30kHz at full rated accuracy. The AD363K is specified for operation over a 0 to $+70°C$ temperature range while the AD363S operates to specification from $-55°C$ to $+125°C$. Processing to MIL-STD-883B is available for the AD363S. Both device grades are guaranteed to have no missing codes over their specified temperature ranges.

Figure 4–15 AD363 data sheets. (From *Data-Aquisition Databook*, 1982, Vol. 1. Courtesy of Analog Devices, Inc., Two Technology Way, Norwood, MA.)

SPECIFICATIONS
(typical @ +25°C, ±15V and +5V with 2000pF hold capacitor as provided unless otherwise noted)

MODEL	AD363K	AD363S
ANALOG INPUTS		
Number of Inputs	16 Single-Ended or 8 Differential (Electronically Selectable)	*
Input Voltage Ranges		
Bipolar	±2.5V, ±5.0V, ±10.0V	*
Unipolar	0 to +5V, 0 to +10V	*
Input (Bias) Current, Per Channel	±50nA max	*
Input Impedance		
On Channel	$10^{10}\,\Omega$, 100pF	*
Off Channel	$10^{10}\,\Omega$, 10pF	*
Input Fault Current (Power Off or On)	20mA, max, Internally Limited	*
Common Mode Rejection		
Differential Mode	70dB min (80dB typ) @ 1kHz, 20V p-p	*
Mux Crosstalk (Interchannel,		
Any Off Channel to Any On Channel)	−80dB max (−90dB typ) @ 1kHz, 20V p-p	*
RESOLUTION	12 BITS	*
ACCURACY		
Gain Error[1]	±0.05% FSR (Adj. to Zero)	*
Unipolar Offset Error	±10mV (Adj to Zero)	*
Bipolar Offset Error	±20mV (Adj to Zero)	*
Linearity Error	±½LSB max	*
Differential Linearity Error	±1LSB max (±½LSB typ)	*
Relative Accuracy	±0.025% FSR	*
Noise Error	1mV p-p, 0 to 1MHz	*
TEMPERATURE COEFFICIENTS		
Gain	±30ppm/°C max (±10ppm/°C typ)	±25ppm/°C max (±15ppm/°C typ)
Offset, ±10V Range	±10ppm/°C max (±5ppm/°C typ)	±8ppm/°C max (±5ppm/°C typ)
Differential Linearity	No Missing Codes Over Temperature Range	*
SIGNAL DYNAMICS		
Conversion Time[2]	25μs max (22μs typ)	*
Throughput Rate, Full Rated Accuracy	25kHz min (30kHz typ)	*
Sample and Hold		
Aperture Delay	100ns max (50ns typ)	*
Aperture Uncertainty	500ps max (100ps typ)	*
Acquisition Time		
To ±0.01% of Final Value for Full Scale Step	18μs max (10μs, typ)	*
Feedthrough	−70dB max (−80dB typ) @ 1kHz	*
Droop Rate	2mV/ms max (1mV/ms typ)	*
DIGITAL INPUT SIGNALS[3]		
Convert Command (to ADC Section, Pin 21)	Positive Pulse, 200ns min Width. Leading Edge ("0" to "1") Resets Register, Trailing Edge ("1" to "0") Starts Conversion. 1TTL Load	*
Input Channel Select (To Analog Input Section, Pins 28-31)	4 Bit Binary, Channel Address. 1LS TTL Load	*
Channel Select Latch (To Analog Input Section, Pin 32)	"1" Latch Transparent "0" Latched 4LS TTL Loads	*

Figure 4–15 (cont.)

MODEL	AD363K	AD363S
DIGITAL INPUT SIGNALS, cont.		
Sample-Hold Command (To Analog Input Section Pin 13 Normally Connected To ADC "Status", Pin 20)	"0" Sample Mode "1" Hold Mode 2LS TTL Loads	* * *
Short Cycle (To ADC Section Pin 14)	Connect to +5V for 12 Bits Resolution. Connect to Output Bit n + 1 For n Bits Resolution. 1TTL Load	* * *
Single Ended/Differential Mode Select (To Analog Input Section, Pin 1)	"0": Single-Ended Mode "1": Differential Mode 3TTL Loads	* * *
DIGITAL OUTPUT SIGNALS[3] (All Codes Positive True) Parallel Data		
Unipolar Code	Binary	*
Bipolar Code	Offset Binary/Two's Complement	*
Output Drive	2TTL Loads	*
Serial Data (NRZ Format)		
Unipolar Code	Binary	*
Bipolar Code	Offset Binary	*
Output Drive	2TTL Loads	*
Status ($\overline{Status}$)	Logic "1" ("0") During Conversion	*
Output Drive	2TTL Loads	*
Internal Clock		
Output Drive	2TTL Loads	*
Frequency	500kHz	*
INTERNAL REFERENCE VOLTAGE	+10.00V, ±10mV	*
Max External Current	±1mA	*
Voltage Temp. Coefficient	±20ppm/°C, max	*
POWER REQUIREMENTS		
Supply Voltages/Currents	+15V, ±5% @ +45mA max (+38mA typ) −15V, ±5% @ −45mA max (−38mA typ) +5V, ±5% @ +136mA max (+113mA typ)	* * *
Total Power Dissipation	2 watts max (1.7 watts typ)	*
TEMPERATURE RANGE		
Specification	0 to +70°C	−55°C to +125°C
Storage	−55°C to +85°C[4]	−55°C to +150°C

NOTES:
[1] With 50Ω, 1% fixed resistor in place of Gain Adjust pot; see Figures 7 and 8.
[2] Conversion time of ADC Section.
[3] One TTL Load is defined as I_{IL} = −1.6mA max @ V_{IL} = 0.4V, I_{IH} = 40μA max @ V_{IH} = 2.4V.
One LS TTL Load is defined as I_{IL} = −0.36mA max @ V_{IL} = 0.4V, I_{IH} = 20μA max @ V_{IH} = 2.7V.
[4] AD363K External Hold Capacitor is limited to +85°C; Analog Input Section and ADC Section may be stored at up to +150°C.
*Specifications same as AD363K.
Specifications subject to change without notice.

ABSOLUTE MAXIMUM RATINGS (ALL MODELS)	
+V, Digital Supply	+5.5V
+V, Analog Supply	+16V
−V, Analog Supply	−16V
V_{IN}, Signal	±V, Analog Supply
V_{IN}, Digital	0 to +V, Digital Supply
A_{GND} to D_{GND}	±1V

Figure 4–15 (cont.)

Sec. 4.4 Data Acquisition ICs

PIN FUNCTION DESCRIPTION

ANALOG INPUT SECTION		ANALOG TO DIGITAL CONVERTER SECTION	
Pin Number	Function	Pin Number	Function
1	Single-End/Differential Mode Select	1	Data Bit 12 (Least Significant Bit) Out
	"0": Single-Ended Mode	2	Data Bit 11 Out
	"1": Differential Mode	3	Data Bit 10 Out
2	Digital Ground	4	Data Bit 9 Out
3	Positive Digital Power Supply, +5V	5	Data Bit 8 Out
4	"High" Analog Input, Channel 7	6	Data Bit 7 Out
5	"High" Analog Input, Channel 6	7	Data Bit 6 Out
6	"High" Analog Input, Channel 5	8	Data Bit 5 Out
7	"High" Analog Input, Channel 4	9	Data Bit 4 Out
8	"High" Analog Input, Channel 3	10	Data Bit 3 Out
9	"High" Analog Input, Channel 2	11	Data Bit 2 Out
10	"High" Analog Input, Channel 1	12	Data Bit 1 (Most Significant Bit) Out
11	"High" Analog Input, Channel 0	13	Data Bit $\overline{1}$ ($\overline{MSB}$) Out
12	Hold Capacitor (Provided, See Figure 1)	14	Short Cycle Control
13	Sample-Hold Command		Connect to +5V for 12 Bits
	"0": Sample Mode		Connect to Bit (n+1) Out for n Bits
	"1": Hold Mode	15	Digital Ground
	Normally Connected to ADC Pin 20	16	Positive Digital Power Supply, +5V
14	Offset Adjust (See Figure 6)	17	$\overline{Status}$ Out
15	Offset Adjust (See Figure 6)		"0": Conversion in Progress
16	Analog Output		(Parallel Data Not Valid)
	Normally Connected to ADC		"1": Conversion Complete
	"Analog In" (See Figure 1)		(Parallel Data Valid)
17	Analog Ground	18	+10Volt Reference Out (See Figures 3, 7, 8, 11)
18	"High" ("Low") Analog Input, Channel 15 (7)	19	Clock Out (Runs During Conversion)
19	"High" ("Low") Analog Input, Channel 14 (6)	20	Status Out
20	Negative Analog Power Supply, -15V		"0": Conversion Complete
21	Positive Analog Power Supply, +15V		(Parallel Data Valid)
22	"High" ("Low") Analog Input, Channel 13 (5)		"1": Conversion in Progress
23	"High" ("Low") Analog Input, Channel 12 (4)		(Parallel Data Not Valid)
24	"High" ("Low") Analog Input, Channel 11 (3)	21	Convert Start In
25	"High" ("Low") Analog Input, Channel 10 (2)		Reset Logic :
26	"High" ("Low") Analog Input, Channel 9 (1)		Start Convert :
27	"High" ("Low") Analog Input, Channel 8 (0)	22	Comparator In (See Figures 3, 7, 8)
28	Input Channel Select, Address Bit AE	23	Bipolar Offset
29	Input Channel Select, Address Bit A0		Open for Unipolar Inputs
30	Input Channel Select, Address Bit A1		Connect to ADC Pin 22 for
31	Input Channel Select, Address Bit A2		Bipolar Inputs
32	Input Channel Select Latch		(See Figure 8)
	"0": Latched	24	10V Span R In (See Figure 7)
	"1": Latch "Transparent"	25	20V Span R In (See Figure 8)
		26	Analog Ground
		27	Gain Adjust (See Figures 7 and 8)
		28	Positive Analog Power Supply, +15V
		29	Buffer Out (For External Use)
		30	Buffer In (For External Use)
		31	Negative Analog Power Supply, -15V
		32	Serial Data Out
			Each Bit Valid On Trailing ( )
			Edge Clock Out, ADC Pin 19

Figure 4–15 (cont.)

AD363 Description

AD363 DESIGN

Concept

The AD363 consists of two separate functional blocks as shown in Figure 1.

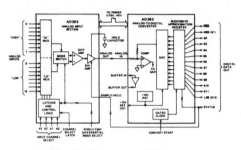

Figure 1. AD363 Functional Block Diagram

The Analog Input Section contains multiplexers, a differential amplifier, a sample-and-hold, a channel address register and control logic. Analog-to-digital conversion is provided by a 12 bit, 25 microsecond "ADC" which is also available separately as the AD572.

By dividing the data acquisition task into two sections, several important advantages are realized. Performance of each design is optimized for its specific function. Production yields are increased thus decreasing costs. Furthermore, the standard configuration 32 pin packages plug into standard sockets and are easier to handle than larger packages with higher pin counts.

Analog Input Section Design

Figure 2 is a block diagram of the AD363 Analog Input Section (AIS).

The AIS consists of two 8-channel multiplexers, a differential amplifier, a sample-and-hold, channel address latches and control logic. The multiplexers can be connected to the differential amplifier in either an 8-channel differential or 16-channel single-ended configuration. A unique feature of the AD363 is an internal analog switch controlled by a digital input that performs switching between single-ended and differential modes. This feature allows a single product to perform in either mode without external hard-wire interconnections. Of more significance is the ability to serve a mixture of both single-ended and differential sources with a single AD363 by dynamically switching the input mode control.

Multiplexer channel address inputs are interfaced through a level-triggered ("transparent") input register. With a Logic "1" at the Channel Select Latch input, the address signals feed through the register to directly select the appropriate input channel. This address information can be held in the register by placing a Logic "0" on the Channel Select Latch input. Internal logic monitors the status of the Single-Ended/Differential Mode input and addresses the multiplexers accordingly.

A differential amplifier buffers the multiplexer outputs while providing high input impedance in both differential and single-ended modes. Amplifier gain and common mode rejection are actively laser-trimmed.

The sample-and-hold is a high speed monolithic device that can also function as a gated operational amplifier. Its uncommitted differential inputs allow it to serve a second role as the output subtractor in the differential amplifier. This eliminates one amplifier and decreases drift, settling time and power consumption. A Logic "1" on the Sample-and-Hold Command input will cause the sample-and-hold to "freeze" the analog signal while the ADC performs the conversion. Normally the Sample-and-Hold Command is connected to the ADC Status output which is at Logic "1" during conversion and Logic "0" between conversions. For slowly-changing inputs, throughput speed may be increased by grounding the Sample-and-Hold Command input instead of connecting it to the ADC status.

A Polystyrene hold capacitor is provided with each commercial temperature range system (AD363K) while a Teflon capacitor is provided with units intended for operation at temperatures up to 125°C (AD363S). Use of an external capacitor allows the user to make his own speed/accuracy tradeoff; a smaller capacitor will allow faster sample-and-hold response but will decrease accuracy while a larger capacitor will increase accuracy at slower conversion rates.

The Analog Input Section is constructed on a substrate that includes thick-film resistors for non-critical applications such as input protection and biasing. A separately-mounted laser-trimmed thin-film resistor network is used to establish accurate gain and high common-mode rejection.

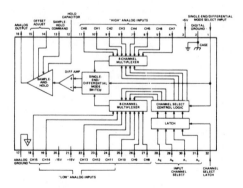

Figure 2. AD363 Analog Input Section Functional Block Diagram and Pinout

Figure 4–15 (cont.)

Sec. 4.4 Data Acquisition ICs

Analog-to-Digital Converter Design

Figure 3 is a block diagram of the Analog-to-Digital Converter Section (ADC) of the AD363.

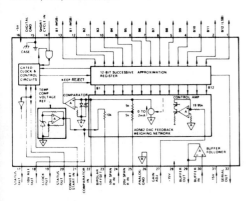

Figure 3. AD363 ADC Section (AD572) Functional Diagram and Pinout

Available separately as the AD572, the ADC is a 12-bit, 25 microsecond device that includes an internal clock, reference, comparator and buffer amplifier.

The +10V reference is derived from a low T.C. zener reference diode which has its zener voltage amplified and buffered by an op amp. The reference voltage is calibrated to +10V, ±10mV by active laser-trimming of the thin-film resistors which determine the closed-loop gain of the op amp. 1mA of current is available for external use.

The DAC chip uses 12 precision, high speed bipolar current steering switches, control amplifier and a laser-trimmed thin film resistor network to produce a very fast, high accuracy analog output current. This ladder network is active laser-trimmed to calibrate all bit ratio scale factors to a precision of 0.005% of FSR (full-scale range) to guarantee no missing codes over the operating temperature range. The design of the ADC includes scaling resistors that provide user-selectable analog input signal ranges of ±2.5, ±5, ±10, 0 to +5, or 0 to +10 volts.

Other useful features include true binary output for unipolar inputs, offset binary and two's complement output for bipolar inputs, serial output, short-cycle capability for lower resolution, higher speed measurements, and an available high input impedance buffer amplifier which may be used elsewhere in the system.

THEORY OF OPERATION

System Timing

Figure 4 is a timing diagram for the AD363 connected as shown shown in Figure 1 and operating at maximum conversion rate.

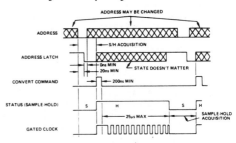

Figure 4. AD363 Timing Diagram

The normal sequence of events is as follows:

1. The appropriate Channel Select Address is latched into the address register. Time is allowed for the multiplexers to settle.
2. A Convert Start command is issued to the ADC which indicates that it is "busy" by placing a Logic "1" on its Status line.
3. The ADC Status controls the sample-and-hold. When the ADC is "busy" the sample-and-hold is in the hold mode.
4. The ADC goes into its 25 microsecond conversion routine. Since the sample-and-hold is holding the proper analog value, the address may be updated during conversion. Thus multiplexer settling time can coincide with conversion and need not effect throughput rate.
5. The ADC indicates completion of its conversion by returning Status to Logic "0". The sample-and-hold returns to the sample mode.
6. If the input signal has changed full-scale (different channels may have widely-varying data) the sample-and-hold will typically require 10 microseconds to "acquire" the next input to sufficient accuracy for 12-bit conversion.

After allowing a suitable interval for the sample-and-hold to stabilize at its new value, another Convert Start command may be issued to the ADC.

ADC Operation

On receipt of a Convert Start command, the analog-to-digital converter converts the voltage at its analog input into an equivalent 12-bit binary number. This conversion is accomplished as follows:

The 12-bit successive-approximation register (SAR) has its 12-bit outputs connected both to the respective device bit output pins and to the corresponding bit inputs of the feedback DAC.

Figure 4–15 (*cont.*)

Applying the AD363

The analog input is successively compared to the feedback DAC output, one bit at a time (MSB first, LSB last). The decision to keep or reject each bit is then made at the completion of each bit comparison period, depending on the state of the comparator at that time.

The versatility and completeness of the AD363 concept results in a large number of user-selectable configurations. This allows optimization of most systems applications.

Single-Ended/Differential Mode Control

The 363 features an internal analog switch that configures the Analog Input Section in either a 16-channel single-ended or 8-channel differential mode. This switch is controlled by a TTL logic input applied to pin 1 of the Analog Input Section:

"0": Single-Ended (16 channels)
"1": Differential (8 channels)

When in the differential mode, a differential source may be applied between corresponding "High" and "Low" analog input channels.

It is possible to mix SE and DIFF inputs by using the mode control to command the appropriate mode. Figure 11 illustrates an example of a "mixed" application. In this case, four microseconds must be allowed for the output of the Analog Input Section to settle to within ±0.01% of its final value, but if the mode is switched concurrent with changing the channel address, no significant additional delay is introduced. The effect of this delay may be eliminated by changing modes while a conversion is in progress (with the sample-and-hold in the "hold mode"). When SE and DIFF signals are being processed concurrently, the DIFF signals must be applied between corresponding "High" and "Low" analog input channels. Another application of this feature is the capability of measuring 16 sources individually and/or measuring differences between pairs of those sources.

Input Channel Addressing

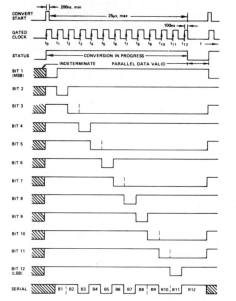

Figure 5. ADC Timing Diagram (Binary Code 110101011001)

The timing diagram is shown in Figure 5. Receipt of a Convert Start signal sets the Status flag, indicating conversion in progress. This, in turn, removes the inhibit applied to the gated clock, permitting it to run through 13 cycles. All SAR parallel bit and Status flip-flops are initialized on the leading edge, and the gated clock inhibit signal removed on the trailing edge of the Convert Start signal. At time t0, B1 is reset and B2-B12 are set unconditionally. At t1 the Bit 1 decision is made (keep) and Bit 2 is unconditionally reset. At t2, the Bit 2 decision is made (keep) and Bit 3 is reset unconditionally. This sequence continues until the Bit 12 (LSB) decision (keep) is made at t12. After 100ns delay period, the Status flag is reset, indicating that the conversion is complete and that the parallel output data is valid. Resetting the Status flag restores the gated clock inhibit signal, forcing the clock output to the Logic "0" state.

Corresponding serial and parallel data bits become valid on the same positive-going clock edge. Serial data does not change and is guaranteed valid on negative-going clock edges, however; serial data can be transferred quite simply by clocking it into a receiving shift register on these edges.

Incorporation of the 100ns delay period guarantees that the parallel (and serial) data are valid at the Logic "1" to "0" transition of the Status flag, permitting parallel data transfer to be initiated by the trailing edge of the Status signal.

Table 1 is the truth table for input channel addressing in both the single-ended and differential modes. The 16 single-ended channels may be addressed by applying the corresponding digital number to the four Input Channel Select address bits, AE, A0, A1, A2 (Analog Input Section, pins 28–31). In the differential mode, the eight channels are addressed by applying the appropriate digital code to A0, A1 and A2; AE must be enabled with a Logic "1". Internal logic monitors the status of the SE/DIFF Mode input and addresses the multiplexes singly or in pairs as required.

ADDRESS				ON CHANNEL (Pin Number)		
					Differential	
AE	A2	A1	A0	Single Ended	"Hi"	"Lo"
0	0	0	0	0 (11)	None	
0	0	0	1	1 (10)	None	
0	0	1	0	2 (9)	None	
0	0	1	1	3 (8)	None	
0	1	0	0	4 (7)	None	
0	1	0	1	5 (6)	None	
0	1	1	0	6 (5)	None	
0	1	1	1	7 (4)	None	
1	0	0	0	8 (27)	0 (11)	0 (27)
1	0	0	1	9 (26)	1 (10)	1 (26)
1	0	1	0	10 (25)	2 (9)	2 (25)
1	0	1	1	11 (24)	3 (8)	3 (24)
1	1	0	0	12 (23)	4 (7)	5 (23)
1	1	0	1	13 (22)	5 (6)	5 (22)
1	1	1	0	14 (19)	6 (5)	6 (19)
1	1	1	1	15 (18)	7 (4)	7 (18)

Table 1. Input Channel Addressing Truth Table

Figure 4–15 *(cont.)*

Sec. 4.4 Data Acquisition ICs

When the channel address is changed, six microseconds must be allowed for the Analog Input Section to settle to within ±0.01% of its final output (including settling times of all elements in the signal path). The effect of this delay may be eliminated by performing the address change while a conversion is in progress (with the sample-and-hold in the "hold" mode).

Input Channel Address Latch

The AD363 is equipped with a latch for the Input Channel Select address bits. If the Latch Control pin (pin 32 of the Analog Input Section) is at Logic "1", input channel select address information is passed through to the multiplexers. A Logic "0" "freezes" the input channel address present at the inputs at the time of the "1" to "0" transition.

This feature is useful when input channel address information is provided from an address, data or control bus that may be required to service many devices. The ability to latch an address is helpful whenever the user has no control of when address information may change.

Sample-and-Hold Mode Control

The Sample-and-Hold Mode Control input (Analog Input Section, pin 13) is normally connected to the Status output (pin 20) from the ADC section. When a conversion is initiated by applying a Convert Start command to the ADC (pin 21), Status goes to Logic "1", putting the sample-and-hold into the "hold" mode. This "freezes" the information to be digitized for the period of conversion. When the conversion is complete, Status returns to Logic "0" and the sample-and-hold returns to the sample mode. Eighteen microseconds must be allowed for the sample-and-hold to acquire ("catch up" to) the analog input to within ±0.01% of the final value before a new Convert Start command is issued.

The purpose of a sample-and-hold is to "stop" fast changing input signals long enough to be converted. In this application, it also allows the user to change channels and/or SE/DIFF mode while a conversion is in progress thus eliminating the effects of multiplexer, analog switch and differential amplifier settling times. If maximum throughput rate is required for slowly changing signals, the Sample-and-Hold Mode Control may be wired to ground (Logic "0") rather than to ADC Status thus leaving the sample-and-hold in a continuous sample mode.

Hold Capacitor

A 2000pF capacitor is provided with each AD363. One side of this capacitor is wired to the Analog Input Section pin 12, the other to analog ground as close to pin 17 as possible. The capacitor provided with the AD363K is Polystyrene while the wider operating temperature range of the AD363S demands a Teflon capacitor (supplied).

Larger capacitors may be substituted to minimize noise, but acquisition time of the sample-and-hold will be extended. If less than 12 bits of accuracy is required, a smaller capacitor may be used. This will shorten the S/H acquisition time. In all cases, the proper capacitor dielectric must be used; i.e., Polystyrene (AD363K only) or Teflon (AD363K or S). Other types of capacitors may have higher dielectric absorption (memory) and will cause errors. CAUTION: Polystyrene capacitors will be destroyed if subjected to temperatures above +85°C. No capacitor is required if the sample-and-hold is not used.

Short Cycle Control

A Short Cycle Control (ADC Section, pin 14) permits the timing cycle shown in Figure 5 to be terminated after any number of desired bits has been converted, permitting somewhat shorter conversion times in applications not requiring full 12-bit resolution. When 12-bit resolution is required, pin 14 is connected to +5V (ADC Section, pin 10). When 10-bit resolution is desired, pin 14 is connected to Bit 11 output pin 2. The conversion cycle then terminates, and the Status flag resets after the Bit 10 decision (t10 + 100ns in timing diagram of Figure 2). Short Cycle pin connections and associated maximum 12-, 10- and 8-bit conversion times are summarized in Table 2.

Connect Short Cycle Pin 14 to Pin:	Bits	Resolution (% FSR)	Maximum Conversion Time (μs)	Status Flag Reset at: (Figure 5)
16	12	0.024	25	t_{12} +100ns
2	10	0.10	21	t_{10} +100ns
4	8	0.39	17	t_8 +100ns

Table 2. Short Cycle Connections

One should note that the calibration voltages listed in Table 4 are for 12-bit resolution only, and are not those corresponding to the center of each discrete quantization interval at reduced bit resolution.

Digital Output Data Format

Both parallel and serial data are in positive-true form and outputted from TTL storage registers. Parallel data output coding is binary for unipolar ranges and either offset binary or two's complement binary, depending on whether Bit 1 (ADC Section pin 12) or its logical inverse $\overline{\text{Bit}}$ 1 (pin 13) is used as the MSB. Parallel data becomes valid approximately 200ns before the Status flag returns to Logic "0", permitting parallel data transfer to be clocked on the "1" to "0" transition of the Status flag.

Serial data coding is binary for unipolar input ranges and offset binary for bipolar input ranges. Serial output is by bit (MSB first, LSB last) in NRZ (non-return-to-zero) format. Serial and parallel data outputs change state on positive-going clock edges. Serial data is guaranteed valid on all negative-going clock edges, permitting serial data to be clocked directly into a receiving register on these edges. There are 13 negative-going clock edges in the complete 12-bit conversion cycle, as shown in Figure 5. The first edge shifts an invalid bit into the register, which is shifted out on the 13th negative-going clock edge. All serial data bits will have been correctly transferred at the completion of the conversion period.

Analog Input Voltage Range Format

The AD363 may be configured for any of 3 bipolar or 2 unipolar input voltage ranges as shown in Table 3.

Range	Connect Analog Input To ADC Pin:	Connect ADC Span Pin:	Connect Bipolar ADC Pin 23 To:
0 to +5V	24	25 to 22	—
0 to +10V	24	—	—
-2.5V to +2.5V	24	25 to 22	22
-5V to +5V	24	—	22
-10V to +10V	25	—	22

Table 3. Analog Input Voltage Range Pin Connections

Figure 4–15 (*cont.*)

Analog Input - Volts (Center of Quantization Interval)			Input Normalized to FSR		Digital Output Code (Binary for Unipolar Ranges; Offset Binary for Bipolar Ranges)	
0 to +10V Range	-5V to +5V Range	-10V to +10V Range	Unipolar Ranges	Bipolar Ranges	B1 (MSB)	B12 (LSB)
+9.9976	+4.9976	+9.9951	+FSR-1 LSB	+½FSR-1 LSB	1 1 1 1 1 1 1 1 1 1 1 1	
+9.9952	+4.9952	+9.9902	+FSR-2 LSB	+½FSR-2 LSB	1 1 1 1 1 1 1 1 1 1 1 0	
:	:	:	:	:		
+5.0024	+0.0024	+0.0049	+½FSR+1 LSB	+1 LSB	1 0 0 0 0 0 0 0 0 0 0 1	
+5.0000	+0.0000	+0.0000	+½FSR	ZERO	1 0 0 0 0 0 0 0 0 0 0 0	
:	:	:	:	:		
+0.0024	-4.9976	-9.9951	+1 LSB	-½FSR+1 LSB	0 0 0 0 0 0 0 0 0 0 0 1	
+0.0000	-5.0000	-10.0000	ZERO	-½FSR	0 0 0 0 0 0 0 0 0 0 0 0	

Table 4. Digital Output Codes vs Analog Input For Unipolar and Bipolar Ranges

The resulting input-output transfer functions are given by Table 4.

<u>Analog Input Section Offset Adjust Circuit</u>

The offset voltage of the AD363 may be adjusted at either the Analog Input Section or the ADC Section. Normally the adjustment is performed at the ADC but in some special applications, it may be helpful to adjust the offset of the Analog Input Section. An example of such a case would be if the input signals were small (<10mV) relative to Analog Input Section voltage offset and gain was inserted between the Analog Input Section and the ADC. To adjust the offset of the Analog Input Section, the circuit shown in Figure 6 is recommended.

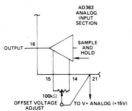

Figure 6. Analog Input Section Offset Voltage Adjustment

Under normal conditions, all calibration is performed at the ADC Section.

<u>ADC Offset Adjust Circuit</u>

Analog and power connections for 0 to +10V unipolar and -10V to +10V bipolar input ranges are shown in Figures 7 and 8, respectively. The Bipolar Offset, ADC pin 23 is open-circuited for all unipolar input ranges, and connected to Comparator input (ADC pin 22) for all bipolar input ranges. The zero adjust circuit consists of a potentiometer connected across ±V_S with its slider connected through a 3.9MΩ resistor to Comparator input (ADC pin 22) for all ranges. The tolerance of this fixed resistor is not critical, and a carbon composition type is generally adequate. Using a carbon composition resistor having a -1200ppm/°C tempco contributes a worst-case offset tempco of $8 \times 244 \times 10^{-6} \times 1200$ ppm/°C = 2.3ppm/°C of FSR, if the OFFSET ADJ potentiometer is set at either end of its adjustment range. Since the maximum offset adjustment required is typically no more than ±4LSB, use of a carbon composition offset summing resistor normally contributes no more than 1ppm/°C of FSR offset tempco.

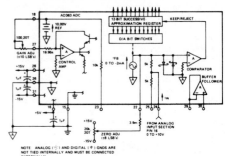

Figure 7. ADC Analog and Power Connections for Unipolar 0 to +10V Input Range

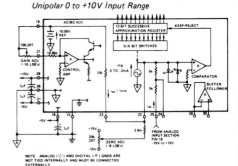

Figure 8. ADC Analog and Power Connections for Bipolar -10V to +10V Input Range

An alternate offset adjust circuit, which contributes negligible offset tempco if metal film resistors (tempco <100 ppm/°C) are used, is shown in Figure 9.

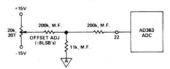

Figure 9. Low Tempco Zero Adj Circuit

Figure 4–15 *(cont.)*

Sec. 4.4 Data Acquisition ICs

In either zero adjust circuit, the fixed resistor connected to ADC pin 22 should be located close to this pin to keep the connection runs short, since the Comparator input (ADC pin 22) is quite sensitive to external noise pick-up.

Gain Adjust

The gain adjust circuit consists of a 100Ω potentiometer connected between +10V Reference Output pin 18 and Gain Adjust Input (ADC pin 27) for all ranges. Both GAIN and ZERO ADJ potentiometers should be multi-turn, low tempco types; 20T cermet (tempco = 100ppm/°C max) types are recommended. If the 100Ω GAIN ADJ potentiometer is replaced by a fixed 50Ω resistor, absolute gain calibration to ±0.1% of FSR is guaranteed.

Calibration

Calibration of the AD363 consists of adjusting offset and gain. Relative accuracy (linearity) is not affected by these adjustments, so if absolute zero and gain error is not important in a given application, or if system intelligence can correct for such errors, calibration may be unnecessary.

External ZERO ADJ and GAIN ADJ potentiometers, connected as shown in Figures 7, 8, and 9, are used for device calibration. To prevent interaction of these two adjustments, Zero is always adjusted first and then Gain. Zero is adjusted with the analog input near the most negative end of the analog range (0 for unipolar and -½FSR for bipolar input ranges). Gain is adjusted with the analog input near the most positive end of the analog range.

0 to +10V Range: Set analog input to +1LSB = +0.0024V. Adjust Zero for digital output = 000000000001; Zero is now calibrated. Set analog input to +FSR -2LSB = +9.9952V. Adjust Gain for 111111111110 digital output code; full-scale (Gain) is now calibrated. Half-scale calibration check: set analog input to +5.0000V; digital output code should be 100000000000.

-10V to +10V Range: Set analog input to -9.9951V; adjust Zero for 000000000001 digital output (offset binary) code. Set analog input to +9.9902V; adjust Gain for 111111111110 digital output (offset binary) code. Half-scale calibration check: set analog input to 0.0000V; digital output (offset binary) code should be 100000000000.

Other Ranges: Representative digital coding for 0 to +10V, -5V to +5V, and -10V to +10V ranges is shown in Table 4. Coding relationships are calibration points for 0 to +5V and -2.5V to +2.5V ranges can be found by halving the corresponding code equivalents listed for the 0 to +10V and -5V to +5V ranges, respectively.

Zero and full-scale calibration can be accomplished to a precision of approximately ±¼LSB using the static adjustment procedure described above. By summing a small sine or triangular-wave voltage with the signal applied to the analog input, the output can be cycled through each of the calibration codes of interest to more accurately determine the center (or end points) of each discrete quantization level. A detailed description of this dynamic calibration technique is presented in "A/D Conversion Notes", D. Sheingold, Analog Devices, Inc., 1977, Part II, Chapter II-4.

Other Considerations

Grounding: Analog and digital signal grounds should be kept separate where possible to prevent digital signals from flowing in the analog ground circuit and inducing spurious analog signal noise. Analog Ground (Analog Input Section pin 17, ADC Section pin 26) and Digital Ground (Analog Input Section pin 2 and ADC Section pin 15) are not connected internally; these pins must be connected externally for the system to operate properly. Preferably, this connection is made at only one point, as close to the system as possible. The cases are connected internally to Digital Ground to provide good electrostatic shielding. If the grounds are not tied common on the same card with both system packages, the digital and analog grounds should be connected locally with back-to-back general-purpose diodes as shown in Figure 10. This will protect the AD363 from possible damage caused by voltages in excess of ±1 volt between the ground systems which could occur if the key grounding card should be removed from the overall system. The system will operate properly with as much as ±200mV between grounds, however this difference will be reflected directly as an input offset voltage.

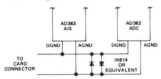

Figure 10. Ground-Fault Protection Diodes

Power Supply Bypassing: The ±15V and +5V power leads should be capacitively bypassed to Analog Ground and Digital Ground respectively for optimum device performance. 1μF tantalum types are recommended; these capacitors should be located close to the system. It is not necessary to shunt these capacitors with disc capacitors to provide additional high frequency power supply decoupling since each power lead is bypassed internally with a 0.039μF ceramic capacitor.

Applications

The AD363 contains several unique features that contribute to its application versatility. The more significant features include a precision +10V reference, an uncommitted buffer amplifier, the dynamic single-ended/differential mode switch and simple, uncommitted digital interfaces.

Transducer Interfacing

The precision +10V reference, buffer amplifier and mode switch can simplify transducer interfacing. Figure 11 illustrates how these features may be used to advantage.

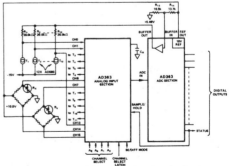

Figure 11. AD363 Transducer Interface Application

Figure 4–15 (cont.)

The AD590 is a temperature transducer that can be considered an ideal two-terminal current source with an output of one microamp per degree Kelvin ($1\mu A/°K$). With an offsetting current of $273\mu A$ sourced from the +5.46 volt buffered reference through $20k\Omega$ resistors (R1-R12) each of the 12 AD590 circuits develop $-20mV/°C$. The outputs are monitored with the AD363 front-end in the single-ended mode (Logic "0" on the Mode Control input). The +5.46 volt reference is derived from the ADC +10 volt precision reference and voltage divider R13, R14. Low output impedance for this +5.46 volt reference is provided by the ADC internal buffer amplifier. (The $10\mu V/°C$ offset voltage drift of the buffer amplifier contributes negligible errors.) At $0°C$, each temperature transducer circuit delivers a 0 volt output. At $125°C$, the output is $-2.5V$; at $-55°C$, the output is $+1.10V$. By using the two's complement ADC output (complemented MSB or sign bit), the negative voltage versus temperature function is inverted and digital reading proportional to temperature in degrees centigrade is provided. Resolution is $0.061°C$ per least significant bit.

The precision +10 volt reference is also used to power several bridge circuits that require differential read-out. When addressing these bridge transducers, a Logic "1" at the mode control input will switch the AD363 to the differential mode. In many cases, this feature will eliminate the requirement for a differential amplifier for each bridge transducer.

Microprocessor Interfacing
Digital interfacing to the AD363 has been deliberately left uncommitted; every processor system and application has different interface requirements and designing for one specific processor could complicate other applications.

The addition of a small amount of hardware will satisfy most interface requirements; an example based on 8080-type architecture is shown in Figure 12.

In this system the data bus is used to transmit multiplexer channel selection and convert and read commands to the AD363. It is also possible to address the AD363 as memory using the address bus to perform channel selection, convert and read operations.

The address lines can be decoded to provide channel selection, ADC convert start, status and ADC data (2 bytes) locations. These are accessed with I/O read/write instructions.

The ADC outputs are buffered with tri-state drivers. Figure 12 shows the 4 most significant ADC data bits and status as one byte.

FF_H	STATUS	—	—	—	B1 (MSB)	B2	B3	B4
	D7							D0

and the 8 least significant ADC data bits as the second byte.

FE_H	B5	B6	B7	B8	B9	B10	B11	B12 (LSB)
	D7							D0

Internal tri-state buffering is not provided because in many applications it would be better to have the first byte contain the 8 most significant bits. To accomodate both left and right justified formats would require more package pins and increase complexity.

The operating sequence for this system is as follows:

1	MVI	80_H	puts the address for channel 0 (including SE/DIFF mode) into accumulator
2	OUT	FF_H	puts 80_H on data bus and FF_H on address bus. Pulses I/O WRITE. OUT FF_H is decoded as a "LOAD ADDRESS" command to the channel select latches.
3	OUT	$F0_H$	puts $F0_H$ on address bus and pulses I/O WRITE. This is decoded to issue a "CONVERSION START" to the ADC. Accumulator contents are of no significance.
4	IN	FF_H	puts FF_H on address bus and pulses I/O READ. This is decoded to enable the appropriate tri-states, thus putting status and the 4 most significant bits on the data bus.

The status may be examined for "0" (conversion complete). In that case, the 4 MSB's would be read.

| 5 | IN | FE_H | puts FE_H on address bus and pulses I/O READ. This is decoded to enable the appropriate tri-states, thus putting the 8 least significant bits on the data bus. |

At this point, the multiplexer channel selection may be changed and another channel processed with the same instruction set (steps 2 through 5).

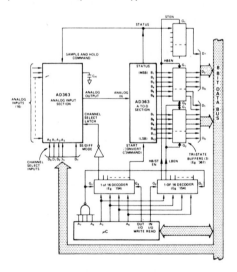

Figure 12. AD363 Microprocessor Interface Application

Figure 4-15 (*cont.*)

CAPACITOR OUTLINE DIMENSIONS
Dimensions shown in inches and (mm).

HOLD CAPACITOR

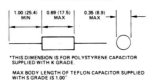

*THIS DIMENSION IS FOR POLYSTYRENE CAPACITOR SUPPLIED WITH K GRADE.

MAX BODY LENGTH OF TEFLON CAPACITOR SUPPLIED WITH S GRADE IS 1.00"

AD363 ORDERING GUIDE

Model	Specification Temp Range	Max Gain T.C.	Max Reference T.C.	Guaranteed Temp Range No Missing Codes	Package Styles[1] Analog Input Section	ADC Section
AD363KD	0 to +70°C	±30ppm/°C	±20ppm/°C	0 to +70°C	HY32D	HY32G
AD363SD	-55°C to +125°C	±25ppm/°C	±20ppm/°C	-55°C to +125°C	HY32D	HY32G
AD363SD/ 883B		Meets all AD363SD specifications after processing to the requirements of MIL-STD-883B, Method 5008.				

NOTE: D Suffix = Dual-In-Line package designator.
[1] See Section 20 for package outline information.

Figure 4–15 (*cont.*)

4.5 SUMMARY

This chapter introduces the ADC units that will allow the analog input portion of the system to interface with the digital processor section. For more information on ADCs, the reader is urged to investigate the literature from the device manufacturers and the References at the end of the chapter.

PROBLEMS

1. For an FSR of 10 V, what is the range of each division for a 10- and 12-bit ADC?
2. For an 8-bit ADC and a bipolar input (±5 V), what are the first 5 ($n = 0, 1, 2, 3, 4$) ADC output values for $s(t) = 5 \sin(\omega t)$ V with $f = 5$ kHz and $T = 10$ μs.
3. Sketch Figure 4–3 for a 4-bit ADC.
4. If $s(t)$ is sampled when it is 1.3 V, what is the nearest binary value for a 4, 8, and 12-bit ADC. Use binary offset with FSR = 10V.
5. Repeat Problem 4 if the ADC is 16-bits.
6. Find the SQNR of an 8-bit ADC.
7. Determine the minimum conversion time required for a TV signal sampled at 12 MHz. Look up several manufacturers' data sheets for ADCs and determine if there is a device available that meets the specification.
8. Using Figure 4–11, with the reference voltage at 5 V, what is t_1 if V_1 is 2.7 V and $t = 100$ μs?

9. Referring to Problem 8, what is the binary output of a 10-bit ADC if the clock rate in Figure 4–10 is 1 MHz? 3MHz?
10. Referring to Figures 4–12 and 4–13, tabulate and compare the values for **(a)** resolution, **(b)** relative accuracy, **(c)** conversion time, and **(d)** power supply values.

REFERENCES

HNATEK, E. R.   *A User's Handbook of D/A and A/D Converters.*   New York: Wiley, 1976.

SHEINGOLD, D. H., ED.   *Analog-Digital Conversion Notes.*   Norwood, Mass.: Analog Devices, 1980.

DISCRETE-TIME SIGNALS AND THE z TRANSFORM

5

The output of an ideal TH circuit with impulse sampling is a sequence of voltages with each value equal to the input signal voltage at the time of sampling, nT. This is shown in Figure 5–1 for an arbitrary signal, $s(t)$.

The continuous-time function $s(t)$ is now transformed into a discrete-time function, $s(nT)$, with values being defined only for $t = nT$, n an integer and T the (ideal) sampling rate. Although the domain of $s(nT)$ is discrete, its range is still continuous: any value of voltage within a given set of limits is possible. It is not until we reach the output of the ADC that we have a discrete range, brought about by quantization of the sampled signal. In either case, we are now concerned with discrete-time systems.

The set of numbers formed by the sampling process is considered a sequence of numbers and will be represented by $\{s(nT)\}$. This is sometimes presented as $\{s(n)\}$, where T is implied or set equal to one second. The value of the sampled function at time nT will be designated as $s(nT)$.

The following sections look at some sequences and their characteristics. The transition will be made from the continuous-time system (CTS) to the discrete-time system (DTS) and a convolution relation will be established along the lines of Chapter 1.

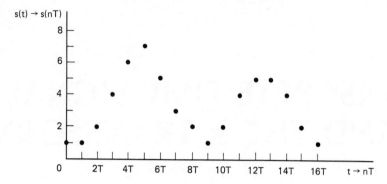

Figure 5–1 Output of an ideal TH circuit with ideal impulse sampling.

5.1 DISCRETE-TIME SIGNALS

As we did for the CTS, we will begin by defining the discrete impulse function, $\delta(nT)$.

$$\delta(nT) = \begin{cases} 1, & nT = 0 \\ 0, & \text{otherwise} \end{cases} \quad (5\text{--}1)$$

Unlike the CTS δ function, $\delta(nT)$ has an amplitude of unity. This is shown in Figure 5–2.

Other DTS signals are the *step function*, $u(nT)$, where

$$u(nT) = \begin{cases} 1, & n \geq 0 \\ 0, & \text{otherwise} \end{cases} \quad (5\text{--}2)$$

the *discrete exponential function*, $\exp(-anT)$, shown in Figure 5–3; the *discrete sinusoidal wave*, $s(nT) = A \sin(2\pi f nT - \theta)$, shown in Figure 5–4; and the *delayed impulse*, $\delta((n-m)T)$, shown in Figure 5–5.

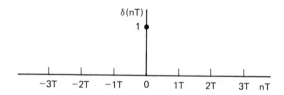

Figure 5–2 Discrete impulse function $\delta(nT)$.

Sec. 5.2 Impulse Response

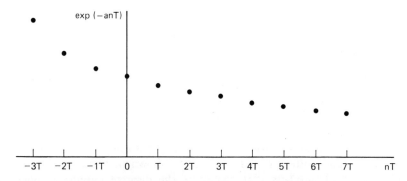

Figure 5–3 Discrete exponential function.

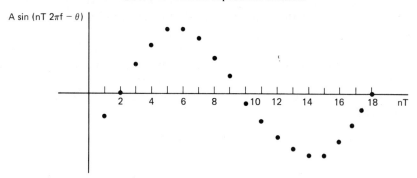

Figure 5–4 Discrete sine wave function.

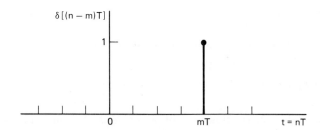

Figure 5–5 Delayed impulse function.

5.2 IMPULSE RESPONSE

If the DTS impulse signal is applied to a linear, discrete-time, shift-invariant, causal system (LDTSICS), the output is the discrete-time impulse response of the system. Figure 5–6 presents the DT impulse response of an arbitrary system.

The meaning of shift invariance is similar to that of time invariance for the CTS and is demonstrated as follows. If we apply the delayed impulse

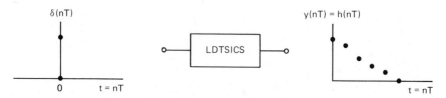

Figure 5-6 Impulse response of the LDTSIC system.

function to a discrete system, then, due to its shift-invariant property, we obtain a delayed (or time-shifted) response. This is shown in Figure 5-7. $\{h(nT)\}$ is the sequence of numbers that comprises the discrete impulse response of the system and its transfer function. $h(nT)$ or $h((n-m)T)$ will represent the individual terms of the sequence. In discribing the discrete systems, we will use the term *causal* if $h(nT) = 0$, for $nT < 0$; if $h(nT) = 0$ for all $nT > NT$ (N finite), the system has a finite impulse response (FIR); if there is no value of N for which $h(nT) = 0$, the system has an infinite impulse response (IIR). These terms will be used to describe the discrete systems of Chapter 6. In addition, the discrete systems, whether FIR or IIR, will be linear, time invariant, and causal.

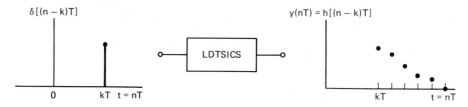

Figure 5-7 Delayed impulse response of the LDTSICS system.

5.3 INPUT SEQUENCE

If the impulse function is multiplied by a constant, its amplitude is changed. We can represent it as $x(nT) = a\delta(nT)$, where a is the constant multiplier. The signal applied to a discrete system yields the constant, a, times the impulse response, as shown in Figure 5-8, where each term of $h(nT)$ is multiplied by a.

In a similar manner, a constant times the delayed δ function will again produce the delayed response, but with each output term multiplied by the constant as shown in Figure 5-9, where the constant value is $x(mT)$, nT being the discrete variable, and mT is fixed in value.

As the system is linear, the sum of two impulses, each multiplied by a constant, will produce the sum of individual outputs. This is given in Equation 5-3.

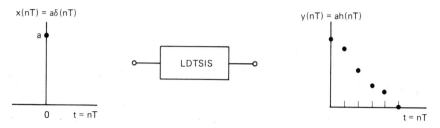

Figure 5–8 Impulse response of $a\delta(nT) = ah(nT)$.

$$x(kT)\delta((n-k)T) + x(mT)\delta((n-m)T) \longrightarrow$$
$$x(kT)h((n-k)T) + x(mT)h((n-m)T) \quad (5\text{-}3)$$

As we did for the CTS, we can represent a specific sample value, $x(nT)$, as the sum of products that consist of the delayed impulse function and the values of the sample at parameter kT. This is given in Equation 5–4, where nT is the sample time of interest and kT is the parameter.

$$x(nT) = \sum_{k=-\infty}^{\infty} x(kT)\delta((n-k)T) \quad (5\text{-}4)$$

This is recognized as the DTS equivalent of the sifting integral (Equation 1–18) for the CTS δ function and it allows us to represent $x(nT)$ as an infinite sum of impulse functions. When $x(nT)$ is applied to the DT system, we will obtain

$$\sum_{k=-\infty}^{\infty} x(kT)\delta((n-k)T) = x(nT) \longrightarrow y(nT)$$
$$= \sum_{j=-\infty}^{\infty} x(jT)h((n-j)T) \quad (5\text{-}5)$$

The term on the right hand side is the convolution equation for the discrete-time system and states that the output response for a given value of nT is the sum of the impulse responses for all values of input. This is represented in Figure 5–10.

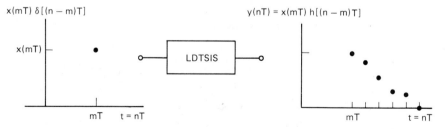

Figure 5–9 Impulse response of $x(mT)$, $\delta[(n-m)T]$.

154   Chap. 5   Discrete-Time Signals and the *z* Transform

$$x(nT) = \sum_k x(kT)\,\delta[(n-k)T] \quad \boxed{\text{LDTSICS}} \quad y(nT) = \sum_k x(kT)\,h[(n-k)T]$$

Figure 5–10 I/O representation of the LDTSIC system.

Equation 5–6 presents the discrete-time convolution relation.

$$y(nT) = x(nT) * h(nT) = \sum_{k=-\infty}^{\infty} x(kT)h((n-k)T) \qquad (5\text{-}6)$$

By letting $r = (n - k)$, we obtain

$$y(nT) = x(nT) * h(nT) = \sum_{r=-\infty}^{\infty} x((n-r)T)h(kT) \qquad (5\text{-}7)$$

5.4 DISCRETE CONVOLUTION EXAMPLE

We will demonstrate the use of Equation 5–7 by an example. We consider the finite duration pulse of unit height and length N, $(u(nT) - u((n - N)T)$, as the input to a DTS whose impulse response, $h(nT)$, is $u(nT)\exp(-anT)$. These are shown in Figure 5–11.

To find the output sequence, $y(nT)$, we will use the graphical approach.

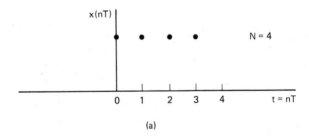

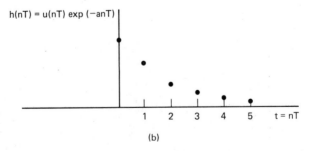

Figure 5–11 Input and impulse response for an example of convolution: (a) finite duration pulse input; (b) impulse response of LDTSIC system.

Sec. 5.4 Discrete Convolution Example

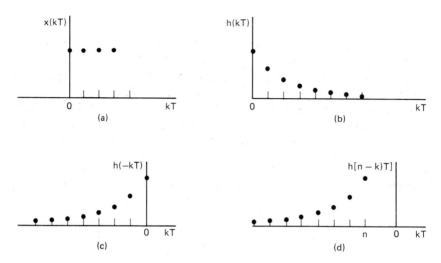

Figure 5-12 Development of convolution sum: (a) $x(k)$ signal; (b) $h(k)$ impulse response; (c) $h(-k)$ impulse response; (d) $h(n - k)$ impulse response.

This involves finding $x(kT)$, $h(kT)$, $h(-kT)$, and finally $h((n - k)T)$, where n is fixed in value and k assumes the role of variable. This is shown in Figure 5-12.

The next step is to find the sample values that coincide (overlap) for different values of (fixed) n, and then take the product of each term, $x(kT)h((n - k)T)$, adding them to obtain $y(nT)$.

For $n < 0$, there is no overlap; that is, each product term in the sum is zero, as one or the other factor is zero for all values of k. For $n > 0$, there is some overlap, which implies that there will be several nonzero product terms in the sum. This is shown in Figure 5-13 for $n = 2$.

The terms for $k = 0, 1,$ and 2 will yield

$$x(0T)h((2-0)T) + x(1T)h((2-1)T) + x(2T)h((2-2)T) = y(2T) \quad (5\text{-}8a)$$

$$1 \exp(-2aT) + 1 \exp(-aT) + 1 \exp(-0aT) = y(2T) \quad (5\text{-}8b)$$

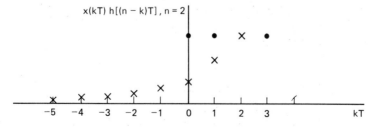

Figure 5-13 Convolution sum for $n = 2$.

This can be rewritten as

$$y(2T) = \exp(-2aT) + \exp(-aT) + 1 \qquad (5\text{–}9)$$

This yields the value of $y(nT = 2T)$ in the output sum. Values of $y(nT)$ for other values of nT can be obtained in a similar manner.

This section has developed the discrete convolution relation. We will see that this is a special case of a general form of the system I/O which will allow us to develop the digital system relation. Like the CTS, a transform technique is available that generates an equation similar to that of Equation 1–26. For the DTS, the transform used is the z transform, and the relation that is developed is

$$Y(z) = X(z)H(z) \qquad (5\text{–}10)$$

where the z transform is given by

$$X(z) = \sum_{n=-\infty}^{\infty} x(nT) z^{-n} \qquad (5\text{–}11)$$

This will be developed further in Section 5.8.

5.5 TRANSFORM CONCEPTS

It was mentioned that continuous-time (CT) signals can be transformed to the (complex) frequency domain by the Fourier and Laplace transforms. The result of the transformation usually provides us with additional insight into the opera-

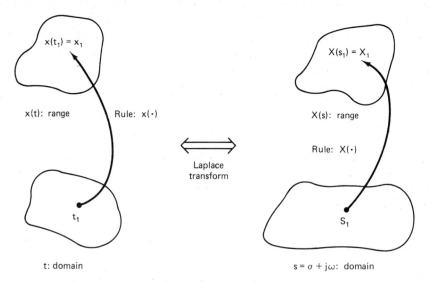

Figure 5–14 Representation of the Laplace transform.

tion of the system. For some signals that do not meet the conditions required for the FT, the Laplace transform is used, which involves a transformation from the time domain to the complex s domain, where s is given as $\sigma + j\omega$. This is shown in Figure 5–14.

The Fourier transform is a special case of the Laplace transform for $\sigma = 0$, that is, for input signals that can be represented by sinusoidal waveforms. It will be shown that the discrete-time (DT) signals can also be represented by a FT; however, instead of the Laplace transform providing the more general relationship, use is made of the z transform.

5.6 FREQUENCY RESPONSE OF A DISCRETE-TIME SYSTEM

If a sampled sinusoidal signal is applied to a DT system, the output, $y(nT)$, is also a sampled sinusoid of the same frequency, but with a modified amplitude and phase. This is analogous to the CTS, discussed in Section 1.15, in which a sine wave (or exponential wave) signal at the input yielded a modified sine wave (or exponential wave) at the output. This is shown in Equation 5–12.

$$y(t) = H(2\pi f)x(t) \tag{5-12a}$$

where $H(2\pi f)$ is the transfer function of the CT system in the frequency domain.

$$h(t) \iff H(2\pi f) \tag{5-12b}$$

If $x(t) = A \sin 2\pi ft$, then $y(t) = A|H(2\pi f)| \sin (2\pi ft + \theta)$, where θ is the phase of $H(2\pi f)$.

A similar relation was shown to exist for the Laplace transform:

$$y(t) = H(s)x(t) \tag{5-13a}$$

where $H(s)$ is the Laplace transform of $h(t)$:

$$h(t) \iff H(s) \tag{5-13b}$$

As we did for the CTS in Section 1.15, we apply the discrete version of the exponential input to our discrete-time system: $x(nT) = \exp(j2\pi fnT)$. This yields the discrete output, $y(nT)$:

$$y(nT) = \sum_{k=-\infty}^{\infty} h(kT) \exp[j2\pi f(n-k)T] \tag{5-14a}$$

$$= \sum_{k=-\infty}^{\infty} h(kT) \exp(j2\pi fnT) \exp(-j2\pi fkT) \tag{5-14b}$$

$$= \exp(j2\pi fnT) \sum_{k=-\infty}^{\infty} h(kT) \exp(-j2\pi fkT) \tag{5-14c}$$

Upon examination, we see that Equation 5–14c represents $y(nT)$ as

$$y(nT) = x(nT)\left[\sum_{k=-\infty}^{\infty} h(kT) \exp(-j2\pi fkT)\right] \quad (5\text{–}15a)$$

As the output is the input signal times a constant complex number, the similarity between Equations 5–12 and 5–15 leads us to define the term in brackets as the frequency-domain representation of the DT transfer function.

$$H(2\pi f) \equiv \sum_{k=-\infty}^{\infty} h(kT) \exp(-j2\pi fkT) \quad (5\text{–}16a)$$

$H(2\pi f)$ is termed the *frequency response* of the DTS. Equation 5–15a becomes

$$y(nT) = H(2\pi f)x(nT) \quad (5\text{–}15b)$$

Note that even though $h(kT)$ is discrete, its frequency response is continuous. This occurs as the frequency in question is that of the analog input signal prior to sampling. It is to this frequency that the discrete-time system will respond: the output of a discrete system will vary if the original (analog) signal's frequency varies.

Example 5–1

For the sequence $h(0T) = h(1T) = 1$, all other values of $h(nT) = 0$, what is the system frequency response?

Solution Using Equation 5–16a, we have

$$H(2\pi f) = \sum_{k=-\infty}^{\infty} h(kT) \exp(-j2\pi fkT)$$
$$= h(0T) \exp(-j2\pi f \cdot 0 \cdot T) + h(1T) \exp(-j2\pi f \cdot 1 \cdot T)$$
$$= 1 \cdot 1 + 1 \exp(-j2\pi fT)$$
$$= [1 + \exp(-j2\pi fT)] \exp\left(+j\frac{2\pi fT}{2}\right) \exp\left(-j\frac{2\pi fT}{2}\right) \quad (5\text{–}16a)$$
$$= \left[\exp\left(j\frac{2\pi fT}{2}\right) + \exp\left(-j\frac{2\pi fT}{2}\right)\right] \exp\left(-j\frac{2\pi fT}{2}\right)$$
$$= 2\cos\pi fT \exp(-j\pi fT)$$

The first factor in $H(2\pi f)$ is the magnitude and the second factor is the phase angle, with both terms being functions of frequency.

Example 5–2

If $s_1(t) = 10 \sin[2\pi(100 \text{ Hz})t]$ and $s_2(t) = 10 \sin[2\pi(500 \text{ Hz})t]$, what is $y(nT)$ for each of the inputs above and for the impulse response of Example 5–1 if the sampling frequency is 1500 Hz?

Solution Using the result of Example 5–1 and Equation 5–15b, we have

$$y_1(nT) = x_1(nT)H(2\pi f)$$
$$= 10 \sin[2\pi(100 \text{ Hz})nT] \, 2\cos(\pi fT) \exp(-j\pi fT)$$

Sec. 5.6 Frequency Response of a Discrete-Time System

$$= 10 \sin (0.419nT) \cdot 2 \cos (0.209) \exp(-j0.209)$$
$$= 19.65 \sin (0.419nT) \exp(-j0.209)$$
$$y_2(nT) = x_2(nT) H(2\pi f)$$
$$= 10 \sin [2\pi(500 \text{ Hz})nT] \, 2 \cos (\pi fT) \exp(-j\pi fT)$$
$$= 10 \sin (2.09nT) \, 2 \cos (1.046) \exp(-j1.076)$$
$$= 10 \sin (2.09nT) \exp(-j1.046)$$

This example clearly shows that the discrete system outputs are different in amplitude and phase due to the variation in the system's response for different input frequencies.

If we replace f with $f + nf_{sa}$, where n is an integer, in Equation 5–16a we obtain

$$H(2\pi(f + f_{sa})) = \sum_{k=-\infty}^{\infty} h(kT) \exp(-j2\pi(f + f_{sa})kT) \quad (5\text{–}16\text{b})$$

$$= \sum_{k=-\infty}^{\infty} h(kT) \exp(-j2\pi fkT) \exp(-j2\pi f_{sa}kT) \quad (5\text{–}16\text{c})$$

Since $f_{sa}T = 1$, the second exponential in Equation 5–16c is equal to unity for all k. We then have

$$H(2\pi(f + f_{sa})) = \sum_{k=-\infty}^{\infty} h(kT) \exp(-j2\pi fkT) = H(2\pi f) \quad (5\text{–}16\text{d})$$

Therefore, $H(2\pi f)$ is a periodic function with the period equal to the sampling frequency. This can be demonstrated by replacing f with $f + nf_{sa}$ in the solution to Problem 5–1. As $H(2\pi f)$ is periodic, it must have a Fourier series representation. This was given by Equation 5–16a where the Fourier coefficients are the values of the impulse response, $h(kT)$.

As we did with the CT system Fourier series, the coefficients of the exponential term in Equation 5–16a are given by

$$h(nT) = \frac{1}{2\pi} \int_{-\pi}^{\pi} H(2\pi f) \exp(j2\pi fnT) d\omega \quad (5\text{–}17)$$

Equations 5–16a and 5–17 form a discrete-time transform pair:

$$h(nT) \iff H(2\pi f) \quad (5\text{–}18)$$

where $H(2\pi f)$ is the continuous and periodic frequency response of the system. We can then establish a transform pair for discrete-time signals:

$$x(nT) \iff X(2\pi f) \quad (5\text{–}19\text{a})$$
$$y(nT) \iff Y(2\pi f) \quad (5\text{–}19\text{b})$$

where the function in the frequency domain represents the frequency content of the discrete signal and is (1) continuous, (2) periodic, and (3) doubly infinite.

The question arises as to whether or not there is any relationship between the frequency response of a continuous and a discrete system, and if there is any relationship between the frequency content of a continuous and discrete signal. Using the results of Section 3.4.1, Ideal Sampling, Equation 3–18c, we see that the discrete system has a frequency response or content which is the same as that of the continuous system, only replicated about each integer multiple of the sampling frequency. In both cases, we must insure that sampling occurs at a sufficiently high rate to prevent aliasing from taking place.

5.7 CONVOLUTION IN THE FREQUENCY DOMAIN

We begin with the DT convolution relation.

$$y(nT) = \sum_{k=-\infty}^{\infty} x(kT)h((n-k)T) \tag{5-20}$$

If we multiply both sides of Equation 5–20 by $\exp(-j2\pi fnT)$ and sum over n, we have

$$\sum_{n=-\infty}^{\infty} y(nT)\exp(-j2\pi fnT) = \sum_{n=-\infty}^{\infty} \exp(-j2\pi fnT) \sum_{k=-\infty}^{\infty} x(kT)h((n-k)T) \tag{5-21}$$

The left-hand side is $Y(2\pi f)$. We can exchange summation signs on the right-hand side to obtain

$$Y(2\pi f) = \sum_{k=-\infty}^{\infty} x(kT) \sum_{n=-\infty}^{\infty} h((n-k)T)\exp(-j2\pi fnT) \tag{5-22}$$

If we insert the term $1 = \exp(-j2\pi fkT)\exp(+j2\pi fkT)$ on the right-hand side of Equation 5–22, we have

$$Y(2\pi f) = \sum_{k=-\infty}^{\infty} x(kT)\exp(-j2\pi fkT) \sum_{n=-\infty}^{\infty} h((n-k)T)\exp[-j2\pi f(n-k)T] \tag{5-23}$$

Letting $(n-k)T = rT$, Equation 5–23 becomes

$$Y(2\pi f) = \left[\sum_{k=-\infty}^{\infty} x(kT)\exp(-j2\pi kT)\right]\left[\sum_{r=-\infty}^{\infty} h(rT)\exp(-j2\pi rT)\right] \tag{5-24}$$

The right-hand side of Equation 5–24 is recognized as the product of the FTs of $x(kT)$ and $h(rT)$.

$$Y(2\pi f) = X(2\pi f)H(2\pi f) \tag{5-25}$$

Sec. 5.8 z Transform

which is similar to Equation 1–65 for the CTS. Equation 5–25 states that the FT of the response $y(nT)$ of a DT system to an input $x(nT)$ is the product of the FTs of the input and transfer function of the system, $h(nT)$. This implies, as it did for the CTS, that convolution in the frequency domain is a multiplication process and that $y(nT)$ can be obtained from $Y(2\pi f)$ using Equation 5–17.

5.8 z TRANSFORM

If we choose $x(nT) = \exp(snT) = \exp[(\sigma + j2\pi f)nT]$ as an input signal, the response of the system will be similar in form to Equation 5–14c.

$$y(nT) = \exp(snT) \sum_{k=-\infty}^{\infty} h(kT) \exp(-skT) \quad (5\text{--}26)$$
$$= \exp(snT)H(s)$$

where $H(s)$ is a more general form of the transform function $H(2\pi f)$. This is in line with the use of a Laplace transform as a more general form of the Fourier transform. $H(s)$ will also be periodic in the complex frequency, s.

With $H(s)$ given by

$$H(s) \equiv \sum_{k=-\infty}^{\infty} h(kT) \exp[-skT] \quad (5\text{--}27)$$

we can introduce the z transform by setting up the relation between the complex s plane and the complex z plane:

$$z \equiv \exp(sT) \quad (5\text{--}28)$$

This function is a mapping of points in the s plane to points in the z plane and is shown in Figure 5–15.

The relation of Equation 5–27 becomes

$$H(z) \equiv \sum_{k=-\infty}^{\infty} h(kT)z^{-k}) \quad (5\text{--}29)$$

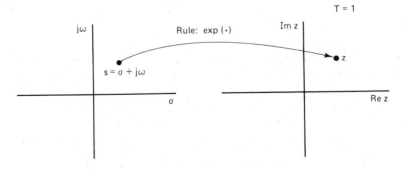

Figure 5–15 Mapping of the s plane into the z plane.

This is generalized to include any sequence of numbers, $\{x(nT)\}$.

$$x(z) \equiv \sum_{k=-\infty}^{\infty} x(kT)z^{-k} \qquad (5\text{-}30)$$

which is the definition of the z transform.

5.9 ROLE OF THE z PLANE

In circuit analysis and control theory, values of s that caused the system transfer function to become zero (zeros) or infinity (poles) provided information about the system's response to signals with given frequency values. These zeros and poles have a similar role in DT systems and provide similar information.

The mapping of the s plane into the z plane is as follows. The $j\omega$ axis of the s plane ($\sigma = 0$) corresponds to $z = \exp(-j\omega T)$. This is a circle of unit radius in the z plane. As ωT varies from $-\pi$ to $+\pi$, this generates a circular path in the z plane from $z = -1$ ($\angle -180°$) to $z = -1$ ($\angle 180°$) (i.e., "almost a full circle"). The z transform of Equation 5–28 takes a strip in the s plane between $(-\pi/T) < \omega < (+\pi/T)$ and maps it into the unit circle of the z plane.

Example 5–3
What values of z correspond to $s = j(75 \text{ rad/s})$ if $\omega_{sa} = 225$ rad/s? $= 2\pi f$
Solution Using Equation 5–28, we have
$$T = \frac{1}{f} = \frac{2\pi}{225}$$

$$\begin{aligned}
z &\equiv \exp(sT) \\
&= \exp(j\omega T) \\
&= \exp\left(j \cdot \frac{75 \text{ rad}}{s} \cdot 0.02790 s\right) \qquad (5\text{-}28) \\
&= \exp(j2.09 \text{ rad}) \\
&= \cos(2.09 \text{ rad}) + j\sin(2.09 \text{ rad}) \\
&= -0.5 + j0.866 \\
&= +1\ \angle 120°
\end{aligned}$$

$= (6.28)(.00444)$
$= .02790$

The general mapping is shown in Figure 5–16. The point $s = 0$, $(\sigma, \omega) = (0, 0)$, maps into the point $z = 1$.

For values of $\sigma < 0$, $|z| < 1$. This implies that the left half of the s plane with its poles and zeros is mapped inside the unit circle of the z plane. Correspondingly, the right-half plane ($\sigma > 0$, $|z| > 1$) is mapped to points exterior to the unit circle. This is shown in Figure 5–17.

If an analog system has poles only in the left-hand s plane, it is stable. As these poles will map inside the unit circle in the z plane, we see that a discrete system will also be stable if all its poles are inside the unit circle.

Sec. 5.10 Examples of the z Transform   163

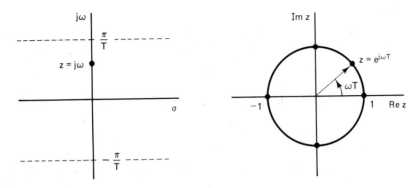

Figure 5–16 Mapping of the $j\omega$ axis in the s plane onto a unit circle ($|z| = 1$) in the z plane.

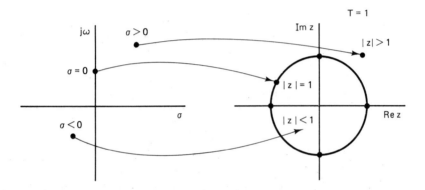

Figure 5–17 Mapping of the s plane into the z plane.

We will use the z transform to establish the DT system transfer function (as we used the Laplace transform for CT systems).

5.10 EXAMPLES OF THE z TRANSFORM

In the section that follows, several sequences are considered and their transforms obtained.

5.10.1 Finite Impulse Sequence

Figure 5–18 shows a finite impulse sequence $x(nT)$. Its z transform is obtained using Equation 5–30.

$$X(z) = \sum_{k=-\infty}^{\infty} x(kT)z^{-k} \qquad (5\text{--}30)$$

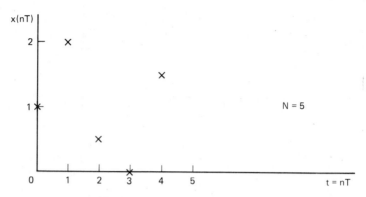

Figure 5–18 Finite impulse sequence.

We obtain

$$X(z) = x(0T)z^{-0} + x(T)z^{-1} + x(2T)z^{-2} + x(3T)z^{-3} + x(4T)z^{-4} \quad (5\text{–}31)$$

Using the values for $x(nT)$, we have

$$X(z) = 1 + 2z^{-1} + 0.5z^{-2} + 0z^{-3} + 1.5z^{-4} \quad (5\text{–}32)$$

Equation 5–32 becomes

$$X(z) = (1 + 2z^{-1} + 0.5z^{-2} + 1.5z^{-4}) \quad (5\text{–}33)$$

Upon investigation of Equation 5–33, we observe that $X(z)$ will become infinite only when $z = 0$. This allows us to introduce the *region of convergence* of $X(z)$. It is defined as the set of values of z for which $X(z)$ remains finite. For the example given, $X(z)$ is finite for all z except $z = 0$. This is stated in Equation 5–34.

$$0 < |z| \leq \infty = \text{region of convergence} \quad (5\text{–}34a)$$

The region of convergence can also be determined by examining the magnitude of all the poles of $H(z)$. If the largest magnitude is $= a$, the region of convergence is the set of all $|z|$ greater than the maximum pole. This is given in Equation 5–34b

$$\max\{|z_i|\} = a < |z| \leq \infty = \text{region of convergence} \quad (5\text{–}34b)$$

where $|z_i|$ represents the set of poles of $H(z)$.

This information allows us to determine if the system we are working with is causal and/or stable. If the region of convergence includes $|z| = \infty$, as in Equation 5–34a, the system is causal; if the unit circle is inside the region of convergence, the system is stable. This last fact implies that all system poles are inside the unit circle, the condition for stability.

Sec. 5.10   Examples of the z Transform   **165**

5.10.2 Causal Sequence

If $x(nT) = 0$ for $n < 0$, the resulting sequence is causal. As another example, consider the sequence $x(nT) = u(nT)\, a^{nT}$. This is shown in Figure 5–19 for $a < 1$. The z transform for this sequence is

$$X(z) = \sum_{k=-\infty}^{\infty} x(kT) z^{-k} = \sum_{k=-\infty}^{\infty} a^{kT} z^{-k} \qquad (5\text{–}35\text{a})$$

$$= a^0 z^{-0} + a^T z^{-1} + \cdots \qquad (5\text{–}35\text{b})$$

This can be written as

$$X(z) = \sum_{k=0}^{\infty} (a^T z^{-1})^k \qquad (5\text{–}36)$$

This equation is in the form of a geometric progression. The closed form representation of a general geometric series is given by Equation 5–37a

$$\sum_{j=p}^{N} (\gamma)^j = \frac{\gamma^p - \gamma^{N+1}}{1 - \gamma} \qquad (5\text{–}37\text{a})$$

For a sum of N terms, Equation 5–36 becomes

$$X(z) = \sum_{k=0}^{N-1} (a^T z^{-1})^k = \frac{1 - (a^T z^{-1})^N}{1 - (a^T z^{-1})} \qquad (5\text{–}37\text{b})$$

If $a^T z^{-1} < 1$, the limit as N goes to infinity of the sum is

$$\lim_{N \to \infty} X(z) = \sum_{k=0}^{\infty} (a^T z^{-1})^k = \frac{1}{1 - a^T z^{-1}} \qquad (5\text{–}38)$$

Equation 5–38 will not converge, that is, it diverges (or blows up, or becomes infinite) at $|z| = a^T$, which is then a pole of $X(z)$. Using this and the information

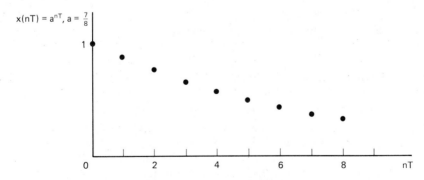

Figure 5–19   Causal exponential signal.

above, we determine that $X(z)$ has $|z| > a^T$, as a region of convergence. For values of $a^T > 1$, the sequence represents an unstable system. Also, as the region of convergence includes $|z| = \infty$, the system is causal.

5.11 FOURIER TRANSFORM OF THE DTS FROM THE z TRANSFORM

If the z transform of the impulse response of a DTS is evaluated on the unit circle, $z = \exp(j\omega T)$, then the z transform becomes the Fourier transform (i.e., the frequency response) of the DTS. The frequency content of a sampled sequence can also be determined by evaluating $X(z)$ on the unit circle. This is shown in Equation 5–39.

$$H(z)\bigg|_{z=\exp(j\omega T)} = H(\exp(j2\pi f)) = H(2\pi f) \qquad (5\text{--}39)$$

The z transform of a sequence evaluated on the unit circle provides us with the frequency response of the system.

$$H(z)\bigg|_{z=\exp(j\omega T)} = H(2\pi f) = \sum_{k=-\infty}^{\infty} h(kT)\exp(j2\pi fkt) \qquad (5\text{--}40)$$

Using Equations 5–33 and 5–38, we obtain the frequency response of the sequences by evaluating them on the unit circle: $z = \exp(j\omega T)$

$$X(z)\bigg|_{z=e^{j\omega T}} = 1 + 2e^{-j\omega T} + 0.5e^{-j3\omega T} + 1.5e^{-j4\omega T} \qquad (5\text{--}41a)$$

$$X(z)\bigg|_{z=e^{j\omega T}} = \frac{1}{1 - a^T e^{-j\omega T}} \qquad (5\text{--}41b)$$

Both equations are functions of ω, periodic, and doubly infinite which indicates that they have a Fourier series representation whose coefficients are the original sequence. The magnitude and phase of Equations 5–41a and 5–41b can be found using the arithmetic of complex numbers.

5.12 z TRANSFORM PROPERTIES

The mathematics of complex variables is required to interpret some of the properties of the z transform and inverse z transform. This section presents some of the properties with a minimal amount of proof or derivation. Material that is relevant to digital signal processing will receive further discussion as required in the chapter where it is introduced. In addition, several excellent texts on complex variables and digital signal processing are available that expand on the topics covered here.

5.12.1 Delay Property

The FT of a CT signal that was shifted or delayed in time was the original transform times a phase factor. This was presented in Equation 1–63.

$$f(t) \iff F(f)$$
$$f(t - \tau) \iff F(f) \exp(-j2\pi f \tau)$$
(1–63)

In a similar manner, we can obtain the effect on the z-transform of a shift in the sequence.

$$x(nT) \iff X(z) \tag{5–42a}$$
$$x((n-k)T) \iff X(z)z^{-k} \tag{5–42b}$$

As an example, consider the sequence shown in Figure 5–20. Its z transform is given by

$$X(z) = 1 + z^{-1} + z^{-2} + z^{-3} = \sum_{k=0}^{N-1=3} z^{-k} \tag{5–43}$$

Using Equation 5–37, we obtain

$$X(z) = \frac{1 - (z^{-1})^4}{1 - (z^{-1})} = \frac{1 - z^{-4}}{1 - z^{-1}} \tag{5–44}$$

If we shift the sequence 4 units to the right [i.e., $\{x(nT)\} \longrightarrow \{x([n-4]T)\}$], we obtain as its z transform, using Equation 5–42b,

$$X(z) \longrightarrow X(z)z^{-4} \tag{5–45a}$$

$$= \frac{1 - z^{-4}}{1 - z^{-1}} \cdot z^{-4} \tag{5–45b}$$

$$= \frac{z^{-4} - z^{-8}}{1 - z^{-1}} \tag{5–45c}$$

This property will be most useful when we study the difference equations that describe a DT system.

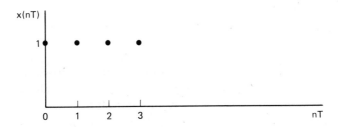

Figure 5–20 Finite unit sequence.

It should be mentioned that if $X(z)$ is a rational function (i.e., the quotient of two polynomials) and is the z transform of a causal sequence, the series form of $X(z)$ can be obtained by dividing the numerator by the denominator term. This yields a series in powers of z^{-1}. The coefficients of z^{-k} will be the $x(kT)$ (i.e., the terms of the DT sequence). For the z transform of Equation 5–44, we have

$$
\begin{array}{r}
1 + z^{-1} + z^{-2} + z^{-3} \\
1 - z^{-1} \overline{) 1 - z^{-4}} \\
\underline{1 - z^{-1}} \\
z^{-1} - z^{-4} \\
\underline{z^{-1} - z^{-2}} \\
z^{-2} - z^{-4} \\
\underline{z^{-2} - z^{-3}} \\
z^{-3} - z^{-4} \\
\underline{z^{-3} - z^{-4}} \\
0
\end{array}
$$

The result of the division is the same as Equation 5–43: The series form of the z transform and the coefficients are all unity.

5.12.2 Convolution

By taking the z transform of the DT convolution relation, Equation 5–20, we obtain

$$Y(z) = X(z)H(z) \qquad (5\text{–}46)$$

where $Y(z)$, $X(z)$, and $H(z)$ are the z transforms of $y(nT)$, $x(nT)$, and $h(nT)$, respectively. From this, we can define $H(z) = Y(z)/X(z)$. If $H(z)$ is causal, we can obtain the DT sequence by division of $Y(z)$ by $X(z)$. If we evaluate Equation 5–46 on the unit circle, we can obtain the frequency form of the convolution equation, Equation 5–25.

5.13 SUMMARY

This chapter has introduced the z transform of a DTS. It will be used when digital filters are discussed. In addition, discrete-time signals were covered, and several relationships between them were introduced.

PROBLEMS

1. Find the first 20 sample values of the following signals.
   (a) $s(t) = 5 \sin [2\pi(1000 \text{ Hz})t]$ V, sampled at 5 kHz.
   (b) A ± 4-V square wave, $f = 300$ Hz, sampled at 1.2 kHz.
   (c) A ± 6-V triangle wave, $f = 2$ kHz, sampled at 5 kHz.

   (d) $s(t) = \begin{cases} +3t, & 0 \le t \le 1.5 \text{ ms, sampled at 10 kHz} \\ 0, & \text{otherwise} \end{cases}$

2. Sketch for $T = 0.1$ ms:
   (a) $\delta((n-7)T)$
   (b) $\delta((n-20)T)$
   (c) $\delta((n+18)T)$

3. Sketch $h(nT)$ if the input to the linear system are the signals from Problem 2.

   $$h(nT) = \begin{cases} 1, & 0 < n < 5 \\ 0, & \text{otherwise} \end{cases}$$

4. Repeat Problem 3 if the input signals of Problem 2 are added together and their sum is applied to the system of Problem 3.

5. Obtain the samples of $s(t)$ at 10 kHz.

   $$s(t) = \begin{cases} 1, & 0 \le t < 1 \text{ ms} \\ -1, & 1 \text{ ms} < t \le 2 \text{ ms} \\ 0, & \text{otherwise} \end{cases}$$

6. Obtain the samples of $h(t)$ at 10 kHz.

   $$h(t) = \begin{cases} -2t^2, & 0 \le t \le 1 \text{ ms} \\ 0, & \text{otherwise} \end{cases}$$

7. Represent the samples of Problem 5 as a constant $s(nT)$ times a shifted impulse function. Then apply each input to the system of Problem 6 and sketch the output. Refer to Equation 5–5, left-hand side.

8. Find the output, $y(nT)$, if $s(nT)$ is applied to the system whose impulse response is $u(nT) e^{-nT}$ as shown in Figure P5–8.

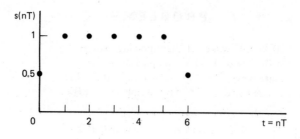

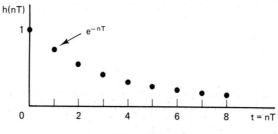

Figure P5–8

9. Find the output for the signal shown in Figure P5–9 if the system impulse response is the sampled sine wave, $v(nT) = \left(0.811 \sin\left(\dfrac{2\pi n}{8} - 3.53\right) + 0.25\right)$ V.

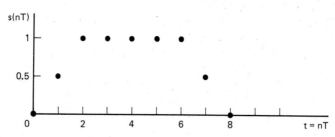

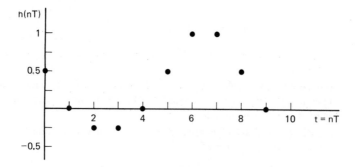

Figure P5–9

Chap. 5 Problems

10. For a sampling rate of 10 kHz, what is the system output?

$$s(t) = \begin{cases} 10 \sin [2\pi(1000 \text{ Hz})t] \text{ V}, & 0 \le t \le 0.5 \text{ ms} \\ 0, & \text{otherwise} \end{cases}$$

$$h(t) = \begin{cases} -(2/\text{ms})t + 2, & 0 \le t \le 0.25 \text{ ms} \\ 0, & \text{otherwise} \end{cases}$$

11. Find the FT of the sequences shown (Figure P5–11).

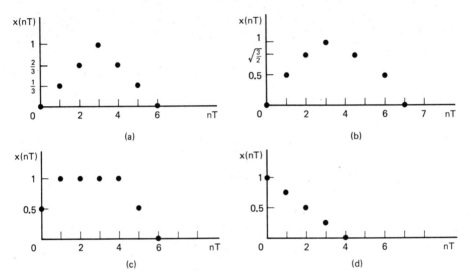

Figure P5–11

12. What is the frequency response of $h(nT)$ if sampling occurs at 1 kHz? Sketch the response for -2 kHz $\le f \le 2$ kHz.

$$h(nT) = \begin{cases} 1, & 0 \le n \le 3 \\ 0, & \text{otherwise} \end{cases}$$

13. For the system of Problem 12, what are the respective outputs if $f_{sa} = 1000$ Hz and the input signals are:
  (a) $2 \sin [2\pi(200 \text{ Hz})t]$?
  (b) $2 \cos [2\pi(280 \text{ Hz})t]$?

14. Find $Y(2\pi f)$ for the system of Problem 12 if the input is

$$s(nT) = \begin{cases} 1, & 0 \le n \le 2 \\ 0, & \text{otherwise} \end{cases}$$

15. Find z for the following values of s if sampling is at 1 kHz.

  (a) $0 + j100$ rad/s   (b) $-2 \pm j300$ rad/s
  (c) $1 \pm j200$ rad/s   (d) $-10 \pm j500$ rad/s
  (e) $0 \pm j1000$ rad/s

16. Map the s-plane poles to the z plane if sampling is at 200 Hz (see Figure P5–16).

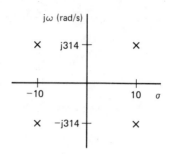

Figure P5–16

17. What is $S(z)$ for the sequence sampled at 10 kHz?

$$s(nT) = \begin{cases} -(2 \text{ V/ms})nT + 2 \text{ V}, & 0 \leq n \leq 5 \\ 0, & \text{otherwise} \end{cases}$$

Expand the series using sin/cos terms.

18. If $T = 1$ ms, $a = 0.75$, what is the closed-form expression for $X(z)$ if $x(nT) = u(nT)(0.5)^{-nT}$?

19. What is the z transform of $h((n-6)T)$ if

$$H(z) = 1 + z^{-2} + 0.7z^{-3} + 0.2z^{-4}?$$

20. Find $h(nT)$ if we have

$$H(z) = \frac{1 + z^{-5}}{1 - z^{-1}}$$

Does $h(nT)$ terminate? Is $h(nT)$ causal? stable?

REFERENCES

CHEN, C.-T. *One-Dimensional Signal Processing.* New York: Marcel Dekker, 1979.

KARWOSKI, R. J. *Introduction to the z Transform and Its Derivation.* TRW publication, La Jolla, 1979.

OPPENHEIM, A. V., and SCHAFER, R. W. *Digital Signal Processing.* Englewood Cliffs, N.J.: Prentice-Hall, 1975.

PELED, A., and LIU, B. *Digital Signal Processing.* New York: Wiley, 1976.

RABINER, L. R., and GOLD, B. *Theory and Application of Digital Signal Processing.* Englewood Cliffs, N.J.: Prentice-Hall, 1975.

STANLEY, W. D. *Digital Signal Processing.* Reston, Va.: Reston, 1975.

TRETTER, S. A. *Introduction to Discrete-Time Signal Processing.* New York: Wiley, 1976.

DIGITAL SYSTEMS

6

The output of the TH circuit was a voltage that could assume any value within a given range (i.e., it was continuous). Its domain was discrete, as it consisted of time values that were integer multiples of the sampling period, T. Each ADC output was a binary number that corresponded to a range (equal to 1 LSB) of input voltages. The effect of quantizing the TH output gave rise to a discrete set of values at the ADC output.

The overall effect of the initial processing was to take a continuous time/continuous voltage signal, the transducer output, and convert it to one that is discrete time/discrete value, the ADC output. In the ideal situation, the sequence of binary numbers at the ADC output would represent the samples of the transducer's output signal in digital form; in the practical case, it represents the sequence of the transducer's output signal together with error terms introduced by noise, aliasing, the TH device, the sampling process, and quantization. We will assume an ideal situation and, where possible, discuss the practical system's effect on the final output.

Having obtained a discrete time/discrete value signal, digital processing techniques can be implemented either in software, hardware, or a combination of both. The sections that follow introduce the techniques used to analyze a digital system and the design of a digital filter.

Chap. 6 Digital Systems

We have indicated above that the ADC output is in digital form and can readily be applied to a processor; however, we can also use the output signal of the TH and construct the system using components that will respond to these sample analog values. Our primary goal is to establish the basic concepts of digital systems. When the fast Fourier transform is introduced, we will see how to construct the system using software rather than hardware. This is the main thrust of DSP.

6.1 DISCRETE LINEAR SYSTEM FUNDAMENTALS

Whether continuous- or discrete-time, the response of a system (e.g., a filter) to a signal at its input can be found from convolution theory. Figures 6–1 and 6–2 show the block diagram of a continuous-time and a discrete-time system, together with their inputs and outputs. $x(nT)$, $h(nT)$, and $y(nT)$ are the discrete-time input, transfer function, and response of a discrete-time system.

Equations 6–1 and 6–2 present both the continuous-time (CT) and discrete-time (DT) convolution relations.

$$y(t) = x(t) * h(t) = \int_{-\infty}^{\infty} x(\tau)h(t-\tau)\, d\tau \qquad (6\text{--}1)$$

$$y(nT) = x(nT) * h(nT) = \sum_{k=-\infty}^{\infty} x(kT)h((n-k)T) \qquad (6\text{--}2)$$

where the DT system convolution relation was derived in Chapter 5. To obtain the desired form of the digital system I/O relationship, several concepts are developed below.

6.1.1 Differential Equation for the CT System I/O Relation

For a linear CT system, it is possible to express the I/O relation of the system as a differential equation involving first- and higher-order derivatives of the input and output voltages, e_{in} and e_{out}, and the circuit component values. This was shown in Section 1.14, and Equation 1–95 is repeated for convenience.

$$\sum_{n=0}^{k} b_n \frac{d^n e_o}{dt^n} = \sum_{r=0}^{j} a_r \frac{d^r e_{\text{in}}}{dt^r} \qquad (1\text{--}95)$$

Equation 1–95 is the general I/O relationship for a CT system.

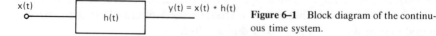

Figure 6–1 Block diagram of the continuous time system.

Sec. 6.1 Discrete Linear System Fundamentals

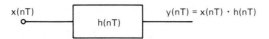

Figure 6–2 Block diagram of the discrete time system.

6.1.2 Difference Equation for a DT System I/O Relation

The reason for introducing the I/O relation for the CT system is to make it more acceptable that a similar relation can be presented for DT systems. The discrete devices with which we will concern ourselves satisfy the I/O relationship in the form of a difference equation.

$$\sum_{k=0}^{N} a_k x((n-k)T) = \sum_{r=0}^{M} b_r y((n-r)T) \tag{6-3}$$

Equation 6–3 is the general DT version of Equation 1–95 (which applies for CT systems) with T as the sampling rate.

We can show that the convolution form (Equation 6–2) can be generated from Equation 6–3 by considering the case for $b_r = 0$, $r > 1$. This yields

$$b_0 y(nT) = \sum_{k=0}^{N} a_k x((n-k)T) \tag{6-4}$$

If $a_k/b_0 \equiv h(kT)$, we have

$$y(nT) = \sum_{k=0}^{N} h(kT) x((n-k)T) \tag{6-5}$$

which is the DT convolution relation. We will assume that all DT systems studied in this text satisfy Equation 6–3.

If we expand Equation 6–3 and examine a few terms, we can obtain some general results.

$$\begin{aligned} a_0 x(nT) + a_1 x((n-1)T) + a_2 x((n-2)T) + \cdots a_N x((n-N)T) \\ = b_0 y(nT) + b_1 y((n-1)T) + b_2 y((n-2)T) + \cdots + b_M y((n-M)T) \end{aligned} \tag{6-6}$$

Solving the equation for $y(nT)$, we have

$$\begin{aligned} y(nT) = \frac{1}{b_0} \{ & a_0 x(nT) + a_1 x((n-1)T) + a_2 x((n-2)T) \\ & + \cdots + a_N x((n-N)T) - b_1 y((n-1)T) + b_2 y((n-2)T) \\ & - \cdots - b_M y((n-M)T) \} \end{aligned} \tag{6-7}$$

This shows that the output of the general DT system, for any value of nT, is determined by the present and previous N inputs and the past M outputs. In addition, for a causal system, if $x(nT) = 0$, for $n < R$, then $y(nT)$ is also zero for $n < R$ where R is an arbitrary integer.

6.1.3 Evaluation of the DT Difference Equation

At this point, we begin to see how we can utilize digital techniques. Equation 6–7 involves only addition and multiplication which can be performed quite readily using digital devices. All this requires is that the values of the $\{a_k\}$ and $\{b_r\}$ be stored in memory and then multiplied with the appropriate values of $x(nT - kT)$ and $y(nT - rT)$. These past inputs and outputs must also be stored in memory so that they will be available to us when they are needed. How this is accomplished will be seen in Chapter 8, as it serves as the basis for using processor devices.

It should be mentioned that the approach discussed above requires $(N + M)$ multiplications and $(N + M)$ additions for each value of n. If we assume that $N = M$, we require $4N$ operations for each point. If there are also N points, there are $4N^2$ required operations to find the full output sequence: A 16-point calculation requires 1024 operations; a 64-point calculation requries 16,384 operations. It becomes obvious that although performing the calculations is straightforward, they will begin to take an excessive amount of execution time as N becomes large. Each value required to determine the sum in Equation 6–7 must be fetched from memory to the processor, an instruction (such as ADD or MULTIPLY) executed, and the sum or product stored in memory. It was this fact that limited the use of digital computers and microprocessors in performing signal processing. We will see in Chapter 7 that the computation time can be reduced from a factor proportional to N^2 to one proportional to $N \log_2 N$, introducing a large savings in time.

Returning to Equation 6–7, several terms can be defined from it. These terms will relate to a general DT system whose I/O relations are satisfied by this equation. If only the present and past inputs determine the present output (i.e., $b_i = 0$, $1 \leq i \leq M$), the DT system is called *nonrecursive*; if, in addition to the past and present inputs, the past outputs also determine the present output, the system is called *recursive*.

6.1.4 System Characteristics

To obtain some insight into the systems that have Equation 6–7 as a representation, let us consider one that is represented by

$$y(nT) = \frac{a_0}{b_0} x(nT) + \frac{a_1}{b_0} x((n-1)T) \qquad (6\text{--}8)$$

From our previous definitions, Equation 6–8 represents a nonrecursive system with $N = 1$. We see that the output is obtained by the present and most recent inputs, $x(nT)$ and $x(nT - T)$ with no past output terms being present. If we want to realize this equation with hardware, it must include (1) two multipliers (e.g., amplifiers or attenuators), (2) a summer circuit, and (3) a device that can store the previous input for a sampling interval so that it can

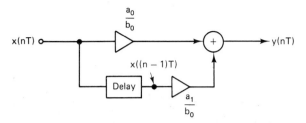

Figure 6–3 Block diagram of the circuit that is obtained from Equation 6–8.

be added to the present one. The first two devices can readily be fabricated from either analog or digital devices; however, as we have gone to the trouble of converting our signal to digital format, we will concentrate on using those devices that can accept the digital signal.

The third item in our circuit must provide storage for one sampling interval [from $(nT - T)$ to nT]. This can be accomplished most readily by a memory or shift register. Figure 6–3 shows the block diagram of a circuit that can be used to realize Equation 6–8. The input signal for a given sampling instant is applied to the system input where it is multiplied by a_0/b_0 and at the same time stored in a delay unit for the next sampling instant. The output of a unit delay for any sampling instant is always the input of the previous sampling instant.

To see how this circuit operates, consider $\{x(nT)\}$ as the sequence shown in Figure 6–4. The circuit parameters are chosen as $a_0/b_0 = 1$ and $a_1/b_0 = \frac{1}{2}$. For $nT = 0$, $x(0) = \frac{1}{2}$ is applied to the system. As $x((n-1)T) = x(nT = -T) = 0$, a zero value was stored in the delay unit at the previous sampling instant $(nT = -T)$. The output is then

$$y(nT) = \frac{a_0}{b_0} x(nT) + \frac{a_1}{b_0} x((n-1)T) \quad (6\text{-}9)$$

For $nT = 0$, this becomes

$$y(0T) = 1 \cdot x(0T) + \frac{1}{2} x(-1T) \quad (6\text{-}10a)$$

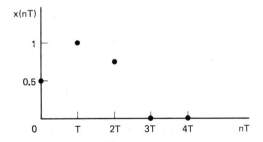

Figure 6–4 Arbitrary input sequence for the circuit of Figure 6–3.

$$y(0T) = 1 \cdot \frac{1}{2} + \frac{1}{2} \cdot 0 \qquad (6\text{-}10\text{b})$$

$$y(0T) = \frac{1}{2} \qquad (6\text{-}10\text{c})$$

At the same time that $x(0) = \frac{1}{2}$ is being multiplied and fed to the output summer, it is being stored in the delay unit, to be used for the next sampling instant.

For $nT = T$, $x(T) = 1$, $x(nT - T) = x(T - T) = x(0) = \frac{1}{2}$, which has been stored from the previous sampling instant. The output is, using Equation 6–9,

$$y(T) = 1 \cdot x(T) + \frac{1}{2} x(T - T) \qquad (6\text{-}11\text{a})$$

$$y(T) = 1 \cdot 1 + \frac{1}{2} \cdot \frac{1}{2} = \frac{5}{4} = 1.25 \qquad (6\text{-}11\text{b})$$

As before, the value of $x(T)$ is stored for use at $nT = 2T$. When $nT = 2T$ arrives, we have $x(2T) = \frac{3}{4}$. $y(2T)$ is then

$$y(2T) = 1 \cdot x(2T) + \frac{1}{2} x(2T - T) \qquad (6\text{-}12\text{a})$$

$$y(2T) = 1 \cdot \frac{3}{4} + \frac{1}{2} \cdot 1 = \frac{5}{4} = 1.25 \qquad (6\text{-}12\text{b})$$

Finally, for $x(3T) = 0$, $y(3T)$ becomes

$$y(3T) = 1 \cdot x(3T) + \frac{1}{2} x(3T - T) \qquad (6\text{-}13\text{a})$$

$$y(3T) = 1 \cdot 0 + \frac{1}{2} \cdot \frac{3}{4} = \frac{3}{8} = 0.375 \qquad (6\text{-}13\text{b})$$

For values of $n > 3$, $x(nT) = 0$, as is $y(nT)$. $y(nT)$ is plotted in Figure 6–5.

Although the example above was straightforward, it sets the ground rules for representing (and realizing) terms in Equation 6–7 in a block diagram and hardware. (We have mentioned that the simple mathematical operations can also be performed with software.) Each term in the equation requires a multiplier and a delay section: the output of a delay, $x(nT - rT)$ was stored at the previous sampling time. The delay units are connected such that each one has its output available so that it can be used as the input to the summer or another delay section.

To demonstrate this, we will represent a nonrecursive system with $N =$

Sec. 6.1   Discrete Linear System Fundamentals   179

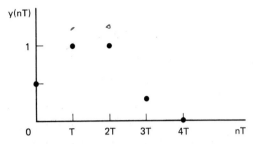

Figure 6-5 Output sequence for the circuit of Figure 6-3 and the input sequence of Figure 6-4.

2 in block diagram form. The equation for this system is given in Equation 6-14.

$$y(nT) = \frac{a_0}{b_0} x(nT) + \frac{a_1}{b_0} x((n-1)T) + \frac{a_2}{b_0} x((n-2)T) \tag{6-14}$$

This is realized by the circuit of Figure 6-6. The present output of each delay unit is the input of the previous sample instant.

If the system includes one or more past inputs, it is recursive. A simple recursive system is one with $N = M = 1$.

$$y(nT) = \frac{a_0}{b_0} x(nT) + \frac{a_1}{b_0} x((n-1)T) - \frac{b_1}{b_0} y((n-1)T) \tag{6-15}$$

The circuit realization uses the same devices as for nonrecursive systems. The circuit that yields Equation 6-15 is given in Figure 6-7. The system output at nT is now dependent on the system output at $nT - T$ as well as the present and most recent past input (i.e., the past output must also be stored).

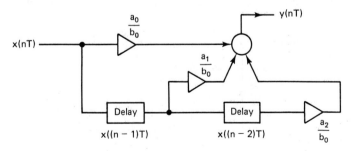

Figure 6-6 Circuit realization of Equation 6-14.

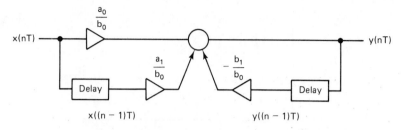

Figure 6–7 Circuit realization of Equation 6–15.

6.1.5 Impulse Response

The transfer function of the circuits shown in Figures 6–3, 6–6, and 6–7 can be obtained by letting $x(nT)$ be the impulse function $\delta(nT)$. For the circuit of Figure 6–6, representing the nonrecursive system, we have

$$h(0T) = y(0T) = \frac{a_0}{b_0} x(0T) + \frac{a_1}{b_0} x(-T) + \frac{a_2}{b_0} x(-2T) \qquad (6\text{–}16a)$$

$$h(T) = y(T) = \frac{a_0}{b_0} x(T) + \frac{a_1}{b_0} x(T-T) + \frac{a_2}{b_0} x(T-2T) \qquad (6\text{–}16b)$$

$$h(2T) = y(2T) = \frac{a_0}{b_0} x(2T) + \frac{a_1}{b_0} x(2T-T) + \frac{a_2}{b_0} x(2T-2T) \qquad (6\text{-}16c)$$

$$h(0T) = \frac{a_0}{b_0} \cdot 1 = \frac{a_0}{b_0} \qquad (6\text{–}17a)$$

$$h(T) = \frac{a_1}{b_0} \cdot 1 = \frac{a_1}{b_0} \qquad (6\text{–}17b)$$

$$h(2T) = \frac{a_2}{b_0} \cdot 1 = \frac{a_2}{b_0} \qquad (6\text{–}17c)$$

for $n > 2$, $y(nT) = 0$. As the output is the result of an impulse input, $y(nT) = h(nT)$, the impulse response. This is shown in Figure 6–8, where we have

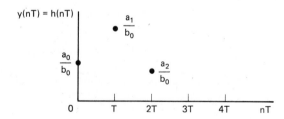

Figure 6–8 Impulse response for the circuit of Figure 6–6.

Sec. 6.1 Discrete Linear System Fundamentals   181

assumed zero values for b_i/b_0, $i > 0$. As the impulse response is zero for $n > 2$, the system is an FIR one. This will be true for all nonrecursive systems with $N = $ finite. In addition, the impulse response of a nonrecursive system is the set of coefficients of the present and past inputs: $h(kT) = a_k/b_0$. These can be obtained from the circuit or system equation by inspection.

To find the impulse response of a recursive system, we apply the impulse to the circuit of Figure 6-7.

$$y(0T) = \frac{a_0}{b_0} x(0T) + \frac{a_1}{b_0} x(-T) - \frac{b_1}{b_0} y(-T) \qquad (6\text{-}18)$$

As the system is causal,

$$y(0T) = \frac{a_0}{b_0} \cdot 1 + \frac{a_1}{b_0} \cdot 0 - \frac{b_1}{b_0} \cdot 0 = \frac{a_0}{b_0} \qquad (6\text{-}19\text{a})$$

$$y(T) = \frac{a_0}{b_0} x(T) + \frac{a_1}{b_0} x(T - T) - \frac{b_1}{b_0} y(T - T) \qquad (6\text{-}19\text{b})$$

$$= \frac{a_0}{b_0} \cdot 0 + \frac{a_1}{b_0} \cdot 1 - \frac{b_1}{b_0} \frac{a_0}{b_0} \qquad (6\text{-}19\text{c})$$

$$y(2T) = \frac{a_0}{b_0} x(2T) + \frac{a_1}{b_0} x(2T - T) - \frac{b_1}{b_0} y(2T - T) \qquad (6\text{-}19\text{d})$$

$$= \frac{a_0}{b_0} \cdot 0 + \frac{a_1}{b_0} \cdot 0 - \frac{b_1}{b_0} \left(\frac{a_1}{b_0} - \frac{b_1 a_0}{b_0^2} \right)$$

$$= -\frac{b_1}{b_0} \left(\frac{a_1}{b_0} - \frac{b_1 a_0}{b_0^2} \right)$$

For $n > 2$, we readily see that all values of $x(nT)$ are zero; however, as the previous output is part of the input (unlike the nonrecursive filter), the recursive output will be nonzero even though the input has gone to zero. For this example, although M is finite, the output does not go to zero. This implies that the recursive system is usually one with an IIR. Note that the impulse response cannot be obtained by inspection and must be calculated for each system.

Based on the above, we see that a system which is causal, linear, discrete-time, and shift invariant can be represented by Equation 6-7. Depending on the form of the equation satisfied by the filter, it will be either recursive (and possibly IIR) or nonrecursive (and FIR for finite N). On the other hand, a system can be designed and realized starting with the difference equation, its characteristics being determined by the equation type and the values of the coefficients, $\{a_i\}$ and $\{b_i\}$. This is discussed below.

6.2 z TRANSFORM OF A DIFFERENCE EQUATION

Equation 6–3 provides the difference equation that we will use to represent a linear DT system with filter characteristics. The initial step is to obtain its z transform. We will make use of the delay property of the z transform as presented in Chapter 5.

$$x(nT) \iff X(z) \qquad (6\text{--}20\text{a})$$
$$x((n-r)T) \iff X(z)z^{-r} \qquad (6\text{--}20\text{b})$$

The z transform of Equation 6–3 is

$$\sum_{k=0}^{N} a_k X(z) z^{-k} = \sum_{r=0}^{M} b_r Y(z) z^{-r} \qquad (6\text{--}21)$$

This becomes

$$X(z) \sum_{k=0}^{N} a_k z^{-k} = Y(z) \sum_{r=0}^{M} b_r z^{-r} \qquad (6\text{--}22)$$

6.3 SYSTEM TRANSFER FUNCTION, H(z)

To find the system transfer function, $H(z)$, we solve Equation 6–22 for $Y(z)/X(z)$. This yields

$$H(z) = \frac{Y(z)}{X(z)} = \frac{\sum_{k=0}^{N} a_k z^{-k}}{\sum_{r=0}^{M} b_r z^{-r}} \qquad (6\text{--}23)$$

Equation 6–23 shows that if the system is described by a difference equation with constant coefficients, its transfer function, $H(z)$, is the quotient of two polynominals in z^{-1} [i.e., $H(z)$ is a rational function]. Equation 6–23 is equivalent in form to the CT system's Laplace transfer function, $H(s) = Y(s)/X(s)$, and will be used in a similar manner.

6.3.1 Nonrecursive System

For values of $b_0 = 1$ and $b_i = 0$, for all $i > 0$, the system under consideration is nonrecursive. Its transfer function is given as

$$H(z) = \sum_{k=0}^{N} a_k z^{-k} \qquad (6\text{--}24)$$

For finite N, the system response is of finite duration (FIR). An example of a FIR system was given in Equation 6–8.

Sec. 6.3 System Transfer Function, $H(z)$

$$y(nT) = a_0 x(nT) + a_1 x((n-1)T) \quad (6\text{–}8)$$

where $b_0 = 1$. The transfer function of this system is given by Equation 6–25.

$$H(z) = a_0 z^{-0} + a_1 z^{-1} \quad (6\text{–}25a)$$

$$H(z) = a_0 + a_1 z^{-1} \quad (6\text{–}25b)$$

In representing the system with a block diagram, each delay block is identified by z^{-1}, which represents a delay (or store) of its inputs at $(nT - T)$ until nT. The block diagram of Equation 6–25b is shown in Figure 6–9. This should be compared with Figure 6–3.

Upon examination of Equation 6–25b, we see that $H(z)$ is zero for $z^{-1} = -(a_0/a_1)$ and infinite for $z = 0$. This is plotted in the pole–zero diagram given in Figure 6–10, where we have assumed that a_0 and a_1 are real and positive. The region of convergence (ROC) for $H(z)$ is $0 < |z| \leq \infty$, which includes the unit circle, $|z| = 1$. As the interior of the unit circle corresponds to the left-half of the s plane, we can infer that the system represented by Equation 6–25b is stable. This can be generalized by stating that a stable system will have all its poles inside the unit circle (i.e., the ROC will include the unit circle). Also, as the region of convergence includes $|z| = \infty$, the system is causal. Finally, we can state that all nonrecursive systems are stable as their regions of convergence will always include the unit circle: all their poles are located at the origin, $|z| = 0$.

As the system of Equation 6–25b has all values of the $b_i = 0$, $i > 0$, its difference equation form is the same as the convolution equation

$$y(nT) = \sum_{r=0}^{\infty} h(rT) x((n-r)T) \quad (6\text{–}26)$$

Figure 6–9 Circuit realization of Equation 6–25b.

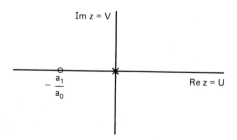

Figure 6–10 Pole-zero diagram of Equation 6–25b.

Therefore, the values of $h(rT)$ are just the coefficients of z^{-r} in $H(z)$. We can then state that for nonrecursive systems, the coefficients of z^{-r} are the terms in the impulse response sequence of the system and vice-versa. The sequence corresponding to the transfer function of Equation 6–25b is

$$h(0) = a_0 \tag{6–27a}$$
$$h(T) = a_1 \tag{6–27b}$$
$$h(nT) = 0, \quad n > 1 \tag{6–27c}$$

This is represented in Figure 6–11.

As the ROC includes the unit circle, we can obtain the Fourier transform of $\{h(nT)\}$ (i.e., its frequency response) by evaluating $H(z)$ on the unit circle, $z = \exp(j2\pi fT)$.

$$H(z)\Big|_{z=\exp(j\omega T)} = H(f) = \sum_{n=-\infty}^{\infty} h(nT) \exp(-j2\pi fnT) \tag{6–28}$$

For the system of Equation 6–25b, this becomes

$$H(f) = a_0 + a_1 \exp(-j2\pi fT) \tag{6–29}$$

$|H(f)|$ is then found to be

$$|H(f)| = \sqrt{H(f)H^*(f)} = \sqrt{a_0^2 + a_1^2 + 2a_1 a_0 \cos(2\pi fT)} \tag{6–30}$$

The frequency response of the system is seen to be periodic in f^{sa}; that is, $|H(f)|$ repeats for $f \to f + fsa$. We will use this periodic relationship later in the text.

6.3.2 Recursive System

If the transfer function of the system includes terms in which some of the difference equation coefficients b_i are not zero, the system is recursive. We will return to Equation 6–15 as an example. For this system, the present output (at $t = nT$) requires the value of its most recent output [at $t = (n-1)T$]. The difference equation form of Equation 6–15 is written as

$$a_0 x(nT) + a_1 x((n-1)T) = b_0 y(nT) + b_1 y((n-1)T) \tag{6–31}$$

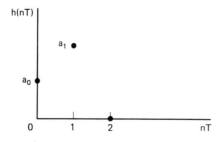

Figure 6–11 Impulse response of equation 6–25b.

Sec. 6.3 System Transfer Function, $H(z)$

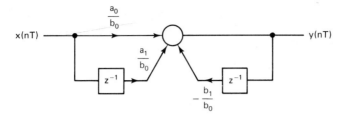

Figure 6–12 Circuit realization of Equation 6–32.

Taking its z transform and finding $Y(z)/X(z)$, we obtain

$$H(z) = \frac{Y(z)}{X(z)} = \frac{a_0 + a_1 z^{-1}}{b_0 + b_1 z^{-1}} \qquad (6\text{–}32)$$

Block diagram representation. The block diagram representation of the recursive filter will be similar to the one we constructed in Figure 6–7. In the present representation, we will use the z^{-1} block to symbolize the unit delay. The recursive system block diagram of Equation 6–32 is presented in Figure 6–12. This will be the standard format used to represent digital systems.

Poles and zeros. If we multiply $H(z)$ (Equation 6–32) by z/z, we obtain

$$H(z) = \frac{a_0 z + a_1}{b_0 z + b_1} \qquad (6\text{–}33)$$

This can be put in the form

$$H(z) = \frac{a_0(z + a_1/a_0)}{b_0(z + b_1/b_0)} \qquad (6\text{–}34)$$

In this form we can readily establish the poles and zeros of $H(z)$. For this system, the zero is at $z = -(a_1/a_0)$ and the pole is at $z = -b_1/b_0$. This is shown in Figure 6–13. In the figure we have assumed that a_0, a_1, b_0, and b_1 are positive and real. This may not always be the situation.

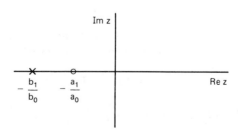

Figure 6–13 Pole-zero diagram of Equation 6–34.

Region of convergence. The region of convergence for the system described by Equation 6–34 is determined by the location of the poles, $z = -b_1/b_0$. If $|-b_1/b_0| < 1$, the pole is inside the unit circle, $|z| = 1$, which indicates that the system is stable; if $|-b_1/b_0| \geq 1$, the pole is either on the unit circle or outside it. Both situations are indicative of an unstable system: $h(nT) \to \infty$ as $n \to \infty$.

If all poles are inside a circle of radius a, the sequence $\{h(nT)\}$ will be right-sided: $h(nT) = 0$ for all $n < -N$. If $H(z) \to$ constant value as $z \to \infty$, the system is causal.

Determination of $\{h(nT)\}$ values. If there are no poles outside or on the unit circle (e.g., $|-b_1/b_0| < 1$ in the example above), $H(z)$ is said to be analytic in this region. The $h(nT)$ values can be determined if $H(z)$ is analytic for $|z| \geq 1$ by dividing the denominator polynomial into the numerator. We will examine the system under consideration (Equation 6–32) for $a_0 = 1$, $a_1 = 1$, $b_0 = 1$, and $b_1 = 0.5$ putting a pole at $z = -\frac{1}{2}$. This yields

$$
\begin{array}{r}
1 + \tfrac{1}{2}z^{-1} - \tfrac{1}{4}z^{-2} + \cdots \\
1 + \tfrac{z^{-1}}{2} \overline{)\, 1 + z^{-1} } \\
1 + \tfrac{1}{2}z^{-1} \\
\tfrac{1}{2}z^{-1} \\
\tfrac{1}{2}z^{-1} + \tfrac{1}{4}z^{-2} \\
-\tfrac{1}{4}z^{-2} \\
-\tfrac{1}{4}z^{-2} - \tfrac{1}{8}z^{-3} \\
\cdots
\end{array}
\qquad (6\text{–}35)
$$

Upon examination of Equation 6–35, we see that $H(z)$ can be written as in Equation 6–36.

$$H(z) = 1 + \frac{1}{2}z^{-1} - \frac{1}{4}z^{-2} + \frac{1}{8}z^{-3} - \frac{1}{16}z^{-4} + \cdots \qquad (6\text{–}36a)$$

The coefficients of powers of z^{-n} are the $h(nT)$. This gives us

$$h(nT) = \begin{cases} 1, & n = 0, \\ (-1)^{n+1}\left(\dfrac{1}{2}\right)^n, & 1 \leq n \leq \infty \\ 0, & \text{otherwise} \end{cases} \qquad (6\text{–}36b)$$

The system is causal, stable, right-sided, and with infinite impulse response. This transfer function can be represented as

$$h(0) = 1, \qquad h(nT) = (-1)^{n+1}\left(\frac{1}{2}\right)^n u(nT - T) \qquad (6\text{–}36c)$$

The impulse response can also be obtained by applying $\delta(nT)$ to the system input and setting $y(nT) = h(nT)$. Additional techniques for obtaining the

Sec. 6.3 System Transfer Function, $H(z)$

$h(nT)$ from $H(z)$ depend on the form of $H(z)$ and its region of convergence. These techniques are (1) the residue theorem and (2) partial-fraction expansion. The former technique depends on a knowledge of complex variables and will not be included here; the latter should be familiar to those who have worked with circuits where the time response of a CT system was found from $H(s)$ using this technique. $H(z)$ is expressed as

$$H(z) = \sum_{i=1}^{N} \frac{a_i}{1 - p_i z^{-1}} \tag{6-37a}$$

where p_i is a pole of $H(z)$. The a_i are formed from Equation 6–37b.

$$a_i = (1 - p_i z^{-1}) H(z) \Big|_{z = p_i} \tag{6-37b}$$

The sequence for $h(nT)$ is then the inverse z transform of each term in Equation 6–37a and given by

$$h(nT) = \sum_{i=1}^{N} a_i (p_i)^n \tag{6-38}$$

Example 6–1

Find $\{h(nT)\}$ for $H(z)$ given as

$$H(z) = \frac{1 + z^{-1}}{1 + 1.2 z^{-1} + 0.35 z^{-2}}$$

Solution We express $H(z)$ in partial-fraction form:

$$H(z) = \frac{a_1}{1 + 0.7 z^{-1}} + \frac{a_2}{1 + 0.5 z^{-1}}$$

Next we find a_1 and a_2:

$$a_1 = (1 + 0.7 z^{-1}) H(z) \Big|_{z = -0.7}$$

$$= (1 + 0.7 z^{-1}) \left[\frac{1 + z^{-1}}{(1 + 0.7 z^{-1})(1 + 0.5 z^{-1})} \right] \Big|_{z = -0.7}$$

$$= \frac{1 + z^{-1}}{1 + 0.5 z^{-1}} \Big|_{z = -0.7} = \frac{1 + (-0.7)^{-1}}{1 + 0.5(-0.7)^{-1}}$$

$$= \frac{1 - 1.43}{1 + 0.5(-1.43)} = -1.5$$

Similarly, $a_2 = 2.5$. Finally, we find the first three terms of sequence as

$$h(nT) = \sum_{i=1}^{2} a_i (p_i)^n$$

$$h(0T) = a_1 (p_1)^0 + a_2 (p_2)^0 = 1.0$$

$$h(T) = a_1(p_1)^1 + a_2(p_2)^1 = -0.2$$
$$h(2T) = a_1(p_1)^2 + a_2(p_2)^2 = -0.11$$

In Example 6–1, $H(z)$ was assumed to be stable and causal. The reader should verify this.

It is also possible to find $h(nT)$ by setting $H(z) = Y(z)/X(z)$, taking the inverse z transform (using the relations presented), and then applying the impulse signal to the system; the system output, $y(nT)$, is $h(nT)$.

6.3.3 General Form of the System Transfer Function

Equation 6–23 presented the general form of a system function represented by a difference equation (Equation 6–3). As $H(z)$ is the ratio of two polynomials in z^{-1}, we can factor each of them into products of terms involving their roots. This is shown in Equation 6–39.

$$H(z) = \frac{K \prod_{n=1}^{N} (1 - 0_n z^{-1})}{\prod_{r=1}^{N} (1 - p_r z^{-1})} \tag{6–39}$$

where 0_n is a zero of $H(z)$ and p_r is a pole. In addition, when $z = 0$, each term in the numerator provides a pole of order n, while each term in the denominator provides a zero.

We will assume that we are dealing with real-world systems that are stable and causal. The region of convergence will be all values of z that lie exterior to a circle whose radius is equal to the value of the maximum valued pole.

$$\text{ROC} = |z| > \max \{p_r\} \tag{6–40}$$

This is shown in Figure 6–14. The ROC in the figure is the set of all z such that

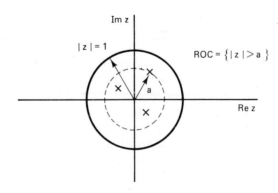

Figure 6–14 Region of convergence exterior to a circle of radius a.

6.4 BLOCK DIAGRAM REPRESENTATION OF A DIGITAL SYSTEM

$$\max \{\|p_r\|\} = a < |z| \leq \infty \tag{6-41}$$

A linear DT system is represented by either a difference equation (Equation 6–3) or an impulse transfer function in the z plane (Equation 6–23). Each can be obtained from the other by using the z transform and the inverse z transform. Using these equations as a starting point, several system structures can be generated.

6.4.1 Direct Form: Type 1

If the DT system under consideration is a filter, then $b_0 = 1$. This allows us to write Equation 6–3 as

$$y(nT) = \sum_{k=0}^{N} a_k x((n-k)T) - \sum_{r=1}^{M} b_r y((n-r)T) \tag{6-42}$$

This states that the nth output of the filter is the sum of its past and present inputs as well as its past M outputs. This equation can be realized by the circuit of Figure 6–15. This circuit requires $M + N$ delay elements for its realization and is called a direct form, type I circuit.

6.4.2 Direct Form: Type 2

If Equation 6–23 is written as the product of two terms, one consisting of the numerator and the other of the denominator, we obtain

$$H(z) = H_1(z) H_2(z) \tag{6-43a}$$

$$= \left(\sum_{k=0}^{N} a_k z^{-k} \right) \frac{1}{1 + \sum_{r=1}^{M} b_r z^{-1}} \tag{6-43b}$$

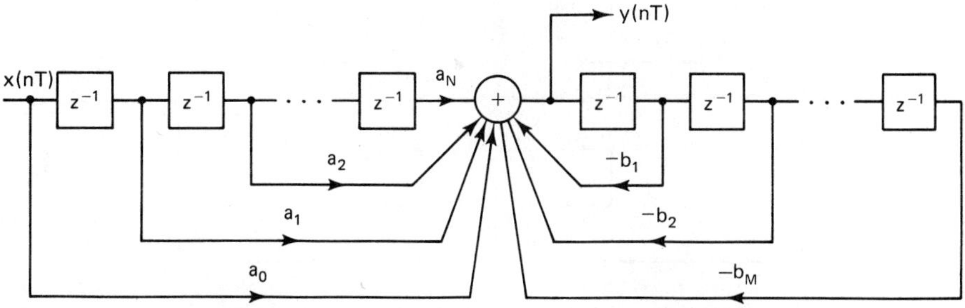

Figure 6–15 Direct form 1 realization of a digital filter.

The circuit is obtained by considering each factor separately and introducing an internal factor, $W(z)$. This is shown in Equation 6–44.

$$H(z) = H_1(z)H_2(z) = \frac{Y(z)}{X(z)} \tag{6-44a}$$

$$= \frac{Y(z)W(z)}{X(z)W(z)} \tag{6-44b}$$

$$H_1(z) \equiv \frac{Y(z)}{W(z)}, \qquad H_2(z) \equiv \frac{W(z)}{X(z)} \tag{6-44c}$$

This is written as

$$H_1(z) = \frac{Y(z)}{W(z)} = \sum_{k=0}^{N} a_k z^{-k} \tag{6-45a}$$

$$H_2(z) = \frac{W(z)}{X(z)} = \frac{1}{1 + \sum_{r=1}^{M} b_r z^{-r}} \tag{6-45b}$$

By cross-multiplying both equations and finding their inverse z transform, we have

$$y(nT) = \sum_{k=0}^{N} a_k w((n-k)T) \tag{6-46a}$$

$$w(nT) = x(nT) - \sum_{r=1}^{M} b_r w((n-r)T) \tag{6-46b}$$

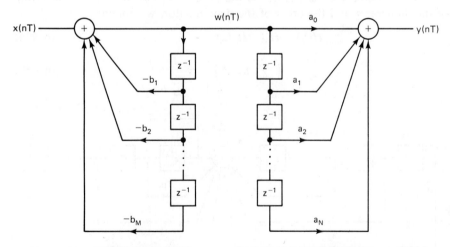

Figure 6–16 Direct form 2 realization of a digital filter.

Sec. 6.4   Block Diagram Representation of a Digital System

Both circuits are constructed from the equations and combined to form the realization of Figure 6–16.

As the sequence $w(nT)$ is internal and undergoes the same amount of delay for both sections, the separate halves can be combined to yield the realization of Figure 6–17. This realization has reduced the number of delays required from $M + N$ to the larger of M or N and is sometimes referred to as the *canonical form*.

6.4.3 Cascade Form

The realization of the transfer function can be obtained using several simpler functions rather than a single complex one. $H(z)$ is decomposed into a product of first- and second-order functions as represented in Equation 6–47.

$$H(z) = KH_1(z)H_2(z) \cdots H_m(z) \qquad (6\text{–}47\text{a})$$

$$= K \prod_{i=1}^{m} H_i(z) \qquad (6\text{–}47\text{b})$$

where the $H_i(z)$ are either first- or second-order terms:

$$H_i(z) = \frac{1 + a_i z^{-1}}{1 + b_i z^{-1}} \qquad (6\text{–}48)$$

is a first-order term, while

$$H_j(z) = \frac{1 + a_{1j}z^{-1} + a_{2j}z^{-2}}{1 + b_{1j}z^{-1} + a_{2j}z^{-2}} \qquad (6\text{–}49)$$

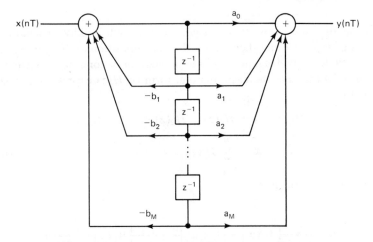

Figure 6–17 Canonical form realization of a digital filter.

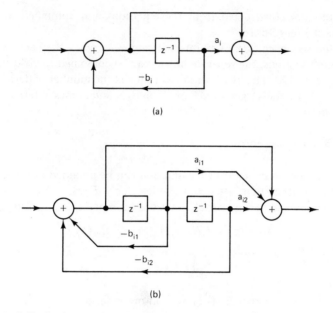

Figure 6–18 Sections used in the cascade form of a digital filter: (a) first order; (b) second order.

is a second-order one. Each term is realized by the canonical form circuits given in Figure 6–18. These circuits are arranged in series (cascade) as shown in Figure 6–19 to realize Equation 6–47a.

6.4.4 Parallel Form

If Equation 6–23 is written as a sum of terms obtained by a partial-fraction expansion of $H(z)$, we can then obtain the decomposition of $H(z)$ in its parallel form. Each term will again be either a first- or second-order term in z^{-1}. Equation 6–50 presents the basic parallel form of $H(z)$ expressed as a sum of individual terms.

$$H(z) = K + H_1(z) + H_2(z) + \cdots + H_N(z) \qquad (6\text{–}50\text{a})$$

$$= K + \sum_{j=1}^{N} H_j(z) \qquad (6\text{–}50\text{b})$$

Each term is of the form

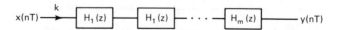

Figure 6–19 Block diagram of the cascade realization of a digital filter.

Sec. 6.4 Block Diagram Representation of a Digital System

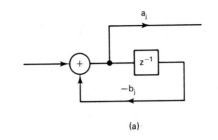

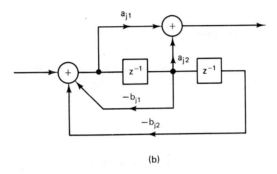

Figure 6–20 Sections used in the parallel form realization of a digital filter: (a) first order; (b) second order.

$$H_j(z) = \frac{a_j}{1 + b_j z^{-1}} \qquad (6\text{-}51)$$

for first-order, or

$$H_j(z) = \frac{a_{1j} + a_{2j}z^{-1}}{1 + b_{1j}z^{-1} + b_{2j}z^{-2}} \qquad (6\text{-}52)$$

for second-order. Their canonical realizations are given in Figure 6–20. Figure 6–21 is the parallel form realization of $H(z)$, Equation 6–50a.

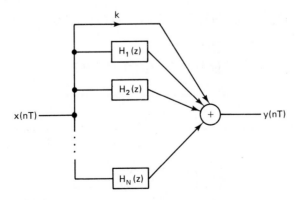

Figure 6–21 Parallel form realization of a digital filter.

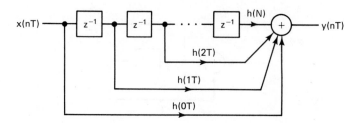

Figure 6–22 Realization of a nonrecursive (all-zero) digital filter.

6.4.5 All-Zero System

The nonrecursive system has its output determined by the convolution of the input sequence and the impulse response of the system. This is given in Equation 6–53.

$$y(nT) = \sum_{k=0}^{N} h(kT)x((n-k)T) \qquad (6\text{--}53)$$

The realization of this equation is given in Figure 6–22. The circuit is also called a transversal, $(N + 1)$-tap, or a smoothing system.

6.5 DISCRETE SYSTEM CIRCUIT ANALYSIS

Given a circuit that represents a discrete system, we would like to know its frequency response. The procedure for doing this is presented in this section. The opposite viewpoint of system synthesis is covered in the following section.

We are given the circuit of Figure 6–23 as an example of the technique used to analyze a DT system. The first step is to determine its difference equation by inspection.

$$y(nT) = 0.63x(nT) - 1.3x((n-1)T) + 0.78y((n-1)T) \qquad (6\text{--}54)$$

The z transform of this equation is

$$Y(z) = 0.63X(z) - 1.3X(z)z^{-1} + 0.78\,Y(z)z^{-1} \qquad (6\text{--}55)$$

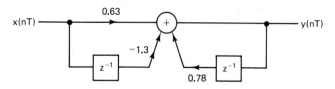

Figure 6–23 Recursive filter.

Sec. 6.5 Discrete System Circuit Analysis

This is rewritten to provide $H(z)$:

$$H(z) = \frac{Y(z)}{X(z)} = \frac{0.63 - 1.3z^{-1}}{1 - 0.78z^{-1}} \quad (6\text{–}56)$$

We can see that $H(z)$ has a zero at $z = 1.3/0.63$ and a pole at $z = 0.78$, which is inside the unit circle, indicating that $H(z)$ is stable.

To find the frequency response of the system, $H(z)$ is evaluated on the unit circle.

$$H(z)\bigg|_{z=\exp(j\omega T)} = H(\omega T) = \frac{0.63 - 1.3e^{-j\omega T}}{1 - 0.78e^{-j\omega T}} \quad (6\text{–}57)$$

This is the same as the Fourier transform of $h(nT)$. After some algebra, we obtain the magnitude and phase of $H(\omega T)$.

$$|H(\omega T)| = \sqrt{\frac{(0.63 - 1.3\cos\omega T)^2 + (1.3\sin\omega T)^2}{(1 - 0.78\cos\omega T)^2 + (0.78\sin\omega T)^2}} \quad (6\text{–}58a)$$

$$\phi(\omega T) = \tan^{-1}\frac{0.81\sin\omega T}{1.64 - 1.79\cos\omega T} \quad (6\text{–}58b)$$

From previous work, we know that $H(\omega T)$ is periodic in f (or ω) with period f_{sa} (or ω_{sa}).

At this point, we must consider the term ωT to obtain some insight into the system frequency response with the Nyquist sampling frequency in mind. We have

$$\omega T = 2\pi f T = 2\pi \frac{f}{f_{sa}} \equiv 2\pi\Omega \quad (6\text{–}59)$$

where f_{sa} is the sampling frequency. The term f/f_{sa} is usually defined as the normalized frequency, Ω. Since $f_{sa} \geq 2f_{max}$, where f_{max} is the maximum frequency component of the sampled signal, we see that

$$0 \leq \Omega \leq 0.5 \quad (6\text{–}60)$$

which can also be given as

$$0 \leq \frac{f}{f_{sa}} \leq \frac{1}{2} \quad (6\text{–}61)$$

This allows us to present the frequency response of a discrete system in terms of Ω, the normalized frequency. Equation 6–58a becomes

$$|H(\omega T)| \longrightarrow |H(\Omega)| = \sqrt{\frac{(0.63 - 1.3\cos 2\pi\Omega)^2 + (1.3\sin 2\pi\Omega)^2}{(1 - 0.78\cos 2\pi\Omega)^2 + (0.78\sin 2\pi\Omega)^2}} \quad (6\text{–}62)$$

where we indicate that the function $H(\omega T)$ is equivalent to the function $H(\Omega)$ as their arguments are functionally related. This is an example of the use of

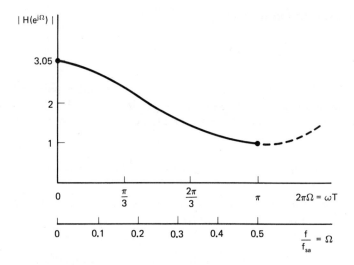

Figure 6–24 Frequency response of the circuit of Figure 6–23.

implicit functions covered in Section 1.1.4. The response is plotted in Figure 6–24. This is the frequency response of the digital system of Figure 6–23, which shows that a CT signal with frequency $f \leq f_{sa}/2$, sampled at a frequency f_{sa}, will be attenuated by an amount $|H(\exp(j2\pi\Omega))|$ as it passes through the system. In addition, there will also be a phase shift, $\phi(2\pi\Omega)$. The system of Figure 6–24 shows the characteristics similar to a LP filter. In this figure, both the ω and Ω axes are presented. In addition, we need show only one segment of the response due to its periodicity and the fact that we assume the maximum frequency of any input signal to be less than $f_{sa}/2$.

6.6 DIGITAL FILTER SYNTHESIS

Given a set of desired filter characteristics, what circuit can be used to realize it? This section will consider the design of the digital filter where its realization can be in the form of an analog or digital hardware circuit or computer software. The techniques used to generate the filter parameters depend on whether the filter is IIR or FIR and will be covered below.

6.6.1 IIR Filter

If an impulse signal is applied to an IIR filter, there is usually an output present for large values of nT. This can be stated in terms of the impulse response as $|h(nT)| > 0$ for n going to infinity. The IIR filter will be the recursive type, requiring past outputs to determine the present output. Also,

Sec. 6.6   Digital Filter Synthesis

as the IIR system is capable of instability, we must determine that all filter poles are inside the unit circle.

In the sections that follow, we look at several techniques used to design IIR filters. An additional characteristic of the IIR filter is the highly nonlinear phase shift in the passband. If a linear phase response is desired, an FIR filter must be used.

Location of poles and zeros in the z plane. In Equation 6–39, the transfer function was expressed as a rational product of terms involving the roots of the numerator and denominator polynomials.

$$H(z) = \frac{\prod_{k=1}^{N} (1 - 0_k z^{-1})}{\prod_{r=1}^{M} (1 - p_r z^{-1})} \tag{6-63}$$

where 0_k and p_r are the roots of the polynomials that form $H(z)$. As an example, for $N = M = 2$, we can express Equation 6-63 as

$$H(z) = \frac{(1 - 0_1 z^{-1})(1 - 0_2 z^{-1})}{(1 - p_1 z^{-1})(1 - p_2 z^{-1})} \tag{6-64}$$

which can be shown in its expanded form as

$$H(z) = \frac{1 - (0_1 + 0_2)z^{-1} + 0_1 0_2 z^{-2}}{1 - (p_1 + p_2)z^{-1} + p_1 p_2 z^{-2}} \tag{6-65a}$$

Arbitrarily choosing the zeros as $0.3 + j0.4$ and $0.3 - j0.4$, and the poles as $-0.4 + j0.3$ and $-0.4 - j0.3$, Equation 6-65a becomes

$$H(z) = \frac{1 - 0.6z^{-1} + 0.25z^{-2}}{1 + 0.8z^{-1} - 0.25z^{-2}} \tag{6-65b}$$

The pole–zero diagram for this equation is given in Figure 6–25. We will consider only systems with poles inside the unit circle (i.e., stable systems).

The frequency response of $H(z)$ will be determined to provide us with some insight into the design approach. This is done by obtaining the magnitude of $H(z)$ in Equation 6–64. Before we do this, the general form of the response magnitude is obtained from Equation 6–63.

$$|H(z)| = \frac{\prod_{k=1}^{N} M_k^0}{\prod_{r=1}^{M} M_r^p} \tag{6-66a}$$

$$M_k^0 \equiv |1 - 0_k z^{-1}| \tag{6-66b}$$

$$M_r^p \equiv |1 - p_r z^{-1}| \tag{6-66c}$$

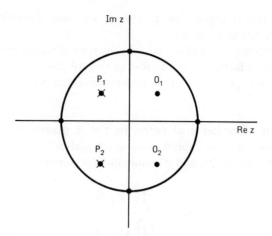

Figure 6–25 Typical pole/zero diagram for Equation 6–65.

The phase angle of $H(z)$ is given by

$$\phi(H(z)) = \frac{\sum_{k=1}^{N} \angle\Theta_k^0}{\sum_{r=1}^{M} \angle\Theta_r^p} \tag{6-67a}$$

$$\angle\Theta_k^0 \equiv \tan^{-1}\frac{\text{Im}\,(1 - O_k z^{-1})}{\text{Re}\,(1 - O_k z^{-1})} \tag{6-67b}$$

$$\angle\Theta_r^p \equiv \tan^{-1}\frac{\text{Im}\,(1 - p_r z^{-1})}{\text{Re}\,(1 - p_r z^{-1})} \tag{6-67c}$$

For the transfer function of Equation 6–64, we have

$$|H(z)| = \frac{M_1^0 M_2^0}{M_1^p M_2^p} \tag{6-68a}$$

$$\phi(H(z)) = \frac{\angle\Theta_1^0 + \angle\Theta_2^0}{\angle\Theta_1^p + \angle\Theta_2^p} \tag{6-68b}$$

Treating each term in Equation 6–64 as a vector, we can obtain the value of $|H(z)|$ and $\phi(H(z))$ for any value of Ω. This is shown in Figure 6–26. By measuring the length of the vector between the point on the unit circle and the corresponding pole/zero and inserting it into Equation 6–68a, we can determine the amplitude of $|H(z)|$ at that point. The angle each vector makes with the Re z axis will yield the phase angle of $H(z)$ when used in Equation 6–68b. For the pole–zero plot of Figure 6–26, the variation of $|H(z)|$ with Ω from $-\frac{1}{2}$ to $+\frac{1}{2}$ will be similar to that of a bandpass filter. The approximate response is given in Figure 6–27.

Sec. 6.6 Digital Filter Synthesis

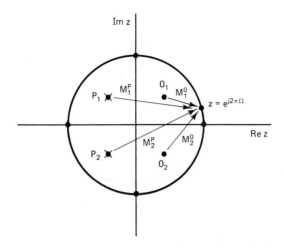

Figure 6–26 Diagram for determining the magnitude response of $|H(\exp j2\pi\Omega)|$.

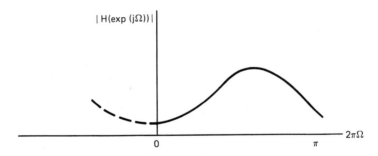

Figure 6–27 Approximate response of the pole–zero diagram of Figure 6–26.

Based on the discussion above, we can begin to see how the frequency response of the transfer function can be changed to meet desired specifications by moving the poles and zeros about the z plane: to reduce the gain at either Ω equal to zero or $\frac{1}{2}$, the zeros of the system can be moved close to $+1$ and -1; choosing the poles closer to the unit circle will cause the gain to roll off more steeply. Figure 6–28 shows the pole–zero diagram of Figure 6–26, but with the poles and zeros moved to provide a more narrow bandwidth. The resulting frequency response of $|H(z)|$ is given in Figure 6–29 with the curve of Figure 6–27 superimposed for comparison.

Once the desired transfer characteristic is obtained, Equation 6–39 (or Equation 6–64 for example) is then expressed as the ratio of two polynomials. By multiplying all terms in Equation 6–64, we obtained

$$H(z) = \frac{1 - (0_1 + 0_2)z^{-1} + 0_1 0_2 z^{-2}}{1 - (p_1 + p_2)z^{-1} + p_1 p_2 z^{-2}} \qquad (6\text{–}69)$$

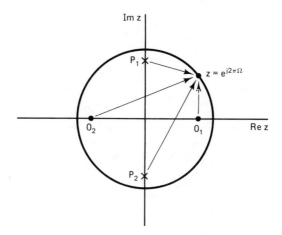

Figure 6-28 Modified pole-zero diagram of Figure 6-26.

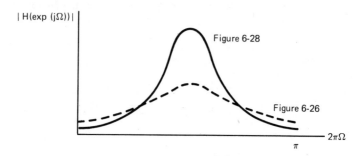

Figure 6-29 Response of $|H(\exp(j\Omega)|$ for pole-zero diagrams of Figures 6-26 and 6-28.

where the zeros and poles are now the ones that yield the desired system response. As this is similar to the second-order section of Equation 6-49, the representation will be similar to that of Figure 6-18b. This is shown in Figure 6-30. This is also the canonical form of realization as given in Figure 6-17.

The difference equation for this circuit is obtained by setting $H(z) = Y(z)/X(z)$ in Equation 6-69 and then performing the inverse z transform:

$$Y(z)(1 - (p_1 + p_2)z^{-1} + p_1 p_2 z^{-2})$$
$$= X(z)(1 - (0_1 + 0_2)z^{-1} + 0_1 0_2 z^{-2}) \quad (6\text{-}70\text{a})$$

$$y(nT) - (p_1 + p_2)y((n-1)T) + p_1 p_2 y((n-2)T)$$
$$= x(nT) - (0_1 + 0_2)x((n-1)T) + 0_1 0_2 x((n-2)T) \quad (6\text{-}70\text{b})$$

$$y(nT) = x(nT) - (0_1 + 0_2)x((n-1)T) + 0_1 0_2 x((n-2)T)$$
$$+ (p_1 + p_2)y((n-1)T) - p_1 p_2 y((n-2)T) \quad (6\text{-}70\text{c})$$

The point to be made is that by judicious and experienced choices of the zero and pole locations, the desired system characteristics can be obtained. With

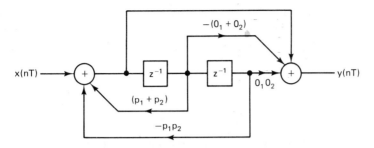

Figure 6–30 Circuit realization of Equation 6–69.

a computer at hand, trial-and-error approaches to the design of filters should take a relatively short time.

Bilinear transformation. The design of an IIR digital filter using the bilinear transformation technique begins with the specifications of an analog filter: (1) its cutoff frequency, f_c, and stopband frequency, f_s; and (2) the characteristics of the pass and stopbands. These specifications are used to generate the analog transfer function, $H(s)$, of a filter, which is then converted to a digital filter transfer function, $H(z)$, using the bilinear transformation. Appendix A is a brief review of analog filter design theory. It is included to serve as a background for the material that follows and should be read at this time.

The transfer function of a LP filter is given in Equation 6–71.

$$H(s) = \frac{Kb_0}{b_0 + b_1 s + b_2 s^2 + \cdots + b_n s^n} \quad (6\text{–}71)$$

This represents an nth order (or pole) filter with gain K at $s = 0$. The choice of $\{b_i\}$ will determine whether it has Butterworth or Chebychev passband characteristics.

The final design for the filter must meet the desired characteristics and provide the proper passband, transition band, and stopband frequencies. Figure 6–31 shows the ideal LP filter characteristics superimposed on the characteristics

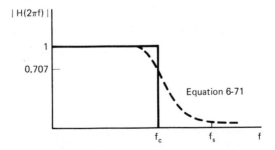

Figure 6–31 Ideal and practical LP filter passband.

obtained from Equation 6–71 for an arbitrary value of n, the filter order.

The passband is defined by $f < f_c$, the stopband by $f > f_s$, and the transition band by $f_c < f < f_s$. Once the analog characteristics are specified and its s plane equation obtained, the IIR filter design can be initiated. This is accomplished by the bilinear transformation (BLT) of Equation 6–72.

$$s = K \frac{1 - z^{-1}}{1 + z^{-1}} \tag{6-72}$$

where $s = \sigma + j\omega$. Depending upon the manner in which the BLT is obtained, K can assume the values of 1, $1/T$, or $2/T$ as a perusal of the references will show. This text will use $K = 1/T$. Solving Equation 6–72 for z, we obtain

$$z = \frac{1 + sT}{1 - sT} = \frac{1 + (\sigma + j\omega)T}{1 - (\sigma + j\omega)T} \tag{6-73a}$$

$$= \frac{(1 + \sigma T) + j\omega T}{(1 - \sigma T) - j\omega T} \tag{6-73b}$$

If we determine the magnitude and phase angle of z, we have

$$|z| = \sqrt{\frac{(1 + \sigma T)^2 + (\omega T)^2}{(1 - \sigma T)^2 + (\omega T)^2}} \tag{6-74a}$$

$$\phi(z) = \tan^{-1} \frac{2\omega T}{1 - (\sigma T)^2 - (\omega T)^2} \tag{6-74b}$$

From Equation 6–74a, we can see that the bilinear transformation maps the left-half s plane ($\sigma < 0$) inside the unit circle, the entire $j\omega$ axis ($\sigma = 0$) onto the unit circle, and the right-half s plane ($\sigma > 0$) outside the unit circle.

This implies that a stable system in the s plane (all poles located in the left-hand side) will map into a stable system in the z plane (all poles inside the unit circle). The technique consists of replacing each s in $H(s)$ with Equation 6–72. This is shown below.

Consider an s-domain transfer function of the form

$$H(s) = \frac{Kb_0}{b_0 + b_1 s + b_2 s^2} \tag{6-75}$$

which represents a second-order filter. Replacing each s that appears in Equation 6–75 with Equation 6–72, we obtain

$$H\left(\frac{1}{T}\frac{1 - z^{-1}}{1 + z^{-1}}\right) \iff H(z) \tag{6-76}$$

$$= \frac{Kb_0}{b_0 + b_1 \left(\frac{1}{T}\frac{1 - z^{-1}}{1 + z^{-1}}\right) + b_2 \left(\frac{1}{T}\frac{1 - z^{-1}}{1 + z^{-1}}\right)^2}$$

Sec. 6.6 Digital Filter Synthesis

Multiplying Equation 6–76 by $T^2(1+z^{-1})^2/T^2(1+z^{-1})^2$ we obtain

$$H(z) = \frac{Kb_0 T^2(1+2z^{-1}+z^{-2})}{(b_0 T^2 + b_1 T + b_2) + 2(b_0 T^2 - b_2)z^{-1} + (b_0 T^2 - b_1 T + b_2)z^{-2}} \quad (6\text{--}77)$$

which can then be put in the standard form

$$H(z) = \frac{Kb_0 T^2(1+2z^{-1}+z^{-2})}{c_0 + c_1 z^{-1} + c_2 z^{-2}} \quad (6\text{--}78)$$

where $c_0 = b_0 T^2 + b_1 T + b_2$, $c_1 = 2b_0 T^2 - 2b_2$, $c_2 = b_0 T^2 - b_1 T + b_2$. Factoring c_0 from the denominator, we then have

$$H(z) = \frac{(Kb_0 T^2/c_0)(1+2z^{-1}+z^{-2})}{1 + (c_1/c_0)z^{-1} + (c_2/c_0)z^{-2}} = \frac{Y(z)}{X(z)} \quad (6\text{--}79)$$

Equation 6–79 is readily recognizable as the transfer function of a recursive digital filter. Its equation is given by

$$y(nT) = KT^2 \frac{b_0}{c_0} [x(nT) + 2x((n-1)T) + x((n-2)T)]$$
$$- \frac{c_1}{c_0} y((n-1)T) - \frac{c_2}{c_0} y((n-2)T) \quad (6\text{--}80)$$

The circuit realization is presented in Figure 6–32. This circuit is the digital equivalent of the analog filter in Equation 6–75. Its frequency response is obtained as discussed in Section 6.3.1.

In order to obtain some further insight into the characteristics of the filter described above, we will look at the relationship between the value of $\omega = 2\pi f$ in the s plane and the angle of z, $2\pi\Omega$, in the z plane. This relation

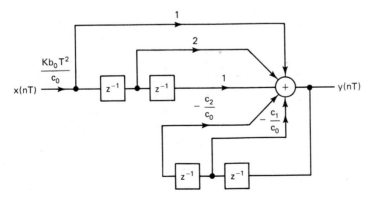

Figure 6–32 Circuit realization of Equation 6–80.

is obtained by evaluating z on the unit circle, $z = \exp(j2\pi\Omega)$, and using it in Equation 6–72.

$$s = \frac{1}{T}\frac{1 - e^{-j2\pi\Omega}}{1 + e^{-j2\pi\Omega}} = \frac{1}{T}\frac{e^{j\pi\Omega} - e^{-j\pi\Omega}}{e^{j\pi\Omega} + e^{-j\pi\Omega}} \qquad (6\text{–}81a)$$

$$= \frac{1}{T}\frac{j\sin\pi\Omega}{\cos\pi\Omega} = \frac{1}{T}j\tan\pi\Omega = \sigma + j\omega \qquad (6\text{–}81b)$$

Comparing the imaginary parts of Equation 6–81b, we have

$$\omega = \frac{1}{T}\tan\pi\Omega \qquad (6\text{–}82)$$

This implies that the infinite-length j axis is mapped (nonlinearly) into the unit circle, introducing a distortion of the frequency response. The distortion is in the form of a relocation of the relevant frequencies: cutoff and stopbands. The general shape of the response is essentially unaffected; that is, the analog low-pass filter will transform into a low-pass digital filter, but with different cutoff and stopband frequencies. This is shown in Figure 6–33. Although the form of the response is that of a LP filter in both cases, the cutoff and stopband frequencies are shifted toward the higher end of the spectrum for the digital filter and not where required by the design.

The shifting of frequencies during the transformation can be corrected by designing the analog filter to account for the warping due to the transformation. Then, when transformed, the digital frequencies of interest will be at their desired locations. The analog filter frequencies are determined by selecting the desired digital filter frequencies and then prewarping them using Equation

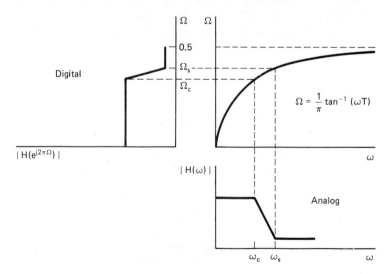

Figure 6–33 Effect of warping on filter frequencies.

Sec. 6.6 Digital Filter Synthesis

6-82. For example, if a digital filter cutoff frequency of $\Omega_c = \frac{1}{4}$ is desired, the analog cutoff frequency required is given by

$$\omega_c = \frac{\tan \pi \Omega_c}{T} \tag{6-83a}$$

$$= \frac{\tan \pi (0.25)}{T} = \frac{1 \text{ rad}}{T} \tag{6-83b}$$

This implies that we will obtain a digital filter whose cutoff frequency is $\Omega_c = \frac{1}{4}$ if we choose the analog filter to have its cutoff frequency at $\omega_c = 1/T$ radians/second and apply the bilinear transformation to the resulting $H(s)$. $\Omega_c = \frac{1}{4}$ indicates that the digital filter's cutoff frequency is $f_{sa}/4$; that is, if $f_{sa} = 8$ kHz, the cutoff frequency will be at 2 kHz.

This technique can be demonstrated by considering the first-order filter

$$H(s) = \frac{b_0}{s + b_0} \tag{6-84}$$

with frequency response magnitude of

$$|H(\omega)| = \frac{b_0}{\sqrt{\omega^2 + b_0^2}} = \frac{1}{\sqrt{(\omega/b_0)^2 + 1}} \tag{6-85}$$

The cutoff frequency occurs at $\omega = b_0$. Applying the bilinear transformation to $H(s)$ in Equation 6-84, we obain

$$H(z) = \frac{b_0}{\left(\frac{1}{T} \cdot \frac{1 - z^{-1}}{1 + z^{-1}}\right) + b_0} = \frac{b_0 T(1 + z^{-1})}{(1 - z^{-1}) + b_0 T(1 + z^{-1})} \tag{6-86a}$$

$$= \frac{b_0 T(1 + z^{-1})}{(1 + b_0 T) - (1 - b_0 T)z^{-1}} \tag{6-86b}$$

Evaluating $H(z)$ for $z = \exp(j\omega T) = \exp(j2\pi \Omega)$, we have

$$H(z) = \frac{b_0 T[1 + \exp(-j2\pi \Omega)]}{(1 + b_0 T) - (1 - b_0 T) \exp[(-j2\pi \Omega)]} \tag{6-87}$$

The magnitude of $H(\exp(j2\pi \Omega))$ is found to be

$$|H(\Omega)| = \sqrt{\frac{[b_0 T(1 + \cos 2\pi \Omega)]^2 + (b_0 T \sin 2\pi \Omega)^2}{[(1 + b_0 T) - (1 - b_0 T) \cos 2\pi \Omega]^2 + [(1 - b_0 T) \sin 2\pi \Omega]^2}} \tag{6-88}$$

When $\Omega = 0$, $|H(\Omega)| = 1$; when $\Omega = (\arctan b_0 T)/\pi$, $|H(\Omega)| = 0.707$, which corresponds to the -3-dB point. The digital filter obtained from Equation 6-86b is shown in Figure 6-34 for $b_0 T = 0.5$. The frequency response of the circuit of Figure 6-34 is shown in Figure 6-35.

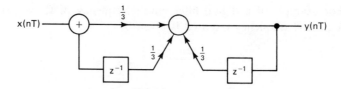

Figure 6-34 Circuit realization of Equation 6-86b.

The general procedure is to start with a desired value for Ω_c and Ω_s, and using prewarping (Equation 6-83), find ω_c and ω_s. These values are used to find $H(s)$ and then $H(z)$ using the bilinear transformation. This is shown in Example 6-2.

Equation 6-85 is a filter with Butterworth characteristics and order 1. The general form for a Butterworth filter of order n is (see Appendix A)

$$|H(\omega)| = \frac{1}{\sqrt{1 + (\omega/\omega_c)^n}} \tag{6-89}$$

Its transfer function is of the form

$$H(s) = \frac{\omega_c^n}{\prod_{k=1}^{n}(s - s_k)} \tag{6-90}$$

where s_k is one of the n poles of $H(s)$.

$$s_k = \omega_c \exp\left[j\left(\frac{\pi}{n}\left(k - \frac{1}{2}\right) + \frac{\pi}{2}\right)\right] \tag{6-91}$$

ω_c is obtained from Equation 6-82.

Example 6-2

Design a LP digital filter with Butterworth characteristics, a cutoff frequency $2\pi\Omega_c = 0.77$ ($\Omega_c = 0.123$), and $f_{sa} = 2$ kHz. Use a two-pole analog filter for the initial design.

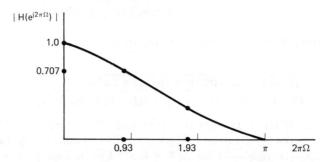

Figure 6-35 Frequency response of the circuit in Figure 6-34.

Sec. 6.6 Digital Filter Synthesis

Solution

$$\omega_c = \frac{\tan \pi \Omega_c}{T} = \frac{\tan(\pi \cdot 0.123)}{0.5 \text{ ms}} \tag{6-82}$$
$$= 813.3 \text{ rad/s}$$

Using Equation 6–91, we find the poles of $H(s)$ for $k = 1, 2$.

$$s_1 = \omega_c \exp\left[j\left(\frac{\pi}{n}\left(k - \frac{1}{2}\right) + \frac{\pi}{2}\right)\right]$$
$$= 813.3 \frac{\text{rad}}{\text{s}} \exp\left[j\left(\frac{\pi}{2}\left(1 - \frac{1}{2}\right) + \frac{\pi}{2}\right)\right] \tag{6-91}$$
$$= 813.3 \frac{\text{rad}}{\text{s}} \exp\left(j\frac{3\pi}{4}\right)$$
$$s_2 = 813.3 \frac{\text{rad}}{\text{s}} \exp\left(j\frac{5\pi}{4}\right)$$

Putting these values into Equation 6–90, we have

$$H(s) = \frac{\omega_c^n}{\prod_{k=1}^{2}(s - s_k)}$$
$$= \frac{\omega_c^2}{(s - s_1)(s - s_2)} \tag{6-90}$$
$$= \frac{(813.3 \text{ rad/s})^2}{\left[s - (813.3)\exp\left(j\frac{3\pi}{4}\right)\right]\left[s - (813.3)\exp\left(j\frac{5\pi}{4}\right)\right]}$$

s_1 and s_2 can be converted to rectangular form:

$$s_1 = 813.3(-0.707 + j0.707) \tag{6-92a}$$
$$s_2 = 813.3(-0.707 - j0.707) \tag{6-92b}$$

where we see that they are complex conjugates: $s_1 = s_2^*$. Equation 6–90 becomes

$$H(s) = \frac{(813.3 \text{ rad/s})^2}{(813.3)^2 + 1150.2s + s^2} \tag{6-93a}$$

which is the same form as Equation 6–75 with $b_0 = \omega_c^2$, $b_1 = \sqrt{2}\omega_c$, and $b_2 = 1$. Applying the bilinear transformation to Equation 6–93, we have (using the form of Equation 6–77), with $\omega_c T = 0.163$,

$$H(z) = \frac{(0.026s)(1 + 2z^{-1} + z^{-2})}{1.257 - 1.947z^{-1} + 0.796z^{-2}} = \frac{Y(z)}{X(z)} \tag{6-93b}$$

This is put into the difference equation format as

$$y(nT) = \frac{0.0265}{1.257}[x(nT) + 2x((n-1)T) + x((n-2)T)]$$
$$+ \frac{1.947}{1.275}y((n-1)T) - \frac{0.796}{1.257}y((n-2)T) \tag{6-94}$$

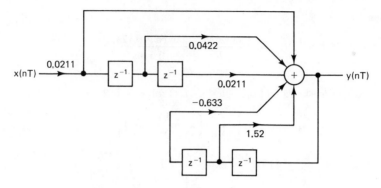

Figure 6-36 Circuit realization of Equation 6-94.

The circuit realization is presented in Figure 6–36.

Example 6–2 has presented the design technique for a digital filter using the bilinear transform. The reader should verify that the resulting digital system meets the requirements stated in the example, specifically, the stopband requirements for $\Omega > \Omega_s$. If not, the order of the analog filter should be increased and the system redesigned.

Direct design of a digital filter. A Butterworth or Chebychev digital filter can be designed more readily if we take their respective analog equations (Appendix A) and replace ω with the prewarp value $(\tan \pi\Omega)/T = \tan(\pi fT)/T$. This yields the digital frequency form of $H(\omega)$:

1. *Butterworth*:

$$|H(\omega)|^2 = |H(\exp(j2\pi fT))|^2 = \frac{1}{1 + \left[\dfrac{(\tan \pi fT)/T}{(\tan \pi f_c T)/T}\right]^{2n}} \quad (9\text{-}95a)$$

$$|H(\Omega)| = \frac{1}{1 + \left[\left(\dfrac{\tan \pi\Omega}{\tan \pi\Omega_c}\right)\right]^{2n}} \quad (6\text{-}95b)$$

2. *Chebychev*:

$$|H(\omega)|^2 = |H(\exp(j2\pi fT))|^2 = \frac{1}{1 + \epsilon^2 C_n^2\left(\dfrac{(\tan \pi fT)/T}{(\tan \pi f_c T)/T}\right)} \quad (6\text{-}95c)$$

$$|H(\Omega)| = \frac{1}{1 + \epsilon^2 C_n^2\left(\dfrac{\tan \pi\Omega}{\tan \pi\Omega_c}\right)} \quad (6\text{-}95d)$$

where c_n is an nth-order Chebychev polynomial. When $\Omega < \Omega_c$, $|H(\Omega)|$ is approximately 1; for $\Omega > \Omega_c$, $|H(\Omega)|$ is approximately zero, which yields the desired LP characteristics.

Sec. 6.6 Digital Filter Synthesis

The stable poles of $H(\Omega)$ lie inside the unit circle of the z plane and are given by:

1. *Butterworth*:

$$z_k = \frac{(1 - \tan^2 \pi \Omega_c) + j(2 \tan \pi \Omega_c \sin(k\pi/N))}{1 - 2 \tan \pi \Omega_c \cos(k\pi/N) + \tan^2 \pi \Omega_c} \quad (6\text{-}96a)$$

where $k = 0, 1, 2, \ldots, 2N - 1$, and $k\pi/N \longrightarrow (2k + 1)\pi/2N$ if N = even.

2. *Chebychev*:

$$z_k = \left[\frac{2(1 - a^- \tan \pi \Omega_c \cos \theta)}{d} - 1 \right] + j2a^+ \tan \pi \Omega_c \sin \theta \quad (6\text{-}96b)$$

where $a^{\pm} = (1/2)\{(\sqrt{1 + 1/\epsilon^2} + 1/\epsilon)^{1/N} \pm (\sqrt{1 + 1/\epsilon^2} + 1/\epsilon)^{1/N}\}$
$\theta = k\pi/N, \; k = 0, 1, 2, \ldots, 2N - 1$ and
$(k\pi/N) \longrightarrow (2k + 1)\pi/2N$ if N = even

Once the poles are found, they can be used to generate the individual second-order sections of a cascade filter (see Section 6.4.3).

Although the equations seem formidable, computer programs can be written that allow the user to generate the required constants.

Impulse invariant design of digital filters. This technique generates a digital filter whose impulse response, $h_d(nT)$, is the same as that of an analog filter, $h_a(t)$, sampled at $t = nT$. The relation between the two forms is given in Equation 6–97 and is shown in Figure 6–37.

$$h_d(nT) = h_a(t = nT) \quad (6\text{-}97)$$

The frequency response of the analog filter whose impulse response is given in Figure 6–37 does not have a sharp cutoff (i.e., it is not band-limited). This causes the transfer function of the digital filter, $H(\exp(j\omega T)$, to introduce distortion in the frequency response due to aliasing. This is shown by presenting the relation between the DT Fourier transform of $h(nT)$ and the FT of $h(t)$. This is given in Equation 6–98.

$$H(e^{j\omega T}) = \frac{1}{T} \sum_{k=-\infty}^{\infty} H_a\left(j\omega - j\frac{2\pi k}{T}\right) \quad (6\text{-}98)$$

where $H(\exp(j\omega T))$ is the frequency response of the digital filter and $H_a(j\omega)$ is that of the analog filter. If $H_a(j\omega)$ is not band-limited, a sufficiently high sampling rate would be required to reduce the amount of overlap. Only if the latter condition can be met will the frequency response of the digital filter closely match that of the analog filter.

Figures 6–38 and 6–39 show the frequency response of a digital system obtained from the impulse invariant technique. In Figure 6–39, the sampling

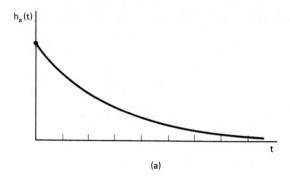

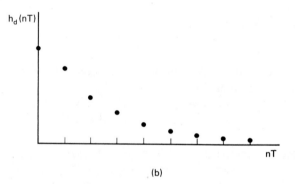

Figure 6–37 CT and DT impulse response: (a) impulse response; (b) sampled impulse response.

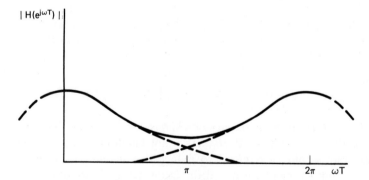

Figure 6–38 Impulse invariant filter frequency response with aliasing.

Sec. 6.6 Digital Filter Synthesis

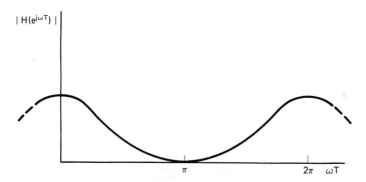

Figure 6-39 Response of Figure 6-38, but with no aliasing.

frequency is increased, bringing about a reduction of the overlap, causing the digital filter's frequency response to match more closely that of the analog filter.

The digital filter's transfer function can be obtained by first expressing the analog filter's transfer function as a partial-fraction sum, using the inverse Laplace transform to find $h_d(t)$, sampling it, and then taking the z transform to obtain $H(z)$. The procedure is outlined below.

The transfer function of the analog filter is given as

$$H_a(s) = \sum_{j=1}^{N} \frac{a_j}{s - s_j} \tag{6-99a}$$

where s_j is the jth pole of $H_a(s)$, and

$$a_j = H_a(s)(s - s_j)\Big|_{s=s_j} \tag{6-99b}$$

Using the inverse Laplace transform, we can find $h_a(t)$.

$$h_a(t) = \sum_{j=1}^{N} a_j e^{+s_j t} u(t) \tag{6-100}$$

Sampling $h_a(t)$ and taking its z transform, we have

$$h_a(nT) = \sum_{j=1}^{N} a_j e^{+s_j nT} u(nT) \tag{6-101a}$$

$$H(z) = \sum_{n=0}^{\infty} h_d(nT) z^{-n} = \sum_{n=0}^{\infty} \sum_{j=1}^{N} a_j e^{+s_j nT} u(nT) z^{-n} \tag{6-101b}$$

Changing the order of summation, we have

$$H(z) = \sum_{j=1}^{N} a_j \sum_{n=0}^{\infty} e^{+s_j nT} z^{-n} \tag{6-102a}$$

$$= \sum_{j=1}^{N} a_j \sum_{n=0}^{\infty} (e^{+s_j T} z^{-1})^n \tag{6-102b}$$

The last summation on the right is a geometric series. This yields

$$H(z) = \sum_{j=1}^{N} a_j \frac{1}{1 - e^{+s_j T} z^{-1}} \qquad (6\text{--}103)$$

The application of this technique is demonstrated in the following example.

Example 6–3

Using the impulse invariance technique, find the transfer function of the analog system given by

$$H_a(s) = \frac{s + 0.5}{s^2 + s + 2.5}$$

Solution Using partial-fraction expansion, we obtain

$$H_a(s) = \frac{0.5}{s - (-0.5 + j1.5)} + \frac{0.5}{s - (-0.5 - j1.5)} \qquad (6\text{--}104)$$

where $s_1 = (-0.5 + j0.75)$ and $s_2 = (-0.5 - j0.75)$.
This yields an impulse response $h_a(t)$:

$$h_a(t) = 0.5 \exp[(-0.5 + j1.5)t]u(t) + 0.5 \exp[(-0.5 - j1.5)t]u(t) \qquad (6\text{--}100)$$

The sampled form is

$$h_d(nT) = 0.5 \exp[(-0.5 + j1.5)nT]u(nT)$$
$$+ 0.5 \exp[(-0.5 - j1.5)nT]u(nT) \qquad (6\text{--}101)$$

We then have $H(z)$ given as

$$H(z) = \frac{0.5}{1 - \exp[(-0.5 + j1.5)T]z^{-1}} + \frac{0.5}{1 - \exp[(-0.5 + j1.5)T]z^{-1}} \qquad (6\text{--}103)$$

Using previous calculational techniques, $H(\exp(j2\pi\Omega))$ can be determined. The frequency response of the resulting digital system will show some distortion, as the frequency response of the analog system used in the example is not band-limited. The digital and analog systems will have approximately the same frequency response, and the impulse response of the digital system will be the sampled version of the impulse response of the analog system. As stated previously, sampling at as high a rate as possible will reduce the distortion due to aliasing. This was not a problem with the previous techniques as the design was performed initially in the discrete time domain.

6.6.2 FIR Filter

If a filter is required with linear-phase shift in the passband, one must use an FIR filter. The characteristic of a linear phase shift in the passband ensures that the phase relations between the input signal's frequency components are maintained at the output. This can result in a minimum of envelope distortion if the gain is flat in the passband. This type of filter can be designed

Sec. 6.6 Digital Filter Synthesis

quite readily using the techniques discussed below. In addition, computer programs are available that allow designs to be realized with a minimum of effort.

In order for the FIR filter to have linear phase, there are conditions placed on the form its impulse response can take: The impulse response must have even or odd symmetry. As the number of samples can be even or odd, this results in four possible formats for the impulse response and the allowed values of $h(n)$.

1. *N odd, even symmetry.* This is given in Figure 6–40a. To show even symmetry, the impulse response with N odd must have the relation

$$h(nT) = h((N - 1 - n)T) \qquad (6\text{–}105)$$

2. *N odd, odd symmetry.* This is presented in Figure 6–40b. The impulse response values satisfy the relation

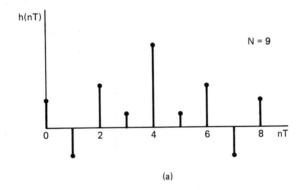

(a)

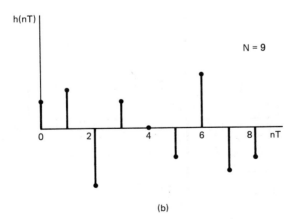

(b)

Figure 6–40 Impulse response for (a) N odd, even symmetry; (b) N odd, odd symmetry.

$$h(nT) = -h((N-1-n)T) \quad (6\text{-}106)$$

3. *N even, odd symmetry.* This is shown in Figure 6–41a. This implies that

$$h(nT) = -h((N-1-n)T) \quad (6\text{-}107)$$

4. *N even, even symmetry.* This is given in Figure 6–41b. We then have

$$h(nT) = h((N-1-n)T) \quad (6\text{-}108)$$

For our purposes, we will assume that the impulse response of the FIR filter will have one of the formats above and therefore linear phase in the passband.

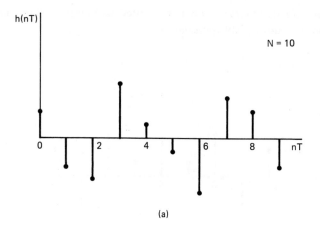

(a)

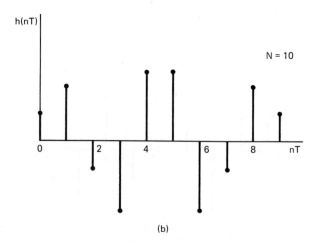

(b)

Figure 6–41 Impulse response for (a) N even, odd symmetry; (b) N even, even symmetry.

Example 6-4

For $\{h(nT)\}$, which has 3-unit pulses for $n = 0, 1,$ and 2, all other values zero, show that the phase is linear.

Solution

$$H(z) = \sum_{n=0}^{2} h(nT)z^{-1} = 1 + z^{-1} + z^{-2}$$

$$H(z)\Big|_{z=\exp(j\omega T)} = 1 + e^{-j\omega T} + e^{-j2\omega T}$$

$$= (e^{j\omega T} + 1 + e^{-j\omega T})e^{-j\omega T}$$

$$= (1 + 2\cos\omega T)e^{-j\omega T}$$

$$H(\Omega) = (1 + 2\cos 2\pi\Omega)e^{-j2\pi\Omega}$$

The phase of $H(z)$ is given by $-\omega T$, which is a linear function of ω. This can be shown for the three other forms of symmetric impulse response.

It is worthwhile to determine the frequency response for each form of linear-phase filter. As this has been done in the References at the end of the chapter, the results will be presented here.

1. *N odd, even symmetry.* The frequency response is symmetric about $\omega = \pi$. Its functional form is given by

$$|H(e^{j\omega T})| = \sum_{n=0}^{(N-1)/2} a(nT)\cos n\omega T \qquad (6\text{-}109\text{a})$$

$$a(0) = h\left(\frac{N-1}{2}T\right) \qquad (6\text{-}109\text{b})$$

$$a(nT) = 2h\left(\left(\frac{N-1}{2} - n\right)T\right) \qquad (6\text{-}109\text{c})$$

where $n = 1, 2, \ldots, (N-1)/2$.

2. *N odd, odd symmetry.* This yields an antisymmetrical frequency response about $\omega = \pi$. This response is

$$|H(e^{j\omega T})| = \sum_{n=1}^{(N-1)/2} c(nT)\sin n\omega T \qquad (6\text{-}110\text{a})$$

$$c(nT) = 2h\left(\left(\frac{N-1}{2} - n\right)T\right), \quad n = 1, 2, \ldots, \frac{N-1}{2} \qquad (6\text{-}110\text{b})$$

3. *N even, odd symmetry.* We have

$$|H(e^{j\omega T})| = \sum_{n=1}^{N/2} d(nT)\sin\left[\left(n - \frac{1}{2}\right)\omega T\right] \qquad (6\text{-}111\text{a})$$

$$d(nT) = 2h\left[\left(\frac{N}{2} - n\right)T\right], \quad n = 1, 2, \ldots, \frac{N}{2} \qquad (6\text{-}111\text{b})$$

4. *N even, even symmetry.* This is

$$|H(e^{j\omega T})| = \sum_{n=1}^{N/2} b(nT) \cos\left[\left(n - \frac{1}{2}\right)\omega T\right] \quad (6\text{–}112a)$$

$$b(nT) = 2h\left[\left(\frac{N}{2} - n\right)T\right], \quad n = 1, 2, \ldots, \frac{N}{2} \quad (6\text{–}112b)$$

where the values of $a(nT)$, $b(nT)$, $c(nT)$, and $d(nT)$ are real.

As an example, we can obtain the result for $N = 5$, with $h(nT)$ having even symmetry. Figure 6–42 shows a typical impulse response that has these characteristics. The frequency response becomes (using Equation 6–109)

$$a(0) = h\left(\frac{N-1}{2}T\right) = h\left(\frac{5-1}{2}T\right) = \frac{1}{2}$$

$$a(T) = 2h\left(\frac{N-1}{2}T - T\right) = 2h\left(\frac{5-1}{2}T - T\right) = 2h(nT) = 2$$

$$a(2T) = 2h\left(\frac{N-1}{2}T - 2T\right) = 2h\left(\frac{5-1}{2}T - 2T\right) = 2h(0) = 1$$

$$|H(e^{j\omega T})| = a(0)\cos(0 \cdot \omega T) + a(T)\cos \omega T + a(2T)\cos 2\omega T$$

$$= \frac{1}{2} + 2\cos \omega T + \cos 2\omega T$$

Figure 6–43 shows the $\{a(nT)\}$ and the resulting frequency response of $h(nT)$ from Figure 6–42. Similar results are obtained for the other forms of $h(nT)$.

With the foregoing characteristics of FIR filters in mind, we can examine some of the design techniques available.

Windowing. The design of FIR digital filters using the window technique begins with the desired ideal frequency response in the digital domain which has $|H(\Omega)| = 1$ in the passbands and $|H(\Omega)| = 0$ in the stopbands. An example of an ideal LP filter is given in Figure 6–44 where Ω_c is the cutoff

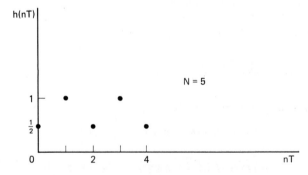

Figure 6–42 Impulse response for $N = 5$, even symmetry.

Sec. 6.6 Digital Filter Synthesis

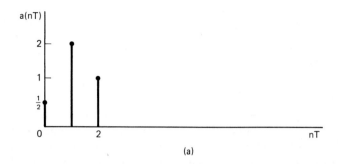

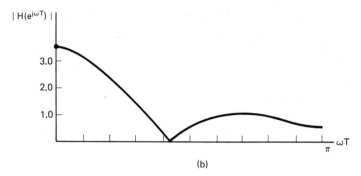

Figure 6–43 Coefficient and response for $h(n)$ of Figure 6–42: (a) values of $a(n)$; (b) frequency response of $h(n)$ of Figure 6–42.

frequency. The impulse response of the ideal filter is calculated below.

As the frequency response is periodic in ω, it can be expressed as a Fourier series (see Chapter 1).

$$H(e^{j\omega T}) = \sum_{n=-\infty}^{\infty} h(nT) \exp(-j\omega nT) \qquad (6\text{–}113)$$

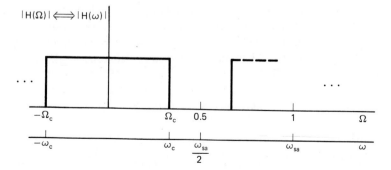

Figure 6–44 Transfer function of ideal LP filter.

Using the inverse FT, we have

$$h(nT) = \frac{1}{2\pi} \int_0^{2\pi} H(e^{j\omega T}) e^{j\omega nT} d\omega \qquad (6\text{--}114)$$

where the $h(nT)$ are the impulse response samples of the filter whose frequency response is $H(\exp(j\omega T))$.

Examination of Equations 6–113 and 6–114 immediately points out a problem: We have indicated that $h(nT)$ is to be causal and finite in extent. The values of $h(nT)$ we obtain from Equation 6–114 are not zero for $n < 0$ and do not go to zero for large n. Therefore, $h(nT)$ is neither causal nor finite in extent.

By its very name, the FIR filter has an impulse response with a finite number of terms. In addition, if it is to be realizable, it must be causal. The filters whose impulse response was given in Figure 6–38, 6–39, 6–40, and 6–41 have these characteristics. The problem is to convert the non-causal, infinite length sequence that represents the ideal filter to one that is finite, causal, and symmetric, where the last condition ensures a linear phase shift in the passband.

The problem is solved by two actions: (1) truncating the series representation of the ideal filter in Equation 6–113 by setting $h(nT) = 0$, for $|n| > A$. This yields

$$H_t(e^{j\omega T}) = \sum_{n=-A}^{A} h(nT) \exp(-jn\omega T) \qquad (6\text{--}115)$$

where $H_t(\exp(j\omega T))$ is the frequency response of the truncated series. This limits the impulse response to finite values of $|n| \leq A$. When we considered the Fourier series of a periodic function in Chapter 1, it was observed that the frequency response was symmetrical or anti-symmetrical with respect to the dc or zero frequency point. For Equations 6–113 and 6–114, this implies that $h(nT)$ will also be either symmetrical or anti-symmetrical about $n = 0$. As the truncation has even symmetry, the symmetry of the ideal impulse response will be maintained if it is truncated. The resulting impulse response is made causal by (2) shifting the impulse response by A samples. This is the same thing as delaying the impulse response by A samples, which is shown in Figure 6–45, where $h_{tc}(nT)$ is the truncated and causal impulse response. The basic idea of the window technique is to use the frequency response of $h_{tc}(nT)$ to approximate the desired ideal frequency response, $H(\exp(j\omega T))$. If we take the z transform of $h_t(nT)$, we have

$$H_t(z) = \sum_{n=-A}^{A} h_t(nt) z^{-n} \qquad (6\text{--}116)$$

From Figure 6–44, we see that the impulse response $h_{tc}(nT)$ is given by

$$h_{tc}(nT) = h_t((n - A)T) \qquad (6\text{--}117)$$

Sec. 6.6   Digital Filter Synthesis   219

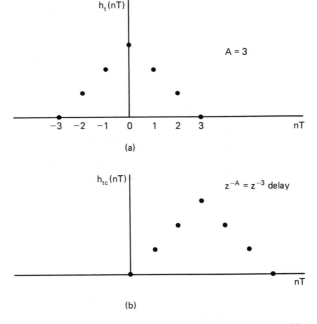

Figure 6–45 Effect of adding delay to the truncated impulse response: (a) truncated impulse response; (b) truncated and causal impulse response.

Multiplying Equation 6–116 by z^{-A}, we obtain the z transform of $h_{tc}(n + A)T$:

$$z^{-A}H_t(z) = H_{tc}(z) = \sum_{n=-A}^{A} h_{tc}((n + A)T)z^{-(N+A)} \quad (6\text{–}118)$$

This yields the difference equation for the approximation to the ideal filter

$$y(mT) = \sum_{n=-A}^{A} h_{tc}((n + A)T) \, x \, ((m - n - A)T) \quad (6\text{–}119)$$

As the $h_{tc}(nT)$ are either symmetric or antisymmetric, the resulting filter will have linear phase.

Although truncating the infinite series that represents the desired filter allows us to obtain an approximation to it, some problems arise, as shown in the following example.

Example 6–5
Design a filter whose frequency response is that of an ideal LP filter, that is, $|H(\exp(j\omega T))| = |H(\Omega)| = 1$, $\Omega < \Omega_c$, 0 otherwise. (See Figure 6–44.) Note that the FIR digital filter is designed directly without the use of an initial analog filter design.

Solution We find the impulse response of the filter using the inverse FT, Equation 6–114.

$$h(nT) = \frac{1}{2\pi} \int_{-\pi}^{\pi} H(e^{j\omega T}) e^{j\omega nT} d\omega$$

$$= \frac{1}{2\pi} \int_{-\omega_c}^{\omega_c} 1 \cdot e^{j\omega nT} d\omega$$

$$= \frac{1}{2\pi} \frac{e^{j\omega_c nT} - e^{-j\omega_c nT}}{jnT} \qquad (6\text{–}114)$$

$$= \frac{1}{\pi} \frac{\sin \omega_c nT}{nT} = 2f_c \frac{\sin \omega_c nT}{2\pi f_c nT}$$

$$= 2f_c \, \text{Sa}(\omega_c nT) = \frac{2\Omega_c}{T} \, \text{Sa}(2\pi n \Omega_c)$$

The impulse values are shown in Figure 6–46, which is symmetric about $\omega = 0$. It is clear that the set $\{h(nT)\}$ does not represent a causal or finite impulse response. If we arbitrarily truncate the series at $|n| = 2$, the frequency response becomes

$$H_t(e^{j\omega T}) = \sum_{n=-2}^{2} h_t(nT) \exp(-jn\omega T)$$

$$= h_t(-2T) \exp(j2\omega T) + h_t(-T) \exp(j\omega T) \qquad (6\text{–}115)$$
$$+ h_t(0) + h_t(T) \exp(-j\omega t) + h_t(2T) \exp(-j2\omega T)$$

where $h(nT) = h(-nT)$. Combining terms, we have

$$H_t(e^{j\omega T}) = h_t(0) + h_t(T)[\exp(+j\omega T) + \exp(-j\omega T)]$$
$$+ h_t(2T) [\exp(+j2\omega T) + \exp(-j2\omega T)]$$
$$= 2f_c + 2f_c \, \text{Sa}(\omega_c T) \, 2 \cos \omega T \qquad (6\text{–}120)$$
$$+ 2f_c \, \text{Sa}(2\omega_c T) 2 \cos 2\omega T$$

This is plotted in Figure 6–47 together with the design goal.

We see that truncation of the infinite series has introduced an error (ripple) in the pass- and stopbands together with an overshoot of the response at the point of transition, ω_c. In addition, the transition band is no longer of zero

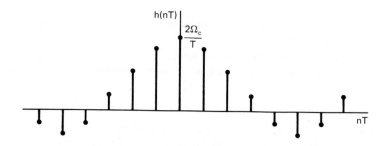

Figure 6–46 Infinite impulse response of Example 6–5.

Sec. 6.6 Digital Filter Synthesis

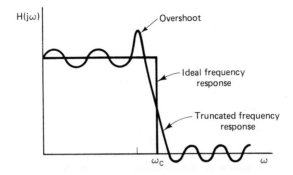

Figure 6-47 Frequency response of a truncated series.

width, but is extended, occurring over a range of frequencies. The z transform of the truncated impulse response is given by Equation 6-116.

$$H_t(z) = \sum_{n=-2}^{2} h_t(n)z^{-n} \tag{6-116}$$

and is not realizable. If we introduce a delay of two samples, we then have

$$H_{tc}(z) = z^{-2}H_c(z) = \sum_{n=-2}^{2} h_{tc}((n+2)T)z^{-(n+2)} \tag{6-118}$$

which is expanded to yield

$$\begin{aligned}H_{tc}(z) &= h_{tc}(0) + h_{tc}(T)z^{-1} + h_{tc}(2T)z^{-2} \\ &\quad + h_{tc}(3T)z^{-3} + h_{tc}(4T)z^{-4} \\ &= 2f_c\,\text{Sa}(2\omega_c T) + 2f_c\,\text{Sa}(\omega_c T)z^{-1} + 2f_c z^{-2} \\ &\quad + 2f_c\,\text{Sa}(\omega_c T)z^{-3} + 2f_c\,\text{Sa}(2\omega_c T)z^{-4}\end{aligned} \tag{6-121}$$

This represents a nonrecursive filter with difference equation

$$y(nT) = 2f_c[\text{Sa}(2\omega_c T)x(nT) + \text{Sa}(\omega_c T)x((n-1)T)$$
$$+ x((n-2)T) + \text{Sa}(\omega_c T)x((n-3)T) + \text{Sa}(2\omega_c T)x((n-4)T)] \tag{6-122}$$

The circuit realization is given in Figure 6-48, which is an FIR filter with LP characteristics and linear-phase shift in the passband.

If we choose a larger number of points (i.e., truncate the series at a higher value of n), the frequency response of the filter will more closely match the ideal response; however, it will still show ripple, overshoot at the transition point, and a nonzero transition width.

The act of truncating the frequency response is termed *windowing* and

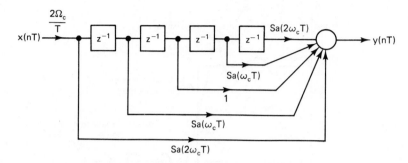

Figure 6–48 Circuit realization of Example 6–5.

is equivalent to multiplying the infinite impulse response by a second, limited, function, $w(nT)$, as shown in Figure 6–49. This produces a new function

$$g_t(nT) = h(nT)w(nT) = \begin{cases} h(nT), & |n| \leq A \\ 0, & |n| > A \end{cases} \quad (6\text{--}123)$$

Just as the product of two functions in the frequency domain was the FT of convolution in the time domain, the product of two functions in the time domain is the inverse FT of convolution in the frequency domain.

$$g_t(nT) = h(nT)w(nT) \Longleftrightarrow H(\omega) * W(\omega) = G_t(\omega) \quad (6\text{--}124)$$

where $H(\omega)$, $W(\omega)$, and $G_t(\omega)$ are the FTs of $h(nT)$, $w(nT)$, and $g_t(nT)$, respectively. This yields

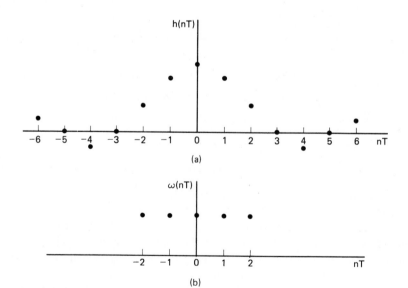

Figure 6–49 Infinite impulse response $h(n)$ and (b) window function $\omega(n)$.

Sec. 6.6 Digital Filter Synthesis

$$G_t(\omega) = \frac{1}{2\pi}\int_{-\infty}^{\infty} W(\beta)H(\omega-\beta)\,d\beta \qquad (6\text{--}125)$$

which is convolution in the frequency domain. $G_t(\omega)$ is the actual frequency response of the FIR filter when the infinite series is truncated. This is what was obtained in Figure 6–47. The large ripple (or overshoot) at ω_c is termed the Gibbs phenomenon. It will cause a large ripple at any sharp transition point in the desired frequency response.

To see the source of the passband error and the overshoot at the transition frequency (Gibbs phenomenon), we can obtain the response of Figure 6–47 through a graphical convolution. To do this, we reverse and slide one of the functions, $W(\omega)$ or $H(\omega)$. For this example, we will slide the response $H(\omega)$ past the FT of the window function, $W(\omega)$.

$$W(\omega) = \text{FT}\{w(nT)\} = \sum_{r=-2}^{2} w(rT)e^{-jr\omega T} \qquad (6\text{--}126)$$

If we factor $\exp(j2\omega T)$, we can obtain a closed form for $W(\omega)$.

$$W(\omega) = \text{FT}\{w(nT)\} = e^{j2\omega T}\sum_{r=-2}^{2} w(rT)e^{-j(2+r)\omega T}$$
$$= e^{j2\omega T}\sum_{k=0}^{4} w((k-2)T)\exp(-jk\omega T) \qquad (6\text{--}127)$$

where $w((k-2)T) = 1$ for $0 \le k \le 4$. As the resulting sum is a geometric series, we have

$$W(\omega) = e^{j2\omega T}\frac{1-(e^{-j5\omega T})}{1-e^{-j\omega T}} \qquad (6\text{--}128)$$

Expanding the exponential terms (and with some algebra), we obtain

$$|W(\omega)| = \frac{\sin(5\omega T/2)}{\sin(\omega T/2)} \qquad (6\text{--}129)$$

which has zeros at $\omega = k\pi/5T$. Figure 6–50 shows a plot of $|W(\omega)|$. The denominator is zero at $\omega = 0$. As the numerator is also zero, the value of $W(\omega)$ is obtained by a limiting process and found to be equal to $N = 5$.

Figure 6–51 represents the convolution of $H(\omega)$ and $W(\omega)$, where β is treated as the variable and ω as the parameter. As ω increases toward zero, the amount of overlap between $H(\omega - \beta)$ and $W(\beta)$ also increases. The ripples in $W(\beta)$ will give rise to the ripples in the passband and stop band of $G_t(\omega)$. As ω passes through $\pm\omega_c$, the overlap of the main lobe of $W(\beta)$ with $H(\omega - \beta)$ will give rise to the overshoot at these points. In addition, the transition width will be increased due to the width of the main lobe.

The discussion above has shown that an FIR filter can be designed using a window function to approximate a specified design with linear phase in the

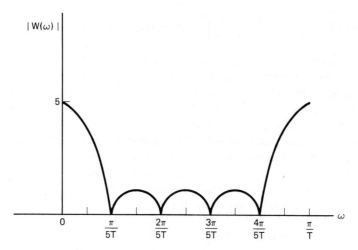

Figure 6–50 Fourier transform of $\omega(w)$ of Figure 6–48.

passband; however, this is achieved at the cost of introducing errors in the passband and at the transition point. We see that, if N is increased, the passband ripples and the transition width can be reduced but the transition overshoot will still be sizable. It can be shown that N must be quite large for a noticeable reduction of these effects to occur.

In an attempt to decrease the amount of ripple in the frequency response, the amount of overshoot, and the transition width, several window functions with tapered sides have been proposed. The FT of these window functions shows a reduction in the amplitude of the side lobes and an increase in the main-lobe width. This has the effect of decreasing the ripple in the pass and stopbands and increasing the transition width.

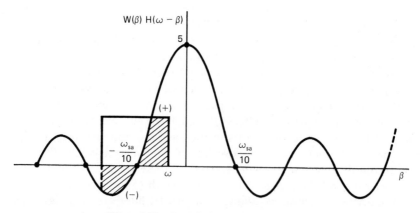

Figure 6–51 Convolution of $H(\omega)$ and $w(\omega)$.

Sec. 6.6 Digital Filter Synthesis

Several functions have been generated that provide an improvement over the rectangular window function. Some of these are listed below.

Bartlett:

$$w(nT) = \begin{cases} \dfrac{2nT}{(N-1)/2}, & -\left(\dfrac{N-1}{2}\right) \le n \le 0 \\ 2 - \dfrac{2nT}{(N-1)/2}, & 0 \le n \le \dfrac{N-1}{2} \end{cases} \quad (6\text{-}130\text{a})$$

Hanning:

$$w(nT) = \left[0.5 - 0.5 \cos \dfrac{2\pi nT}{(N-1)/2}\right], \quad 0 \le |n| \le \dfrac{N-1}{2} \quad (6\text{-}130\text{b})$$

Hamming:

$$w(nT) = \left[0.54 - 0.46 \cos \dfrac{2\pi nT}{(N-1)/2}\right], \quad 0 \le |n| \le \dfrac{N-1}{2} \quad (6\text{-}130\text{c})$$

Blackman:

$$w(nT) = 0.42 - 0.5 \cos \dfrac{2\pi nT}{(N-1)/2}$$

$$+ 0.08 \cos \dfrac{4\pi nT}{(N-1)/2}, \quad 0 \le |n| \le \dfrac{N-1}{2} \quad (6\text{-}130\text{d})$$

Table 6–1 shows the approximate relation between the main-lobe width and the desired amount of signal reduction in the stop band. The values for "Transition Width of Main Lobe" are in terms of ω. To obtain their values in terms of Ω, divide each term by π.

TABLE 6–1 Window Characteristics*

Window	Peak Amplitude of Side Lobe (dB)	Transition Width of Main Lobe	Minimum Stopband Attenuation (dB)
Rectangular	−13	$4\pi/N$	−21
Bartlett	−25	$8\pi/N$	−25
Hanning	−31	$8\pi/N$	−44
Hamming	−41	$8\pi/N$	−53
Blackman	−57	$12\pi/N$	−74

* From *Digital Signal Processing*, A. V. Oppenheim and R. W. Schafer. Copyright © 1975, Prentice-Hall, Inc. Reprinted by permission of Prentice-Hall, Inc.

To indicate how the digital filter specifications influence the choice of a window, the following examples are provided.

Example 6-6

Which window will provide a minimum stopband attenuation of -50dB or more?

Solution Using Table 6-1, we see that the Hamming or Blackman window will provide a minimum stopband attenuation of -53dB and -74dB, respectively, which exceeds the specifications.

Example 6-7

Using the results of Example 6-6, how many sample points are required to provide a normalized transition width of 0.15?

Solution For the Hamming window, we use the value of the transition width of the main lobe given in Table 6-1 to find N.

Hamming:
$$N = (8/0.15) = 53.3$$

Blackman:
$$N = (12/0.15) = 80.0$$

From Example 6-7, we see that the Blackman window requires more sample points in the sequence for $h_t(nT)$ than does the Hamming window. This implies that the design requires a trade-off between filter length (or transition width) and minimum stopband attenuation.

This section has presented a brief discussion of the window method of FIR filter design. Further discussion is available in the References listed at the end of this chapter

The Optimal filter. The material in this section is based on the series of papers written by T. Parks and J. McClellan. They are listed in the References.

As we saw in the preceding section, the approximation to the desired filter response by truncating its infinite Fourier series representation introduced ripple (error) in the pass- and stopbands as well as extending the transition width. Several different window functions were presented that reduced these effects; however, a trade-off was required between the amount of stopband attenuation and the number of sample points required to realize the filters.

The design technique used in this section is to find an approximation to the desired ideal filter in the minimax sense; that is, the maximum error in the pass- and stopbands is minimized. The error is defined as the difference between the desired frequency response and the one obtained from the approximation. The resulting filter is optimal in the sense that the number of samples is a minimum. No better filter (in the minimax sense) can be realized using fewer sample points than the ones provided by this technique.

The design approach minimizes the maximum error (in the stop- and passbands) between the desired ideal filter response, $D(\exp(j\omega T))$, and the

Sec. 6.6 Digital Filter Synthesis

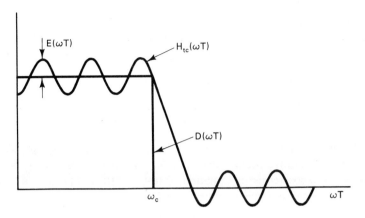

Figure 6–52 Frequency response of ideal and practical LPF.

approximation to it, $H_{tc}(\omega T)$. This is shown in Figure 6–52 where $E(\omega T)$ is the difference (error) between the desired (ideal) response and the practical response $H_{tc}(\omega T)$. The starting point of the analysis that leads to the equation solved by the program is the set of four equations given in Section 6.6.2: Equations 6–109, 6–110, 6–111, and 6–112, which represent the frequency response of the linear-phase FIR filter. If we include the phase response, the general form of the transfer function can be written as

$$H(e^{j\omega T}) = \exp\left(j\frac{N-1}{2}\omega T\right) \exp\left(j\frac{\pi}{2}L\right) H_{tc}(\omega T) \quad (6\text{–}131)$$

where $H_{tc}(\omega T)$ represents the four frequency response functions. If the impulse response is symmetric, $L = 0$; if antisymmetric, $L = 1$.

Each form of $H_{tc}(\omega T)$ can be converted to a format that involves a sum of cosines by relatively simple trigonometric identities. The result can be written as the product of two functions, $Q(\exp(j\omega T))$ and $P(\exp(j\omega T))$.

$$H_{tc}(\omega) = Q(e^{j\omega t})P(e^{j\omega T}) \quad (6\text{–}132)$$

where we have for Q and P,

N odd, symmetric:

$$Q = 1; \qquad P = \sum_{n=0}^{(N-1)/2} \tilde{a}(nT)\cos n\omega T \quad (6\text{–}133a)$$

N even, symmetric:

$$Q = \cos\frac{\omega T}{2}; \qquad P = \sum_{n=0}^{(N/2)-1} \tilde{b}(nT)\cos n\omega T \quad (6\text{–}133b)$$

N odd, asymmetric:

$$Q = \sin \omega T; \qquad P = \sum_{n=0}^{(N-3)/2} \tilde{c}(nT) \cos n\omega T \qquad (6\text{–}133\text{c})$$

N even, asymmetric:

$$Q = \sin \frac{\omega T}{2}; \qquad P = \sum_{n=0}^{(N/2)-1} \tilde{d}(nT) \cos n\omega T \qquad (6\text{–}133\text{d})$$

Depending on the value of N the program calculates one of the sets of coefficients of $\cos \omega nT$ to minimize the maximum error. Their relationship to the coefficients of the sin/cos terms in Equation 6–109, 6–110, 6–111, and 6–112, and therefore, the $h(nT)$ are given below.

$$a(nT) = \tilde{a}(nT) \qquad (6\text{–}134\text{a})$$

$$b(T) = \tilde{b}(0) + \frac{\tilde{b}(T)}{2}$$

$$b(kT) = \frac{1}{2}(\tilde{b}[(k-1)T] + \tilde{b}(kT)),$$

$$k = 2, 3, \ldots, \frac{N}{2} - 1$$

$$b(N/2) = \frac{1}{2}\tilde{b}\left(\left(\frac{N}{2} - 1\right)T\right) \qquad (6\text{–}134\text{b})$$

$$c(T) = \tilde{c}(0) - \frac{\tilde{c}(2T)}{2}$$

$$c(kT) = \frac{1}{2}(\tilde{c}[(k-1)T] - \tilde{c}[(k+1)T]),$$

$$k = 2, 3, \ldots, \frac{N-1}{2} - 2$$

$$c\left(\left[\frac{N-1}{2} - 1\right]T\right) = \frac{1}{2}\tilde{c}\left(\left(\frac{N-1}{2} - 2\right)T\right) \qquad (6\text{–}134\text{c})$$

$$d(T) = \tilde{d}(0) - \frac{\tilde{d}(T)}{2}$$

$$d(kT) = \frac{1}{2}(\tilde{d}[(k-1)T] - \tilde{d}(kT)),$$

$$k = 2, 3, \ldots, \frac{N}{2} - 1$$

$$d\left(\frac{NT}{2}\right) = \frac{1}{2}\tilde{d}\left(\frac{N}{2} - 1\right) \qquad (6\text{–}134\text{d})$$

Sec. 6.6 Digital Filter Synthesis

If we define the error between the desired frequency response, $D(\exp(j\omega T))$, and the approximating frequency response, $H_{tc}(\exp(j\omega T))$, by $E(\exp(j\omega T))$ and introduce a weighting function, $W(\exp(j\omega T))$, which allows the designer to state how much deviation between the desired and approximated values is permitted over the range of frequencies, we have

$$E(e^{j\omega T}) = W(e^{j\omega T})[D(e^{j\omega T}) - H_{tc}(e^{j\omega T})] \quad (6\text{-}135)$$

This becomes

$$E(e^{j\omega T}) = W(e^{j\omega T})[D(e^{j\omega T}) - Q(e^{j\omega T})P(e^{j\omega T})] \quad (6\text{-}136)$$

Factoring out Q and dropping the arguments for clarity, we obtain

$$E = QW\left(\frac{D}{Q} - P\right) \quad (6\text{-}137)$$

The set of constants $\tilde{a}(nT)$, $\tilde{b}(nT)$, $\tilde{c}(nT)$, or $\tilde{d}(nT)$ that minimize the maximum absolute (magnitude) value of $E(\exp(j\omega T))$ for all values of $0 \leq \omega \leq 0.5$ is the solution to Equation 6-137. To see how this solution is obtained, we can consider as an example the case of N odd, and a symmetric impulse response. Equation 6-137 becomes

$$E = W\left(D - \sum_{n=0}^{(N-1)/2} \tilde{a}(nT)\cos n\omega T\right) \quad (6\text{-}138)$$

where $Q(\exp(j\omega T)) = 1$.

If we are attempting to design an approximation to an ideal low-pass filter with cutoff frequency $(\omega_c T/2\pi) = 0.2$, where the passband has a desired value of 1 and the stopband a desired value of 0, we then obtain from Equation 6-135:

$0 \leq (\omega T/2\pi) \leq 0.2$:

$$E = 1\left[1 - \sum_{n=0}^{(N-1)/2} \tilde{a}(nT)\cos n\omega T\right] \quad (6\text{-}139a)$$

$0.2 \leq (\omega T/2\pi) \leq 0.5$:

$$E = 25\left[0 - \sum_{n=0}^{(N-1)/2} \tilde{a}(nT)\cos n\omega T\right] \quad (6\text{-}139b)$$

where the weighting function was chosen with values of 1 and 25, respectively. Typically, the value of the weighting function in the passband is chosen as unity while that of the stopband is much higher. W in the stopband is determined from the system's allowed ripple specifications in both bands. The allowed ripple (error) is given as the amount of deviation from the desired value in dB; δ_1 is the amount of deviation allowed in the passbands while δ_2 is the allowed deviation for the stopband. The values of δ_1 and δ_2 are obtained from Equations 6–139c and 6–139d and shown in Figure 6–53.

$$\text{Passband ripple, dB} = 20 \log(1 + \delta_1) \qquad (6\text{–}139c)$$

$$\text{Stopband ripple, dB} = 20 \log(\delta_2) \qquad (6\text{–}139d)$$

The value of W is given as

$$W = (\delta_1/\delta_2) \qquad (6\text{–}139e)$$

Example 6–8

An LP filter is required with passband ripple of 0.5 dB and stopband ripple of -40 dB. Find W.

Solution Using Equations 6–139c, 6–139d, and 6–139e we obtain

$$\begin{aligned} 0.5 \text{ dB} &= 20 \log(1 + \delta_1) \text{dB} \\ \delta_1 &= \log^{-1}(0.5/20) - 1 \\ &= 0.0593 \end{aligned} \qquad (6\text{–}139c)$$

$$\begin{aligned} -40 \text{ dB} &= 20 \log(\delta_2)\text{dB} \\ \delta_2 &= \log^{-1}(-2) \\ &= 0.01 \end{aligned}$$

$$\begin{aligned} W &= (\delta_1/\delta_2) \\ &= (0.0593/0.01) = 6 \end{aligned} \qquad (6\text{–}139e)$$

Choosing $W = 6$ in the stopband will result in the desired ratio of δ_2 to δ_1; however, the values chosen for N will yield the actual values of δ_2 and δ_1.

Returning to Equations 6–139a and 6–139b, the resulting values of $\tilde{a}(nT)$ must simultaneously minimize the maximum value of $E(\exp(j\omega T))$ over the pass- and stopbands. Once determined, they can generate the impulse response values using Equations 6–134. This allows us to obtain the system difference equation.

The mathematics upon which the program is based is a bit beyond the scope of the text. The reader is referred to the original papers and some additional works listed in the References at the end of the chapter for a more detailed discussion. However, a basic introduction to the concepts involved in the program is introduced as a primer.

The minimax problem is also termed a Chebychev approximation in which a given curve (function) is approximated over a set of values (domain) by Chebychev polynomials which are sums of cosine terms each raised to a specific power. (Note the similarity between this representation of a function and that of the Fourier series.) As with the Fourier series, the approximation is obtained when a set of coefficients is obtained, in this case, the set $\{\tilde{a}(nT)\}$. The solution is

Sec. 6.6 Digital Filter Synthesis

based on the Alternation Theorem, which states that any function having its representation in the form of a sum of r cosines, for example,

$$P(e^{j\omega t}) = \sum_{n=0}^{r-1} \alpha(nt) \cos n\omega t \qquad (6\text{--}140)$$

is the unique best approximation to $D(\exp((j\omega T))/Q) \exp(j\omega T)$ over $0 \le \omega T < \pi$ if $E(\exp(j\omega T))$ has $r + 1$ values of ω at which an extremum (maximum or minimum) occurs. The values of ω are such that consecutive values yield extrema which are alternately above and below the desired value, as shown in Figure 6–53. In addition, the values of $|E(\exp(j\omega T))|$ at the extrema are equal to the maximum value of $E(\exp(j\omega T))$ for all $0 \le \omega T \le \pi$. In the passband, the minimax value of E is δ_1 while in the stopband it is δ_2.

As the computer can manipulate only discrete numbers (in binary form), the continuous domain of $\omega T = 2\pi\Omega$ is replaced with a grid of discrete frequency points. Using this set of points, $D(\exp(j\omega T))$ and $W(\exp(j\omega T))$ are evaluated. This, in turn, enables the computer to evaluate $D(\exp(j\omega T))/Q(\exp(j\omega T))$.

By taking $r + 1$ equally spaced frequency values, an initial guess is made of the extremal frequencies. By iteration, the extremal frequencies are found, yielding the best approximation to $D\exp(j\omega T))$, $P(\exp(j\omega T))$. This last function is evaluated, and using the discrete Fourier transform (DFT) (Chapter 7), the values of the coefficients $\tilde{a}(nT)$, $\tilde{b}(nT)$, $\tilde{c}(nT)$, or $\tilde{d}(nT)$ are found. A second calculation yields the impulse response values, $\{h(nT)\}$. Once these are obtained, the filter that results will be the optimal approximation to the desired response and have linear phase shift in the passband.

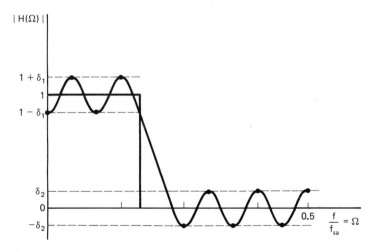

Figure 6–53 Frequency response of an optimal LP filter where the maximum error is minimized to $\pm\delta_1$, and $\pm\delta_2$.

The program of Parks and McClellan finds the coefficients that yield the impulse response of the optimal filter and the impulse response itself. Once the program is loaded into a computer, the user provides the following information either in a card form or directly through an interactive terminal. On the first line, the user enters:

1. The filter length, N
2. The type of filter
   a. Multiple passband/stopband, code 1
   b. Differentiator, code 2
   c. Hilbert transform filter, code 3
3. Number of bands, up to 10
4. Results in card form? (1 = yes, 0 = no)
5. Grid density.

The value of N used in the program is dependent on the values of δ_1 and δ_2, the cutoff and the passband frequencies, and their differences. The equation that ties these terms together can be found in the text by Rabiner and Gold listed in the references. However, if the other terms are known, N is found by providing the program with an arbitrary value and then examining the resulting filter characteristics. If they are within the specifications, repeat the process by using smaller values of N, stopping when the specifications are no longer met; if the original choice did not meet the specifications, increase N until they do.

As an example, a 32-point bandpass filter with three bands, no card punch, and a grid density of 16 would be entered as

$$32,1,3,0,16$$

Additional information required (second line) is:

6. Band edges: lower and upper band edge for each band
7. Desired value for each band

For a three-band, bandpass filter, with the middle band being the passband, the second and third lines could be entered as

$$0, 0.1, 0.2, 0.35, 0.425, 0.5$$
$$0,1,0$$

where the third line indicates that the first and third bands are stopbands, while the second band is a passband. (Entering 1,0,1 would make the filter a band-reject, or notch, filter.) The second line of the input tells us that the first band extends from 0 to 0.1 of the normalized frequency Ω, the second

Sec. 6.6   Digital Filter Synthesis

band from 0.2 to 0.35, and the third band from 0.425 to 0.5. The transition width on both sides of the passband is less than 0.1, which indicates that a very high rate of rolloff can be readily obtained. The upper value of the highest band must always be 0.5.

The last bit of information required (line 4) is the weighting function for each band. The higher this value, the smaller the maximum value of the error in the band will be. As an example, the value for the filter above can be entered as

$$\$10,1,10$$

where the choice of 10 for the stopbands implies that we want the value of the approximation to be very close to the value of the desired function in that

```
************************************************************
                    FINITE IMPULSE RESPONSE (FIR)
                    LINEAR PHASE DIGITAL FILTER DESIGN
                    REMEZ EXCHANGE ALGORITHM
                    BANDPASS FILTER
         FILTER LENGTH =   32
         ***** IMPULSE RESPONSE *****
              H(  1) =  -0.57534121E-02 = H( 32)
              H(  2) =   0.99027198E-03 = H( 31)
              H(  3) =   0.75733545E-02 = H( 30)
              H(  4) =  -0.65141192E-02 = H( 29)
              H(  5) =   0.13960525E-01 = H( 28)
              H(  6) =   0.22951469E-02 = H( 27)
              H(  7) =  -0.19994067E-01 = H( 26)
              H(  8) =   0.71369560E-02 = H( 25)
              H(  9) =  -0.39657363E-01 = H( 24)
              H( 10) =   0.11260114E-01 = H( 23)
              H( 11) =   0.66233643E-01 = H( 22)
              H( 12) =  -0.10497223E-01 = H( 21)
              H( 13) =   0.85136133E-01 = H( 20)
              H( 14) =  -0.12024993E 00 = H( 19)
              H( 15) =  -0.29678577E 00 = H( 18)
              H( 16) =   0.30410917E 00 = H( 17)

                       BAND 1         BAND 2         BAND 3         BAND
LOWER BAND EDGE        0.             0.20000000     0.42500000
UPPER BAND EDGE        0.10000000     0.35000000     0.50000000
DESIRED VALUE          0.             1.00000000     0.
WEIGHTING             10.00000000     1.00000000    10.00000000
DEVIATION              0.00151312     0.01513118     0.00151312
DEVIATION IN DB      -56.40254641   -36.40254641   -56.40254641

EXTREMAL FREQUENCIES
  0.            0.0273437      0.0527344      0.0761719      0.0937500
  0.1000000     0.2000000      0.2195312      0.2527344      0.2839844
  0.3132812     0.3386719      0.3500000      0.4250000      0.4328125
  0.4503906     0.4796875

************************************************************
TIME=      0.8065938 SECONDS
```

(a)

Figure 6–54 Frequency response for a bandpass filter. (From *Theory and Application of Digital Signal Processing*, L. R. Rabiner and B. Gold. Copyright © 1975, Prentice-Hall, Inc. Reprinted by permission of Prentice-Hall, Inc.)

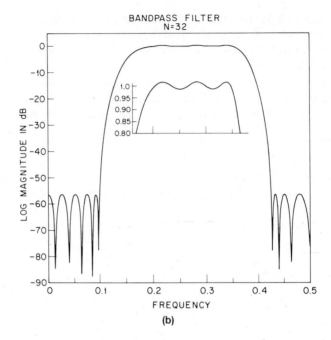

Figure 6-54 (*cont.*)

band; however, the larger this value, the larger the resulting deviation in the passband for a given value of *N*. The program output is the practical filter's impulse response, the extremal frequencies, the deviation, and other relevant information. A printout is shown in Figure 6-54 for the filter design given above along with a plot of its frequency response. Once the impulse values are obtained, the filter can be realized using the circuits provided earlier in this chapter.

6.7 FREQUENCY TRANSFORMATION

As with analog filters, the standard procedure is to design a low-pass (LP) digital filter with desired characteristics and convenient cutoff frequency. Then, by a particular transformation, the LP design is transformed to the desired one.

6.7.1 LP to LP Filter

Given an LP filter with a cutoff frequency, Ω_c, it is desired to have an LP filter with a different cutoff frequency, Ω_c'. This can be accomplished by the transformation given in Equation 6-141.

Sec. 6.7 Frequency Transformation

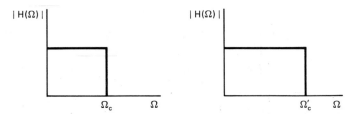

Figure 6–55 LP-to-LP transformation.

$$z^{-1} \longrightarrow \frac{z^{-1} - \alpha}{1 - \alpha z^{-1}} \tag{6-141a}$$

$$\alpha = \frac{\sin\left[\pi(\Omega_c - \Omega_c')\right]}{\sin\left[\pi(\Omega_c + \Omega_c')\right]} \tag{6-141b}$$

This transformation is shown in Figure 6–55.

6.7.2 LP-to-BP Filter

For this situation, a LP filter is converted to a BP filter as shown in Figure 6–56. The transformation is given by

$$\cos 2\pi\Omega_0 = \frac{\cos\left[\pi(\Omega_U + \Omega_L)\right]}{\cos\left[\pi(\Omega_U - \Omega_L)\right]} \tag{6-142a}$$

$$z^{-1} \longrightarrow \frac{z^{-2} - [2\alpha\beta/(\beta+1)]\,z^{-1} + (\beta-1)/(\beta+1)}{[(\beta-1)/(\beta+1)]\,z^{-2} - [2\alpha\beta/(\beta+1)]z^{-1} + 1} \tag{6-142b}$$

$$\alpha = \cos 2\pi\Omega_0 \tag{6-142c}$$

$$\beta = \cot\left[\pi(\Omega_U - \Omega_L)\right] \tan \pi\Omega_c \tag{6-142d}$$

where Ω_0 is the center frequency of the passband.

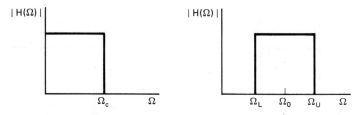

Figure 6–56 LP-to-BP transformation.

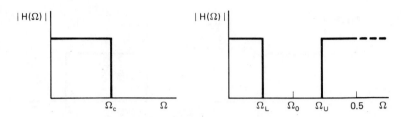

Figure 6–57 LP-to-notch filter transformation.

6.7.3 LP-to-Notch (Band-Reject) Filter

This transformation is shown in Figure 6–57. The equations for the transformation are

$$z^{-1} \longrightarrow \frac{z^{-2} - [2\alpha/(1+k)]\,z^{-1} + (1-k)/(1+k)}{[(1-k)/(1+k)]\,z^{-2} - [2\alpha/(1+k)]\,z^{-1} + 1} \qquad (6\text{–}143\text{a})$$

$$\alpha = \cos 2\pi\Omega_0 = \frac{\cos[\pi(\Omega_U + \Omega_L)]}{\cos[\pi(\Omega_U - \Omega_L)]} \qquad (6\text{–}143\text{b})$$

$$k = \tan[\pi(\Omega_U - \Omega_L)] \tan \pi\Omega_c \qquad (6\text{–}143\text{c})$$

6.7.4 LP-to-HP Filter

The transformation is shown in Figure 6–58 and the equations are presented below.

$$z^{-1} \longrightarrow -\frac{z^{-1} + \alpha}{1 + \alpha z^{-1}} \qquad (6\text{–}144\text{a})$$

$$\alpha = -\frac{\cos[\pi(\Omega_c - \Omega_c')]}{\cos[\pi(\Omega_c + \Omega_c')]} \qquad (6\text{–}144\text{b})$$

An example is provided to show the use of this procedure.

Example 6–9

Given a one-pole LP filter with $\Omega_c = \frac{1}{4}$. If f_{Sa} is 1 kHz, design a HP filter with $\Omega_c' = \frac{1}{3}$.

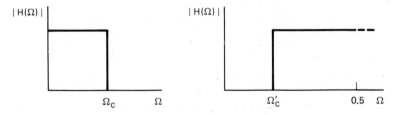

Figure 6–58 LP-to-HP transformation.

Sec. 6.7 Frequency Transformation

Solution A one-pole filter has the analog form

$$H(s) = \frac{b_0}{s + b_0} \tag{6-84}$$

We find ω_c by prewarping Ω_c:

$$\omega_c = \frac{\tan \pi \Omega_c}{T}$$

$$= \frac{\tan (\pi/4)}{T} = \frac{1 \text{ rad}}{T \text{ seconds}}$$

$$= 1 \text{ rad} \left(\frac{1000}{s}\right) = \frac{1 \text{ krad}}{s}$$

$H(s)$ can be written as

$$H(s) = \frac{(1/T) \text{ rad/s}}{s + (1/T) \text{ rad/s}}$$

with $|H(\omega)|$ given as

$$|H(\omega)| = \frac{1/T}{\sqrt{\omega^2 + (1/T)^2}}$$

The bilinear transformation yields

$$H(z) = \frac{1/T}{\left(\dfrac{1}{T}\dfrac{1-z^{-1}}{1+z^{-1}}\right) + \dfrac{1}{T}}$$

$$= \frac{1 + z^{-1}}{1 - z^- + 1 + z^{-1}} = 0.5 + 0.5z^{-1}$$

which is the desired LP filter.
   The transformation to HP is made. First we calculate α.

$$\alpha = -\frac{\cos [\pi(\Omega_c - \Omega_c')]}{\cos [\pi(\Omega_c + \Omega_c')]}$$

$$\alpha = -\frac{\cos [\pi(\tfrac{1}{4} - \tfrac{1}{3})]}{\cos [\pi(\tfrac{1}{4} + \tfrac{1}{3})]} = -\frac{0.9659}{-0.2588}$$

$$= 3.732$$

We then have

$$H_{\text{HP}}(z) = 0.5 + 0.5z^{-1}$$

$$= 0.5 + 0.5\left(-\frac{z^{-1} + \alpha}{1 + \alpha z^{-1}}\right)$$

$$= \frac{0.5(1 + \alpha z^{-1}) - 0.5(z^{-1} + \alpha)}{1 + \alpha z^{-1}}$$

$$= \frac{0.5(1 + 3.732z^{-1}) - 0.5(z^{-1} + 3.732)}{1 + (3.732)z^{-1}}$$

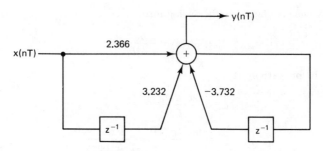

Figure 6–59 Filter design of Example 6–9.

$$= \frac{2.366 + (3.232)z^{-1}}{1 + (3.732)z^{-1}} = \frac{Y(z)}{X(z)}$$

The difference equation is

$$y(nT) = 2.366x(nT) + (3.232)x((n-1)T) - 3.732y((n-1)T)$$

from which the filter design can be implemented. This is shown in Figure 6–59. Once $H(z)$ is obtained for the LP filter, by whatever technique, the proper transformation will yield the desired filter.

6.8 SUMMARY

This chapter has presented the basic concepts in the design of digital filters. Several techniques were presented; however, several others are available in the References at the end of this chapter. In addition, several references are provided that present some of the mathematical concepts in more detail.

PROBLEMS

1. For the following system, draw the block diagram and find the impulse response.
$$y(nT) = 0.6x(nT) + 0.43x[(n-1)T] + 0.9x[(9n-3)T] \\ - 0.3y(n-1) - 0.81y[(n-2)T]$$

2. Find the impulse response for the following systems. Assume $T = 1s$
   (a) $y(n) = 0.6x(n) + 0.43x(n-1) - 0.3y(n-1)$
   (b) $y(n) = -0.2\ x(n) + 0.7x(n-1) + 0.11x(n-2)$
   (c) $y(n) = 0.5x(n) + 1.1x(n-1) - 0.9y(n-1) + 0.6y(n-2)$

3. Find $y(n)$ for the following sequence $x(n) = 10v\ \sin(2\pi 200 H_3 n)$. Find $h(n)$.
$$y(n) = 3x(n) - 2x(n-1) + 2x(n-2)$$

4. Repeat Problem 3 for $y(n) = 3x(n) + x(n-1) - 2y(n-1)$.
5. For the circuit shown, find the frequency response.

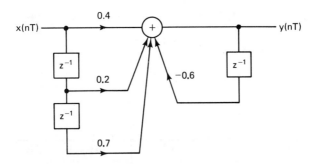

Figure P6–5

6. Using the result of Problem 3, find $H(f)$. Sketch its magnitude as a function of Ω.
7. For the circuit of Problem 5, write the difference equation and determine its transfer function. Is the system stable, FIR, or recursive?
8. Given that $y(n) = 2x(n) - 3x(n-1) + 2x(n-2)$, find the z transform, $H(z)$, $|H(\Omega)|$, and its frequency response. Also find $h(n)$ and draw the block diagram.
9. For the following system, is it stable? Find $H(z)$, $|H(\Omega)|$, $h(n)$, and the block diagram.

$$y(n) = 2x(n) - 3x(n-1) - 0.5y(n-1)$$

10. Repeat Problem 9 for $y(n) = 2x(n) - 3x(n-1) - 0.5y(n-1) + 2y(n-2)$.
11. For the circuit shown, find $H(z)$, $y(n)$, and $h(n)$. Is the system stable? What is $|H(\Omega)|$?

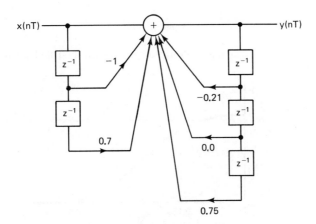

Figure P6–11

12. Transform the pole/zero diagram shown into the z plane. Assume $T = 1$ ms.

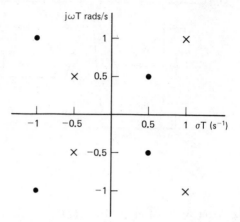

Figure P6–12

13. For the pole/zero diagram shown, sketch the magnitude of the system response.

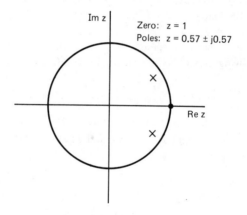

Zero: $z = 1$
Poles: $z = 0.57 \pm j0.57$

Figure P6–13

14. Using trial-and-error techniques, design a system with the LP characteristic shown.

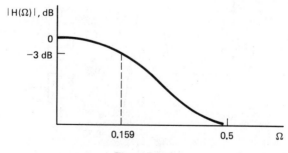

Figure P6–14

Chap. 6 Problems

15. Design an analog low-pass Butterworth filter with $\omega_c = 20$ rad/s. The gain is to be down by 22dB at $\omega_s = 40$ rad/s.
16. Design a four-pole low-pass Chebychev filter with a 1-dB maximum attenuation in the passband (ripple width) and with a cutoff frequency of 50 rad/s.
17. For the following transfer function, use the bilinear transformation to find $H(z)$. If $\omega_c = 1.256$ krad/s and $T = 1$ ms? Sketch $|H(\Omega)|$.

$$H(s) = \frac{1}{(s/\omega_c)^2 + \sqrt{2}(s/\omega_c) + 1}$$

18. Using the bilinear transformation, design a LP filter with $\Omega_c = 0.2$. Use a z-pole Butterworth design with $T = 1$ ms.
19. Repeat Problem 18 for a Chebychev filter with $\Omega_c = 0.135$. The analog design should have a ripple width of 2 dB with $T = 0.5$ ms.
20. Using the analog result of Problem 15, design a digital filter using the impulse invariant technique. Find $|H(\Omega)|$.
21. Design a two-pole Butterworth filter with $\omega_c = 25$ rad/s. Convert it to digital form. Use a transformation to convert the digital design to a LP filter with $\Omega_c = 0.125$ with $T = 10$ ms.
22. Given a LP filter with $\Omega_c = 0.1$. What is the circuit (block diagram) for a HP filter with $\Omega_c = 0.2$? Use the Butterworth two-pole design equation with $T = 10$ ms.
23. For the transfer function shown, find the FIR response for a main-lobe transition width (Δf) of 0.08 and a minimum stopband attenuation of <40 dB.

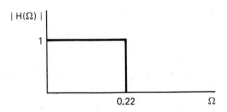

Figure P6–23

24. Repeat Problem 23 for a HP filter with $\Delta f = 0.12$ and a minimum stopband attenuation of <50 dB. $\Omega_c = 0.37$.
25. Design an FIR filter with $\Omega_c = 0.17$ and $T = 0.2$ ms. Use a five-point rectangular window function. Express $H_{tc}(\omega)$ in series form. Draw the circuit.
26. Repeat Problem 25 for a Hamming window.
27. Repeat Problem 25 for a Bartlett window.
28. Plot the window equations for $N = 11$.

29. For the FIR impulse response shown, what is the series form for $H(\omega)$?

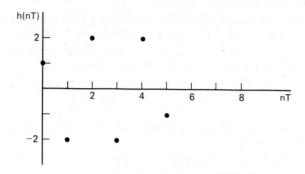

Figure P6-29

30. For $N = 7$, show the impulse response of an FIR filter with even and odd symmetry. Repeat for $N = 8$.

31. For the impulse response shown, what is $H(\Omega)$?

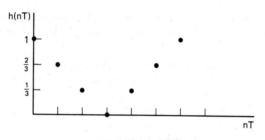

Figure P6-31

32. What would be the computer input to a Parks and McClellan program for the filter function shown? Assume the PB ripple is 1 dB while the stopband ripple is -50 dB down.

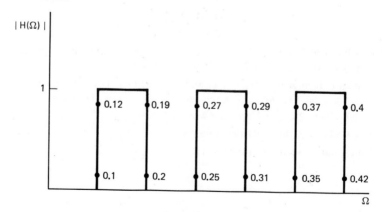

Figure P6-32

33. Using the LP digital filter design of Problem 17, with $\Omega_c = 0.125$, design:
   (a) A LP filter with $\Omega_c = 0.28$.
   (b) A HP filter with $\Omega_c = 0.28$.
   (c) A BP filter with $\Omega_L = 0.2$ and $\Omega_U = 0.35$.

REFERENCES

ACKROYD, M. H. *Digital Filters.* London, England, Butterworth's, 1973.

CHEN, C.-T. *One-Dimensional Digital Signal Processing.* New York: Marcel Dekker, 1979.

KAROWSKI, R. J. *Introduction to the z Transform and Its Derivation.* La Jolla, CA: TRW Publications, 1979.

MCCLELLAN, J. H., and PARKS, T. W. "A Unified Approach to the Design of Optimum FIR Linear-Phase Digital Filters." *IEEE Trans. Circuit Theory,* CT-20 (1973): 697–701.

MCCLELLAN, J. H., PARKS, T. W., and RABINER, L. W. "A Computer Program for Designing Optimum FIR Linear Phase Digital Filters." *IEEE Trans. Audio Electroacoust.,* AU-21 (1973): 506–526.

OPPENHEIM, A. V., and SCHAFER, R. W. *Digital Signal Processing.* Englewood Cliffs, N.J.: Prentice-Hall, 1975.

PARKS, T. W., and MCCLELLAN, J. H. "Chebychev Approximation for Nonrecursive Digital Filters with Linear Phase." *IEEE Trans. Circuit Theory,* CT-19 (1972): 189–194.

PELED, A., and LIU, B. *Digital Signal Processing.* New York: Wiley, 1976.

RABINER, L. R., and GOLD, B. *Theory and Application of Digital Signal Processing.* Englewood Cliffs, N.J.: Prentice-Hall, 1975.

RABINER, L. R., KAISER, J. E., HERMANN, O., and DOLAN, M. T. "Some Comparisons between FIR and IIR Digital Filters." *Bell Syst. Tech. J.,* 53 (1974): 305–331.

RABINER, L. R., MCCLELLAN, J. H., and PARKS, T. W. "FIR Digital Filter Design Using Weighted Chebychev Approximation." *Proc. IEEE* (1975): 595–610.

RADER, C. M. and GOLD, B. "Digital Filter Design Techniques in the Frequency Domain," *Proc. IEEE,* 55, No. 2, (Feb., 1967.): 149–171.

STANLEY, W. D. *Digital Signal Processing.* Reston, Va.: Reston, 1975.

TRETTER, S. A. *Introduction to Discrete-Time Signal Processing.* New York: Wiley, 1976.

THE DISCRETE AND THE FAST FOURIER TRANSFORMS

7

It was shown in Chapter 6 that a computer can be used in the design of a digital filter. In the program developed by Parks and McClellan, the computer found a set of coefficients that yielded the impulse response of a digital filter that was optimal in the Chebychev or minimax sense and with a minimum number of sample points. Once the filter is designed, we would like to use it to select desired and reject undesired frequencies from the frequency content of the input signal. This can be done by convolving the DT input with the impulse response sequence of the digital filter. The possible disadvantage of this approach is the amount of processor time required to perform all the multiplications and additions as indicated in the DT convolution equation.

We will see that this disadvantage can be overcome by performing convolution in the frequency domain. The procedure will require a method of converting the sampled sequences in the time domain to sampled sequences in the frequency domain. This will be accomplished by the discrete Fourier transform (DFT).

Another aspect of signal processing that can be performed digitally in the sampled frequency domain is correlation. This indicates how closely the frequency content of two signals agree and was discussed in Section 1.13. The CT relation for correlation is similar to convolution and is presented in Equation 7–1.

$$x_{12}(\tau) = \int_{-\infty}^{\infty} x_1(t)x_2(\tau + t)\,dt \tag{7-1}$$

where τ is termed the searching parameter.

If two waveforms are similar, but occur at different times (e.g., as with a transmitted and returning radar signal), the value of the integral will be different from zero for specific choices of τ; otherwise, it will be negligibly small or zero. If $x_1(t) = x_2(t)$, then $R_{12}(\tau)$ is termed the *autocorrelation function*.

The discrete form of the correlation equation is similar to that for convolution and is given in Equation 7-2.

$$x_{12}(nT) = \sum_{j=-\infty}^{\infty} x_1(jT)x_2((n+j)T) \tag{7-2}$$

As with convolution, we can work with the DFTs of the signals that are being correlated to reduce the computer time evaluating $x_{12}(nT)$.

In the sections that follow, we introduce the DFT and some of its applications. After that, the fast Fourier transform (FFT) method of evaluating the DFT is presented.

7.1 DISCRETE FOURIER TRANSFORM

We begin our examination of the DFT by considering a causal, finite, nonperiodic sequence, $x(nT)$. Its z transform is given by

$$X(z) = \sum_{n=0}^{\infty} x(nT)z^{-n} \tag{7-3}$$

If we evaluate the z transform on the unit circle, Equation 7-3 becomes

$$\begin{aligned} X(z)\bigg|_{z=\exp(j\omega T)} = X(e^{j\omega T}) &= \sum_{n=0}^{\infty} x(nT)e^{-j\omega nT} \\ &= \sum_{n=0}^{\infty} x(nT)e^{-j2\pi\Omega n} \end{aligned} \tag{7-4}$$

which is a continuous function of Ω. As it is continuous, it is not in a form that can be used by the computer.

Since the input is usually a finite sequence of N samples, we have

$$X(e^{j\omega T}) = \sum_{n=0}^{N-1} x(nT)e^{-j\omega nT} \tag{7-5}$$

As $\omega T = 2\pi fT = 2\pi\Omega$, this can also be written as

$$X(e^{j2\pi\Omega}) = \sum_{n=0}^{N-1} x(nT)e^{-j2\pi\Omega n} \tag{7-6}$$

where $X(\exp(j\omega T))$ is periodic in f_{sa}, the sampling frequency, and $X(\exp(j2\pi\Omega))$ is periodic with period equal to 1. Example 7–1 shows the relevent points of the discussion above.

Example 7–1

Apply the procedures described above to the sequence of Figure 7–1.

Solution Taking the z transform of the sequence in Figure 7–1, we obtain

$$X(z) = \sum_{n=0}^{\infty} x(nT)z^{-n} = \sum_{n=0}^{3} x(nT)z^{-n} \qquad (7\text{–}7)$$

Evaluating the z transform on the unit circle and expanding the sum, we have

$$X(z)\bigg|_{z=\exp(j\omega T)} = X(e^{j\omega T}) = \sum_{n=0}^{3} x(nT)e^{-j\omega nT} \qquad (7\text{–}8a)$$

$$= 1 + \frac{2}{3}e^{-j\omega T} + \frac{1}{3}e^{-j\omega 2T} + 0 \cdot e^{-j\omega 3T}$$

$$= 1 + \frac{2}{3}(\cos \omega T - j\sin \omega T) + \frac{1}{3}(\cos 2\omega T - j\sin 2\omega T) \qquad (7\text{–}8b)$$

$$= 1 + \frac{2}{3}\cos \omega T + \frac{1}{3}\cos 2\omega T - j\left(\frac{2}{3}\sin \omega T + \frac{1}{3}\sin 2\omega T\right)$$

The magnitude of $X(\exp(j\omega T))$ is given as

$$|X(e^{j\omega T})| = \sqrt{\left(1 + \frac{2}{3}\cos \omega T + \frac{1}{3}\cos 2\omega T\right)^2 + \left(\frac{2}{3}\sin \omega T + \frac{1}{3}\sin 2\omega T\right)^2}$$

$$= \sqrt{\left[1 + \frac{2}{3}\cos(2\pi\Omega) + \frac{1}{3}\cos(4\pi\Omega)\right]^2 + \left[\frac{2}{3}\sin(2\pi\Omega) + \frac{1}{3}\sin(4\pi\Omega)\right]^2} \qquad (7\text{–}9)$$

which is continuous and periodic in $\omega T = 2\pi\Omega$, and doubly infinite.

If, given $X(\exp(j\omega T))$, we want to find the sequence it represents, we can use Equation 7–10.

$$x(nT) = \frac{1}{f_{sa}} \int_{f}^{f+f_{sa}} X(e^{j2\pi fT})e^{j2\pi fnT}df \qquad (7\text{–}10)$$

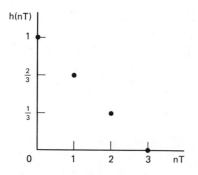

Figure 7–1 Impulse response used in Example 7–1.

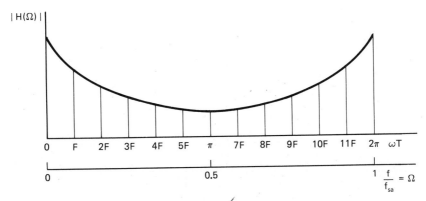

Figure 7-2 Sampled value of $|H(\Omega)|$.

However, if we wish to use a computer to determine $x(nT)$, the continuous nature of the domain of X prevents us from so doing. To change its format, thereby making it acceptable, we treat it as we did the CT signal (i.e., we sample it). This is done in the frequency domain by taking N samples of $X(\exp(j\omega T))$: the sampling frequency is divided into N sections each equal to $(f_{sa}/N) \equiv F$ in length. This allows us to sample the frequency response function every F hertz, which is shown in Figure 7-2. In this manner, the continuous frequency-domain signal is converted to a discrete frequency-domain signal. It should be mentioned that this is also equivalent to sampling the z transform at uniform intervals on the unit circle as shown in Figure 7-3. As with the CT sampling operation, the samples of the frequency-domain function will also contain the same information (i.e., the frequency response of the system).

The sampling of Equation 7-5 yields

$$X(e^{j2\pi fT}) \implies X(e^{j2\pi kFT}) = \sum_{n=0}^{N-1} x(nT)e^{-j2\pi knFT} \qquad (7\text{-}11)$$

As $F = f_{sa}/N$ and $f_{sa} = 1/T$, where T is the sampling interval in the CT domain, we have

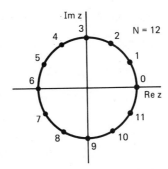

Figure 7-3 Sampled z-transform on the unit circle.

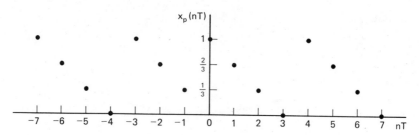

Figure 7-4 Periodic function obtained from sampling.

$$\text{FT} = \frac{1}{N} \tag{7-12}$$

allowing Equation 7-11 to be written as

$$X(e^{j(2\pi/N)k}) = \sum_{n=0}^{N-1} x(nT) e^{-j(2\pi/N)kn} \equiv X(k) \tag{7-13}$$

where $X(k)$ is the discrete Fourier transform, DFT, of $x(nT)$ and the samples of $X(e^{j2\pi fT})$, the frequency content of $x(nT)$.

As always, we can obtain the sequence $x(nT)$ by the inverse operation (IDFT)

$$x(nT) = \frac{1}{N} \sum_{k=0}^{N-1} X(k) e^{j(2\pi/N)kn} \tag{7-14}$$

As there is no limit placed on the integer variable n, we see that $x(nT)$ obtained from Equation 7-14 may be larger than the original sequence. In addition, $x(nT)$ will be periodic. The sampling of $X(\exp(j\omega T))$ yields a periodic DT function which is the original DT function $x(nT)$ (Figure 7-1) continuously repeated as shown in Figure 7-4. As both $x(nT)$ and $X(\exp(j2\pi k/N))$ are periodic, we will use the subscript p when we refer to the periodic sequence formed from the nonperiodic sequence $x(nT)$ when the frequency content $X(\exp(j2\pi fT))$ is sampled.

Example 7-2

Using Equation 7-13, with $N = 4$, find $X(k)$ for $x(nT) = 1$, $n = 0, 1$.

Solution As $N = 4$, $k = 0, 1, 2,$ and 3. For $k = 0$, we have

$$X(k) = \sum_{n=0}^{3} x(nT) \exp(-j[2\pi/N]kn) \tag{7-13}$$

$$X(0) = x(0T) \cdot \exp(0) + x(T) \exp(0) = 2$$

$k = 1$,

$$X(1) = x(0T) \cdot \exp(0) + x(T) \exp\left(-j\frac{2\pi}{4} 1 \cdot 1\right)$$

Sec. 7.1 Discrete Fourier Transform

$$= 1 + 1 \left(\cos \left(\frac{\pi}{2} \right) - j \sin \left(\frac{\pi}{2} \right) \right)$$
$$= 1 - j$$

$k = 2$,

$$X(2) = x(0T) \exp(0) + x(T) \exp \left(-j \frac{2\pi}{4} 2 \cdot 1 \right)$$
$$= 1 + 1 (\cos \pi - j \sin \pi)$$
$$= 1 - 1 - j0 = 0$$

$k = 3$,

$$X(3) = x(0T) \exp(0) + x(T) \exp \left(-j \frac{2\pi}{4} 3 \cdot 1 \right)$$
$$= 1 + 1 \left(\cos \left(\frac{3\pi}{2} \right) - j \sin \left(\frac{3\pi}{2} \right) \right)$$
$$= 1 - j(-1) = 1 + j$$

Even though we found only four values of $X(k)$, it has been shown that $X(\exp(j\omega T))$ is periodic and doubly infinite. The same holds true for $X(k)$.

To demonstrate that $x(nT)$ obtained from $X(k)$ is also periodic, one simply determines two values of $x(nT)$, using Equation 7–14, such that they are N samples apart. This is shown in Example 7–3.

Example 7–3
Demonstrate that $x(nT)$ obtained from $X(k)$, Example 7–1, is periodic.

Solution This will be done by comparing the value of $x(nT)$ and $x((n + N)T)$ for any n using Equation 7–14

$$x(nT) = \frac{1}{N} \sum_{k=0}^{N-1} X(k) \exp \left(j \frac{2\pi}{N} kn \right)$$

$$x((n + N)T) = \frac{1}{N} \sum_{k=0}^{N-1} X(k) \exp \left(j \frac{2\pi}{N} (n + N)k \right)$$
(7–14)

Factoring the exponential term on the right hand side, we obtain

$$x((n + N)T) = \frac{1}{N} \sum_{k=0}^{N-1} X(k) \exp \left(j \frac{2\pi}{N} kn \right) \exp \left(j \frac{2\pi}{N} N \right)$$
$$= x(nT)$$

as the second exponential term on the right is unity for all k.

Equations 7–13 and 7–14 are defined as the discrete Fourier transform (DFT) pair. Equation 7–13 is the DFT and Equation 7–14 is its inverse, IDFT. A program has been written in Pascal to calculate the values of the $X(k)$ given a finite sequence $x(nT)$. It can be used for any number of sample points,

$n = 0, 1, \ldots, N - 1$. This is given in Appendix B together with the results of a 32-point DFT.

Although it may appear that the DFT is applicable only to discrete sequences that are periodic, we will see how it can be used to obtain the correlation and convolution of finite, aperiodic sequences.

7.2 CONVOLUTION USING THE DFT

From Chapter 6 we saw that the discrete convolution of two sequences was given by

$$y(nT) = \sum_{k=0}^{\infty} h(kT)x((n - k)T) \qquad (6-2)$$

If $x(nT)$ and $h(nT)$ are finite-duration sequences, the output, $y(nT)$, is also of finite duration and of length $N_1 + N_2 - 1$, where N_1 and N_2 are the lengths of $x(nT)$ and $h(nT)$, respectively.

In Chapter 5, it was shown that convolution of two discrete time sequences can be obtained by multiplication of their z transforms, Equation 5–46, or their frequency responses, Equation 5–25, and then by taking the respective inverse transform. This leads us to propose that convolution can also be obtained by multiplying the DFTs of the input sequence and the filter's impulse response. This result can be obtained directly by sampling Equation 5–25. This leads to the DFT form of convolution.

$$Y(k) = X(k)H(k) \qquad (7-15)$$

where $Y(k)$ and $X(k)$ are the sampled frequency content of $y(nT)$ and $x(nT)$ respectively, and $H(k)$ is the sampled frequency response of $h(nT)$. As each term in Equation 7–15 represents the periodic form of the finite aperiodic sequence, the resulting application of the IDFT to $Y(k)$ may not yield the expected sequence, $y(nT)$, but rather it will return $y_p(nT)$. This is due to the fact that use of Equation 7–15 is equivalent to convolving the periodic form of $x(nT)$ with the periodic form of $h(nT)$ as shown in Equation 7–16.

$$y_p(nT) = \sum_{k=0}^{\infty} h_p(kT)x_p((n - k)T) \quad \Longleftrightarrow \quad Y_p(k) = H_p(k)X_p(k) \qquad (7-16)$$

This result is termed *circular convolution*, as opposed to linear convolution (Equation 6–2).

In most cases, circular convolution will provide a sequence $y_p(nT)$ that is a distorted, periodic form of $y(nT)$, the linear convolution result of Equation 6–2.

To perform convolution on finite-duration sequences using DFTs and to obtain a sequence $y_p(nT)$ that, though periodic, yields the desired result of linear convolution, each sequence is lengthened by adding zero-value terms to it. This process is termed *zero filling* and will enable us to obtain the results

Sec. 7.2 Convolution Using the DFT

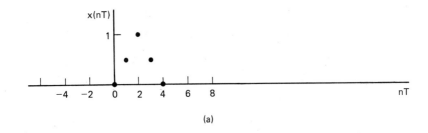

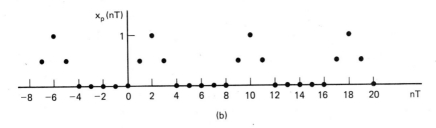

Figure 7–5 Periodic signal obtained from sampled input: (a) sampled input signal; (b) periodic, zero-extended input signal.

of linear convolution. The number of zeros to be added to each sequence is such that each one will have a final length of at least the sum of the two original sequences less 1, $N_1 + N_2 - 1$. They are then made periodic by repeating them for positive and negative n. This is shown in Figure 7–5 where the finite sequence $x(nT)$ is zero-filled and replicated.

As the output sequence of a linear convolution will have a length $N_1 + N_2 - 1$, it is necessary that the periodic sequence formed from $x(nT)$ have a period of at least that length. In addition, $h_p(nT)$, formed from $h(nT)$, must also be of length $N_1 + N_2 - 1$ or greater. This is shown in Figure 7–6 for $x(nT)$ with $N_1 = 5$ and $h(nT)$ with $N_2 = 4$. From $x(nT)$, we formed $x_p(nT)$ by adding three zeros and repeating $x(nT)$; to form $h_p(nT)$, four zeros were added before repeating. In each case, more zeros could have been added if required. The reader is urged to use the program of Appendix B to see the effect of performing the DFT with and without zero filling.

When convolution using Equation 7–15 is performed, the result of the IDFT of $Y(k)$ for $0 \leq n \leq (N_1 + N_2 - 1)$ will be the desired result. This is shown in the following example.

Example 7–4
Find the convolution of the sequences given in Figure 7–6.

Solution As the output will have a period of $N_1 + N_2 - 1 = 8$, the sequences are zero-extended by adding three zeros to $x(nT)$ and four zeros to $h(nT)$, as shown in Figure 7–6. The DFT of the output, $Y_p(k)$, is obtained by the product of the individual DFTs of $x_p(nT)$ and $h_p(nT)$ and then taking the IDFT of $Y_p(k)$. The results for

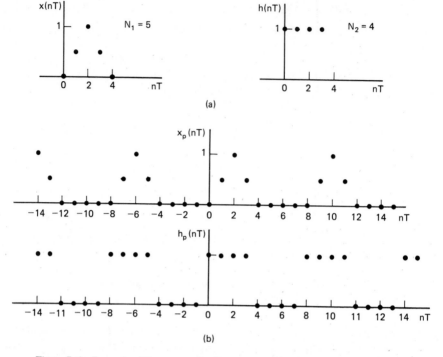

Figure 7-6 Example of "zero extending" sequences: (a) sampled input and impulse response; (b) zero extended periodic input and impulse response.

$0 \leq n \leq (N_1 + N_2 - 1)$ will yield the convolution of $x(nT)$ and $h(nT)$. Using Equation 7-13, we have, with $N = N_1 + N_2 - 1 = 8$,

$$X(e^{j(2\pi/8)k}) \equiv X(k) = \sum_{n=0}^{7} x(nT)e^{-j(2\pi/8)kn} \quad (7\text{-}17a)$$

$$H(e^{j(2\pi/8)k}) \equiv H(k) = \sum_{n=0}^{7} h(nT)e^{-j(2\pi/8)kn} \quad (7\text{-}17b)$$

The product is given by

$$Y(e^{-j(2\pi/8)k}) \equiv Y(k) = X(k)H(k) \quad (7\text{-}17c)$$

which is equivalent to sampling the z transform of each function on the unit circle at $\Omega = k/4$, $0 \leq k \leq 7$. Using Equation 7-14, we obtain

$$y_p(nT) = \frac{1}{N} \sum_{k=0}^{7} Y(k)e^{+j(2\pi/N)nk} = y(nT) \quad (7\text{-}18)$$

By selecting the samples of $y_p(nT)$ for $0 \leq n \leq 7$, we will have the linear convolution of $x(nT)$ and $h(nT)$. The result is shown in Figure 7-7.

Sec. 7.2 Convolution Using the DFT

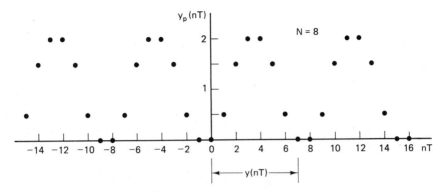

Figure 7-7 Output sequence of Example 7-4.

We can see from the above result that the evaluation of a convolution sum can be accomplished by the computer, as we are now working with discrete values that are to be multiplied and summed. If the sequence consists of real values, for each point n, of $y_p(nT)$, we must perform eight multiplications and additions, or $2N$ operations. As there are N points, we must then perform a total of $2N^2$ operations to obtain the N-point convolution. We will see that the number of operations can be reduced significantly by the use of the FFT.

The program provided in Appendix B not only finds the DFT of a sequence, but, with slight modification, is capable of finding the convolution of two sequences. The reader should use this program as a basis to write a new program that can perform convolution.

An additional use of the DFT (and the program of Appendix B) is to perform spectral analysis of a signal, that is, to determine the frequency content of a complex signal. The output sequence of the DFT will provide the frequency content as long as the input spectrum is band limited to below $f_{sa}/2$. Otherwise, aliasing will occur and a false spectrum will be obtained. The upper half of the sample values (for $n > N/2$) is the mirror image of the lower half set of samples due to the periodicity of the pre-sampled frequency content.

To ensure that all frequency components are sampled, the value of N, the number of sample points, can be increased. This also provides better resolution of components that are close together. Consider a signal $s(t)$ whose components are two sine waves, one at 200 Hz and the other at 250 Hz. If the signal is sampled at 1000 Hz, these components occur at $\Omega = 0.2$ and 0.25 respectively. If N is chosen as 8, then F, the interval between frequency samples is 125 Hz or $\Delta\Omega = 0.125$. This will show the component at $\Omega = 0.25$, but miss the one at $\Omega = 0.2$. Increasing N to 16, reduces the sample spacing to 62.5 Hz or $\Delta\Omega = 0.0625$, yielding a better chance to pick up the component at $= 0.2$. Finally, increasing N to 64 or greater will provide a much better chance of including all spectral components in the output. The reader should determine the value of N that will allow both of the above components to be detected.

7.3 CORRELATION USING THE DFT

Now that we have introduced the DFT, we can apply it to the evaluation of the correlation sum given in Equation 7–2. Using the DFT relation, we obtain, for real sequences,

$$x_{p12}(nT) = \sum_{k=0}^{N-1} x_{p1}(kT)x_{p2}((n+k)T)$$
$$\Longleftrightarrow \quad X_{p12}(k) = X_{p1}(k)X_{p2}^{*}(k) \tag{7-19}$$

where * indicates the complex conjugate. As with convolutions, each DFT is obtained, the pair multiplied, and the result converted back to a sequence by the IDFT. Keep in mind that the signals must be zero-extended to yield satisfactory results with Equation 7–19.

The programs in Appendix B and C can be used to find the FTs of the signals to be correlated; however, prior to using the results in a program that multiplies them, a procedure must be included that forms the complex conjugate of the signal that is being moved past the fixed one. If the signals correlate, the output sequence $x_{p12}(nT)$ will show a large output for some values of n; otherwise the output will be small.

7.4 FILTERING LONG SEQUENCES OF SAMPLES

When obtaining a set of samples over a relatively long interval of time, the amount of sample points becomes quite large. As an example, sampling an EEG signal at 100 hZ will yield, over a five minute interval, 30,000 sample values; if we sample a signal generated by an orchestra playing for 30 minutes, we obtain, assuming a sample rate of 44 KHz, 79,200,000 samples! If we wish to convolve this set of sample points with a 64 point filter we have designed, the DFT must be of length 79,200,063, requiring an excessively large amount of storage space (one for each sample). In addition, Equation 7–14 indicates that all samples of $s(t)$ must be available before we can begin to calculate $S(k)$. In order to make use of these techniques developed above, the input sequence is divided into sections of "reasonable" length, and then processed using the DFT. We will consider an approach to this problem termed the overlap-add technique.

Given an input sequence, $x(nT)$, whose length is large, and a filter, $h(nT)$, with a relatively short finite-duration impulse response (i.e., FIR, of length N_1), it is desired to find the output response $y(nT)$. This is accomplished by

Sec. 7.4 Filtering Long Sequences of Samples

dividing the input sequence into blocks of length N_2. Each section is then convolved with $h(nT)$ forming a section $y_i(nT)$ of length $N_1 + N_2 - 1$. Due to the nature of the cyclic convolution, it is necessary to add the resulting sections in a special manner. This is done by overlapping the $N_1 - 1$ end points of one section with the first $N_1 - 1$ points of the next section, thereby obtaining the desired output. The original sequence can be expressed as

$$x(nT) = \sum_{k=-\infty}^{\infty} x_k(nT) \qquad (7\text{-}20)$$

where $x_k(nT)$ has a length of N_2, which is usually chosen such that $N_1 + N_2 - 1 = 2^\gamma$, with γ an integer. Then

$$y(nT) = h(nT) * x(nT) = h(nT) * \sum_{k=-\infty}^{\infty} x_k(nT) \qquad (7\text{-}21a)$$

$$y(nT) = \sum_{k=-\infty}^{\infty} h(nT) * x_k(nT) = \sum_{k=-\infty}^{\infty} y_k(nT) \qquad (7\text{-}21b)$$

Each section, $y_k(nT)$, is obtained using the IDFT of $Y_k(j)$, where $h(nT)$ and $x_k(nT)$ are zero filled and treated as if they were periodic such that each period is of length $N_1 + N_2 - 1$.

$$Y_k(jT) = H(jT)X_k(jT) \qquad (7\text{-}22a)$$

$$y_k(nT) = \text{IDFT}\{Y_k(jT)\} \qquad (7\text{-}22b)$$

The length of $y_k(nT)$ is again $N_1 + N_2 - 1$. The use of this technique is shown in Example 7–5.

Example 7–5
Find the output sequence if $h(nT)$ and $x(nT)$, as given in Figure 7–8, are convolved.

Solution As $x(nT)$ is much longer than $h(nT)$ we arbitrarily divide it into lengths $N_2 = 6$. Forming the convolution between $h(nT)$ and the first block of $x(nT)$, we obtain the sequence $y_0(nT)$ which is nine-samples long. This is shown in Figure 7–9. Forming the convolution between $h(nT)$ and the second block of $x(nT)$, $x_1(nT)$, we obtain the sequence $y_1(nT)$, shown in Figure 7–10. The sequences $y_0(nT)$ and $y_1(nT)$, together with the other $y_i(nT)$, are added together and form the first 15 points of the output sequence, $y(nT)$. This is shown in Figure 7–11, where the x's indicate overlap-add points ($N_1 - 1 = 3$).

As we expect, the result of convolving the sine wave input with the impulse response of Figure 7–8 is a sine wave. Its amplitude is modified by the value of the filter's gain evaluated at the input frequency, and it is shifted in phase from the input. Observe that Figure 7–11 should include the result of convolution with the block that precedes the zero-block of the input to provide a more accurate plot of the output.

Using the DFTs and a computer, we could have obtained $y(nT)$ much

256   Chap. 7   The Discrete and the Fast Fourier Transforms

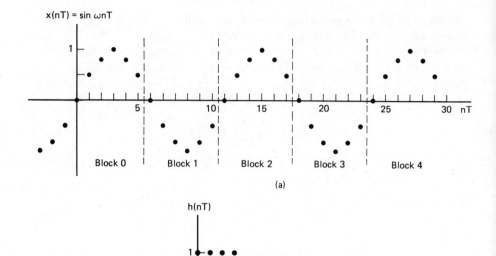

Figure 7–8 (a) Input sequence and (b) impulse response of Example 7–5.

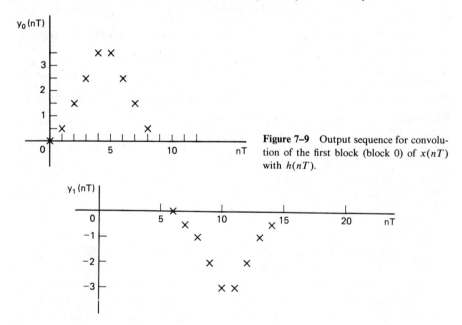

Figure 7–9 Output sequence for convolution of the first block (block 0) of $x(nT)$ with $h(nT)$.

Figure 7–10 Output sequence for convolution of the second block (block 1) of $x(nT)$ with $h(nT)$.

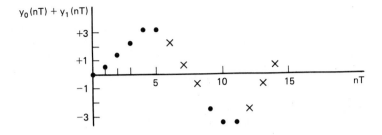

Figure 7–11 Complete output sequence for blocks 1 and 2.

more easily. The purpose of this example was to demonstrate the use of the overlap-add technique.

7.5 FAST FOURIER TRANSFORM

Having obtained the DFT, we find that a large amount of computer time is required for its calculation if N is large. This problem was solved by J. W. Cooley and J. W. Tukey in a paper listed in the References at the end of this chapter. In this section we look at the derivation of the basic flow graph diagram that indicates the steps of the DFT, and then at the procedure of Cooley and Tukey.

7.5.1 Flow Graph Diagram

This procedure begins with an N-point DFT and reduces it by half at each step until we are left with a group of $N/2$ two-point DFTS. The first thing we will do is to introduce some new notation. This is given in Equation 7–23.

$$W_N \equiv \exp\left(-j\frac{2\pi}{N}\right) \qquad (7\text{–}23a)$$

$$kT \longrightarrow k \qquad (7\text{–}23b)$$

$$nT \longrightarrow n \qquad (7\text{–}23c)$$

Equation 7–13 is written as

$$X(W_N^k) = \sum_{n=0}^{N-1} x(n) W_N^{kn}, \qquad k = 0, 1, 2, \ldots, N-1 \qquad (7\text{–}13)$$

As the DFT of $x(nT)$ will depend on k, for a given N we have

$$X(W_N^k) \equiv X(k) \qquad (7\text{–}24)$$

To evaluate the DFT, we are required to perform $2N$ complex multiplications to obtain one term of $X(k)$. Since there are N values of k, the total number of multiplications is $2N^2$. This is shown in Equation 7-25, where the series is expanded for a given value of k.

$$X(k) = x(0)W_N^{0 \cdot k} + x(1)W_N^{1 \cdot k} + \cdots + x(N-1)W_N^{(N-1) \cdot k} \quad (7\text{-}25)$$

As W_N is a complex number and $x(n)$ may also be complex, there are actually $4N^2$ multiplications: one each for the real and imaginary parts of W_N. The factors in each product are usually stored in memory and must be called into the processor. The number of steps required to fetch, multiply, and return to memory any data take a relatively long time and must be reduced to perform the operation as rapidly as possible.

The procedure used takes the DFT equation of the input sequence, $x(nT)$ and continually divides it in half until there are $N/2$ two-point sequences. The approach used to divide the input sequence and combine the results of the DFTs is the algorithm termed the fast Fourier transform (FFT). The end result of this operation is to reduce the number of multiplications to a total of $(N/2) \log_2 N$. The saving in time will be evident for large values of N. If N is 1024 real-valued points, the DFT requires $2N^2$ or 2,097,152 multiplications; for the FFT, only 5120 multiplications are required.

The DFT sum is reordered by first separating the terms of the DFT equation into even and odd values of n. Using Equation 7-13, we have

$$X_0(k) = \sum_{n=0}^{(N/2)-1} x(2n)W_N^{2nk} + \sum_{n=0}^{(N/2)-1} x(2n+1)W_N^{(2n+1)k} \quad (7\text{-}26)$$

From Equation 7-23,

$$W_N^{2nk} = e^{-j(4\pi nk/N)} = e^{-j[2\pi nk/(N/2)]} = W_{N/2}^{nk} \quad (7\text{-}27)$$

Also,

$$W_N^{(2n+1)k} = W_N^{2nk} W_N^k = W_{N/2}^{nk} W_N^k \quad (7\text{-}28)$$

Equation 7-26 becomes

$$X_0(k) = \sum_{n=0}^{(N/2)-1} x(2n)W_{N/2}^{kn} + W_N^k \sum_{n=0}^{(N/2)-1} x(2n+1)W_{N/2}^{nk} \quad (7\text{-}29)$$

where each of the sums corresponds to an $(N/2)$-point DFT over the even and odd terms of the input sequence.

Referring to Equation 7-23, we observe that

$$W_{N/2}^{n(k+N/2)} = \exp\left[-\frac{jn(k+N/2)2\pi}{N/2}\right] \quad (7\text{-}30)$$

$$= \exp\left(-\frac{jnk2\pi}{N/2}\right) \exp\left[-j\frac{n(N/2)2\pi}{N/2}\right]$$

$$= \exp\left(-\frac{jnk2\pi}{N/2}\right) \exp(-j2n\pi)$$

$$= \exp\left(-\frac{jnk2\pi}{N/2}\right) = W_{N/2}^{kn} \qquad (7\text{-}31)$$

Similarly,

$$W_N^{(k+N/2)} = -W_N^k \qquad (7\text{-}32)$$

The labeling of the output sequence for the first half is $k = 0, 1, \ldots, (N/2 - 1)$, while for the second half it is $k = N/2, N/2 + 1, \ldots, N - 1$. The second half can be expressed as $k + N/2$ for $k = 0, 1, 2, \ldots, (N/2 - 1)$. This allows us to write Equation 7-29 in terms of the upper and lower halves of the output sequence as

$$X_0(k) = \sum_{n=0}^{(N/2)-1} x(2n) W_{N/2}^{nk} + W_N^k \sum_{n=0}^{(N/2)-1} x(2n+1) W_{N/2}^{nk} \qquad (7\text{-}33a)$$

$$X_0\left(k + \frac{N}{2}\right) = \sum_{n=0}^{(N/2)-1} x(2n) W_{N/2}^{n(k+N/2)}$$

$$+ W_N^{(k+N/2)} \sum_{n=0}^{(N/2)-1} x(2n+1) W_{N/2}^{n(k+N/2)} \qquad (7\text{-}33b)$$

Using Equations 7-31 and 7-32 in 7-33b, we have

$$X_0\left(k + \frac{N}{2}\right) = \sum_{n=0}^{(N/2)-1} x(2n) W_{N/2}^{kn} - W_N^k \sum_{n=0}^{(N/2)-1} x(2n+1) W_{N/2}^{kn} \qquad (7\text{-}34)$$

that is, the upper-half values differ from the corresponding lower-half values by the minus sign between the sums.

Figure 7-12 shows the representation of the output sequence, $X(k)$, of an N-point DFT. Each block represents an $(N/2)$-point DFT sequence formed from the even and odd terms of the original input sequence, with

$$A(k) = \sum_{n=0}^{N/2-1} x(2n) W_{N/2}^{kn}, \qquad k = 0, 1, 2, \ldots, \frac{N}{2} - 1 \qquad (7\text{-}35a)$$

$$B(k) = \sum_{n=0}^{N/2-1} x(2n+1) W_{N/2}^{kn}, \qquad k = 0, 1, 2, \ldots, \frac{N}{2} - 1 \qquad (7\text{-}35b)$$

Equations 7-33a and 7-34 can now be written in shortened form as

$$X_0(k) = A(k) + B(k) W_N^k \qquad (7\text{-}35c)$$

$$X_0\left(k + \frac{N}{2}\right) = A(k) - B(k) W_N^k \qquad (7\text{-}35d)$$

Upon examining Figure 7-12, we see that the $A(k)$ terms are added, with no multiplications, to other terms to form the output. In this manner, the number

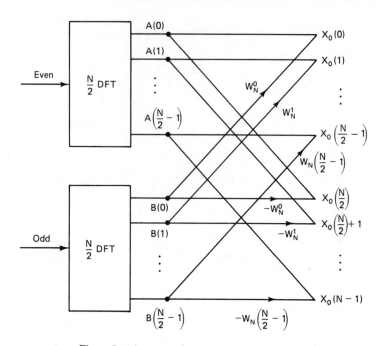

Figure 7-12 Representation of the N-point DFT.

of mathematical operations has been reduced by the FFT. For corresponding pairs of terms in the output sequence (e.g., k, $k + N/2$), we obtain the basic butterfly relationship from Equations 7–35c and 7–35d. This is shown in Figure 7–13.

In the ordering of the DFT discussed above, we initially separated the terms, starting with $n = 0$, by taking terms separated by $2^1 = 2$; for example, if $N = 8$, we have 0, 2, 4, 6; 1, 3, 5, 7. For the next step, we again divide each of the sequences, but now by taking terms separated by $2^2 = 4$; for example, if $N = 8$, the even sequence is divided into the two sequences 0, 4 and 2, 6, while the odd sequence is divided into the sequences 1, 5 and 3, 7. Each of

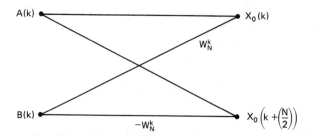

Figure 7-13 Basic butterfly relation of Figure 7-12.

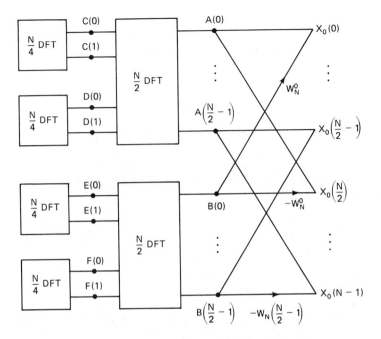

Figure 7-14 Representation of the N-point DFT.

the $(N/2)$-point sequences is separated into two $(N/4)$-point sequences, as indicated in Equations 7-36a through 7-36d and shown in Figure 7-14.

We now apply the procedure followed for separating the $X(k)$ into the $A(k)$ and $B(k)$ terms, $k = 0, 1, 2, \ldots, (N/2) - 1$. This yields Equation 7-36.

$$A(k) = C(k) + W_N^{2k} D(k) \tag{7-36a}$$

$$A\left(k + \frac{N}{4}\right) = C(k) - W_N^{2k} D(k) \tag{7-36b}$$

$$B(k) = E(k) + W_N^{2k} F(k) \tag{7-36c}$$

$$B\left(k + \frac{N}{4}\right) = E(k) - W_N^{2k} F(k) \tag{7-36d}$$

Figure 7-15 shows the flow graph diagram for $N = 8$, based on Equation 7-36. The W_N values have been omitted for clarity.

The technique described above is applied until we reach a point where we have reduced the N-point DFT to $N/2$ two-point DFTs. This is shown in Figure 7-16, where the W have again been omitted for clarity. The output sequence is in order, while the input sequence is in "bit-reversed" order. It is given this term as the binary representation of a given output is the reversed

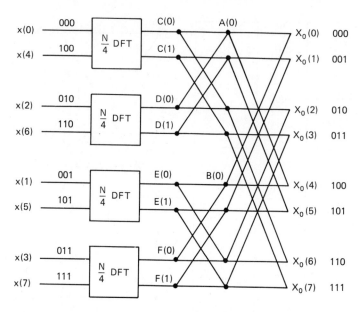

Figure 7-15 Flow graph diagram of Equation 7-36.

binary representation of the corresponding input term [e.g., $X(4) = X(100)$ ··· $x(001) = x(1)$.

The procedure described above is termed *decimation in time* (DIT). If we divide the sequence by breaking it into halves rather than odd-even sequences, it is termed *decimation in frequency* (DIF). We have just seen that the DIT reordering results in an "out-of-order" input sequence, but an in-order output sequence; the DIF is just the reverse: an in-order input and an out-of-order output. The procedure is demonstrated in the following example.

Example 7-6

Using the FFT algorithm, find the DFT of the sequence shown in Figure 7-17.

Solution

1. Using Figure 7-16, we obtain four two-point DFTs: C(0), C(1); D(0), D(1); E(0), E(1); and F(0), F(1), as shown in Figure 7-18.
2. Next, we find the four-point DFTs using Equation 7-36 and Figure 7-16: A(0) through A(3) and B(0) through B(3), shown in Figure 7-18b.
3. Finally, using the results of step 2 and Equation 7-35, we obtain the output sequence $X_0(k)$, Figure 7-18c.

7.5.2 FFT Algorithm of Cooley and Tukey

The algorithm prescribed by Cooley and Tukey is presented in a sequence of steps with an example following each to show its use. The algorithm will be presented for an eight-point FFT. The FFT for higher values of N can also be obtained using this algorithm.

Sec. 7.5   Fast Fourier Transform   263

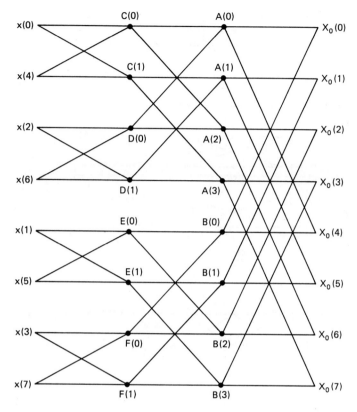

Figure 7–16   Eight-point FFT.

Step. 1.   Starting with Equation 7–13, express the values of k and n as binary numbers. This yields

$$X(4k_2 + 2k_1 + k_0)$$
$$= \sum_{n=0}^{N-1} x(4n_2 + 2n_1 + n_0) W_N^{(4k_2+2k_1+k_0)(4n_2+2n_1+n_0)} \qquad (7\text{–}37)$$

where $\{n_i\}$ and $\{k_i\}$ consist of ones and zeros.

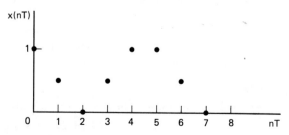

Figure 7–17   Input sequence for Example 7–6.

264 Chap. 7 The Discrete and the Fast Fourier Transforms

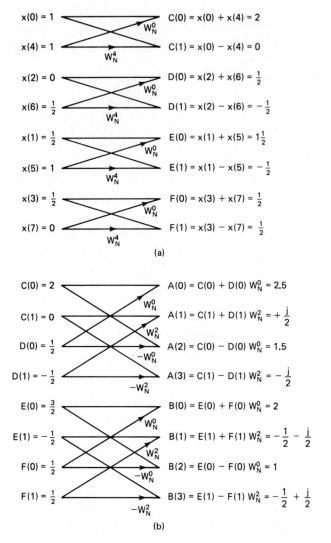

Figure 7–18 (a) Calculation for Example 7–6: C-F; (b) calculation for Example 7–6: A, B; (c) calculation for Example 7–6: X(k).

Step 2. Expand the exponent of W_N such that the terms in k are multiplying each of the n_i. This is shown below.

$$(4k_2 + 2k_1 + k_0)(4n_2 + 2n_1 + n_0)$$
$$= (4k_2 + 2k_1 + k_0)(4n_2) + (4k_2 + 2k_1 + k_0)(2n_1)$$
$$+ (4k_2 + 2k_1 + k_0)(n_0) \qquad (7\text{–}38)$$

Sec. 7.5   Fast Fourier Transform   265

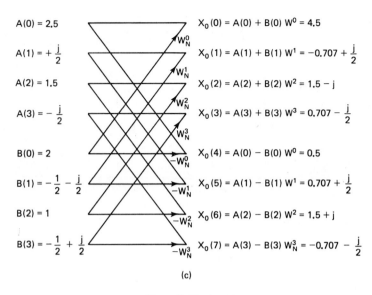

(c)

Figure 7–18 (cont.)

Step 3.   Express the sum over n as a multiple sum over the set $\{n_i\}$.

$$\sum_{n=0}^{N-1} = \sum_{n=0}^{8-1} = \sum_{n_0=0}^{1}\sum_{n_1=0}^{1}\sum_{n_2=0}^{1} \qquad (7\text{-}39)$$

Step 4.   Write the W_N^{kn} as a product of terms using the expansion of the exponent of step 2.

$$\begin{aligned}W_N^{kn} &= W_N^{(4k_2+2k_1+k_0)(4n_2+2n_1+n_0)} \\ &= W_N^{(4k_2+2k_1+k_0)(4n_2)}W_N^{(4k_2+2k_1+k_0)(2n_1)}W_N^{(4k_2+2k_1+k_0)(n_0)} \end{aligned} \qquad (7\text{-}40)$$

Step 5.   Based on the definition of W_N^{kn}, all powers of W_N that are a multiple of 8 must equal 1. This causes the exponent of step 4 to be reduced in the number of terms and appear as

$$W_N^{kn} = W_N^{4k_0n_2}W_N^{(2k_1+k_0)(2n_1)}W_N^{(4k_2+2k_1+k_0)(n_0)} \qquad (7\text{-}41)$$

Step 6.   The equation of step 5 is in a form that can yield the butterfly flow graph. This is done by expressing each of the sums as an intermediate value of X, labeled x_i. This is shown below.

$$\begin{aligned}X(4k_2+2k_1+k_0) = X(k_2k_1k_0) = \sum_{n_0=0}^{1}\Big\{\sum_{n_1=0}^{1}\Big\{\sum_{n_2=0}^{1}\{x_0(n_2n_1n_0) \\ \times W_N^{4k_0n_2}\Big\}W_N^{(2k_1+k_0)(2n_1)}\Big\}W_N^{(4k_2+2k_1+k_0)(n_0)}\Big\}\end{aligned} \qquad (7\text{-}42a)$$

where

$$\sum_{n_2=0}^{1} x_0(n_2 n_1 n_0) W_N^{4k_0 n_2}$$

is defined as $x_1(k_0 n_1 n_0)$;

$$X(k_2 k_1 k_0) = \sum_{n_0=0}^{1} \left\{ \sum_{n_1=0}^{1} \left\{ x_1(k_0 n_1 n_0) W_N^{(2k_1+k_0)(2n_1)} \right\} \right. \\ \left. \times W_N^{(4k_2+2k_1+k_0)n_0} \right\} \quad (7\text{-}42\text{b})$$

where

$$\sum_{n_1=0}^{1} x_1(k_0 n_1 n_0) W_N^{(2k_1+k_0)(2n_1)}$$

is defined as $x_2(k_0 k_1 n_0)$;

$$X(k_2 k_1 k_0) = \sum_{n_0=0}^{1} x_2(k_0 k_1 n_0) W_N^{(4k_2+2k_1+k_0)(n_0)} \quad (7\text{-}42\text{c})$$

$$X(k_2 k_1 k_0) = x_3(k_0 k_1 k_2) \quad (7\text{-}42\text{d})$$

The final result of this alogrithm is the bit-reversed values of the $X(k)$. The resulting flow graph is given in Figure 7–19.

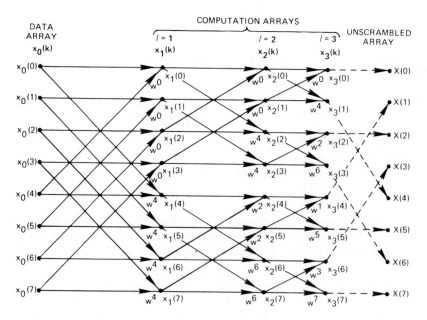

Figure 7–19 FFT flowgraph for $N = 8$. (From *The Fast Fourier Transform*, E. O. Brigham. Copyright © 1974, Prentice-Hall, Inc. Reprinted by permission of Prentice-Hall, Inc.)

Sec. 7.5 Fast Fourier Transform

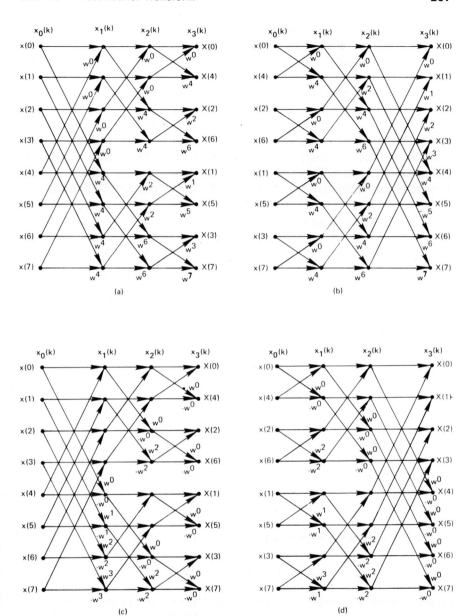

Figure 7-20 Canonical signal flowgraphs, $N = 8$. (From *the Fast Fourier Transform*, E. O. Brigham. Copyright © 1974, Prentice-Hall, Inc. Reprinted by permission of Prentice-Hall, Inc.)

Evaluating a nested sum is the equivalent of moving from x_i to x_{i+1} in the flow diagram. Each dot represents a summation while a multiplication is represented by a factor adjacent to the arrow. In going from x_0 (the input) to x_1, each term in the input is multiplied by either 1 (no multiplication) or W_8^4. This last term has the value -1 which implies a sign change. Therefore, the x_1 are determined without a single multiplication. In going from x_1 to x_2 and from x_2 to $x_3 = X_0$, the same situation occurs but to a lesser extent. The overall result is a reduction in the amount of multiplications required to evaluate the DFT resulting in a faster calculation. The reader is urged to calculate an 8-point DFT and FFT manually to observe the difference in the amount of multiplications required, with the reductions arising from the use of $N = 2^\gamma$ and the manipulations of W_N. Figure 7–19 is similar to Figure 7–16, the difference being the labeling of the intermediate steps. Figure 7–20 is a set of canonical signal flow graphs that are the result of several variations of the Cooley–Tukey algorithm. These graphs show the several I/O relations available for each variation of the algorithm. When writing a program to perform the FFT, the programmer can include a routine that will provide the output values in the desired sequence. Keep in mind that the main purpose of the FFT algorithm is to perform the DFT much more rapidly than a direct calculation (see Appendix B) when the number of points, N, becomes large.

7.6 APPLICATION OF THE FFT

Now that the FFT has been obtained in the DIT form, we are ready to use it. The FFT will not be calculated by hand, as it is much simpler to write a program and let a computer or processor do it. Just as there are many FFT versions available, so are there many programs. In this section the results of a program based on a subroutine by Conte and deBoor are presented. The actual program used is presented in Appendix C and a discussion of it is provided in the References.

7.6.1 Computer Application of the FFT

Using the program given in Appendix C, the FFT of the sequence given in Example 7–6 was determined. The sequence, although real, is entered as though it were a complex number as called for by the program. This will allow us to use the same subroutines to do the IDFT as were used to determine the DFT. Only a slight modification of the data is required rather than having to write a new subroutine in the program. The results are shown in Figure 7–21. Comparing these results with those of Example 7–6, we see that they compare quite favorably. It should be mentioned that the program can be modified to provide the magnitude and phase values of $H(k)$. These results represent the samples of the continuous frequency content of $h(nT)$, taken at

Sec. 7.6 Application of the FFT

Figure 7-21 FFT of an eight-point sequence, Figure 7-17.

```
 4.5000000    0.0000000
-0.7071066    0.5000001
 1.5000000   -0.9999999
 0.7071066   -0.5000000
 0.5000002    0.0000000
 0.7071067    0.5000000
 1.5000000    0.9999999
-0.7071068   -0.4999999
```

intervals of f_{sa}/N. If a more dense frequency sampling is required, a 16-point FFT can be obtained by zero-filling $h(nT)$ (i.e., adding eight zeros to it). This is shown in Figure 7-22. Using the program with $N = 16$, the results are presented in Figure 7-23. These represent samples taken at $f_{sa}/16$. Every other sample (even) is identical to the results of the eight-point FFT.

If finer sampling is desired, zero filling to obtain $N = 32, 64, \ldots$ can be done. Figure 7-24 shows the results of a 32-point FFT, where every fourth sample is the original eight-point sample.

7.6.2 Inverse DFT by FFT

Equation 7-43 presents the IDFT that provides the DT sequence for a given frequency response, $H(e^{j\omega T})$.

$$h(nT) = \frac{1}{N} \sum_{k=0}^{N-1} H(k) W_N^{-km} \qquad (7\text{-}43)$$

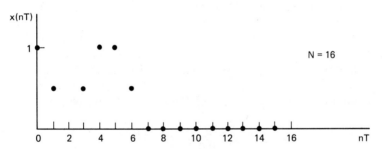

Figure 7-22 Zero filling of an eight-point sequence.

Figure 7-23 Output of a 16-point sequence, zero filled.

```
 4.5000000    0.0000000
 0.9170445   -2.9307144
-0.7071066    0.5000002
 2.0068350    0.7585317
 1.5000000   -1.0000001
 0.7002718   -0.5343612
 0.7071067   -0.5000001
 0.3758482   -0.2236075
 0.5000000    0.0000007
 0.3758491    0.2236078
 0.7071068    0.4999997
 0.7002718    0.5343617
 1.5000006    1.0000000
 2.0068347   -0.7585325
-0.7071068   -0.4999995
 0.9170459    2.9307144
```

```
4.5000000      0.0000000
3.3601463     -2.3758464
0.9170446     -2.9307144
-0.8316420    -1.4790329
-0.7071065     0.5000000
0.7373159      1.3659536
2.0068345      0.7585318
2.1669943     -0.3850608
1.5000000     -0.9999995
0.8645359     -0.8753948
0.7002725     -0.5343615
0.7723502     -0.4309439
0.7071061     -0.5000003
0.4935482     -0.4475026
0.3758483     -0.2236065
0.4367514     -0.0377523
0.5000012      0.0000005
0.4367521      0.0377531
0.3758493      0.2236067
0.4935490      0.4475022
0.7071064      0.4999997
0.7723501      0.4309433
0.7002724      0.5343614
0.8645360      0.8753947
1.5000002      0.9999993
2.1669946      0.3850605
2.0068343     -0.7585321
0.7373156     -1.3659532
-0.7071059    -0.4999994
-0.8316407     1.4790329
0.9170458      2.9307125
3.3601456      2.3758440
```

Figure 7–24 Output of a 32-point sequence, zero filled.

Comparing this equation with Equation 7–13, we see that the right-hand sides differ in that (1) $H(k)$, usually complex, is used, and (2) there is a minus sign in the exponent of W_N^{-kn}. Other than that, the two equations are identical. We can easily determine that the program for the inverse FFT (IFFT) can be obtained from the FFT simply by replacing each plus sign in the exponential with a minus sign, and vice versa. This can be readily done to obtain the IFFT.

Taking the complex conjugate of Equation 7–43, we obtain

$$h^*(nT) = \frac{1}{N} \sum_{k=0}^{N-1} H^*(k) W_N^{+kn} \qquad (7\text{--}44)$$

If $h(nT)$ is real, $h^*(nT) = h(nT)$ allowing us to use the FFT algorithm to generate the DT sequence, using the complex conjugate of the frequency samples as data. This saves the step of having to include a separate subroutine in the program to determine the IDFT.

To determine if this procedure is effective, we take the complex conjugate of the output frequency samples of Figures 7–21 and 7–23 and use them as the complex input to the FFT program. The results of these runs for the 8- and 16-point IFFT are given in Figures 7–25 and 7–26. In each case, we have obtained the original eight values. For the 16-point IFFT, the last eight values are zero, as the original sequence was zero-filled to obtain the 16-point FFT. If the IFFT of the 32-point DFT of Figure 7–24 were obtained, the

Sec. 7.6 Application of the FFT

Figure 7-25 IFFT of the Figure 7-21 sequence, of eight-points.

```
1.0000000    0.0001250
0.4998761    0.0000884
0.0001250    0.0000000
0.5003006   -0.0000884
1.0000000   -0.0001250
0.9996239   -0.0000884
0.4998750    0.0000000
0.0001994    0.0000883
```

Figure 7-26 IFFT of the sequence of Figure 7-23, 16-points.

```
 1.0000002    0.0000001
 0.5000001   -0.0000001
 0.0000000    0.0000001
 0.4999999   -0.0000001
 1.0000001    0.0000001
 1.0000000   -0.0000002
 0.4999998    0.0000001
 0.0000000    0.0000000
-0.0000001   -0.0000001
-0.0000001    0.0000000
 0.0000000    0.0000000
 0.0000000   -0.0000001
 0.0000001   -0.0000002
 0.0000001   -0.0000002
 0.0000000   -0.0000001
 0.0000000    0.0000000
```

last 24 values would be zero to reflect zero filling. In all instances, the first eight values of the IFFT are the original real samples, with any variation due to computer round-off error.

This discussion has shown that we now have a powerful tool in our hands (or computer) for performing calculations on discrete sequences. The following section presents some applications.

7.6.3 Discrete Convolution

Given the discrete transfer function of a system and a discrete input sequence, we would like to determine the system output. This can be readily done by (1) finding the FFT of the input sequence and transfer function, (2) multiplying them point by point to find $Y(k) = H(k)X(k)$, and (3) using the IFFT to find the output sequence, $y(nT)$. This is shown in Example 7-7.

Example 7-7
Using the sequences of Figure 7-27, find the system output $y(nT)$ using the FFT and IFFT.

Solution Using the FFT program as a subroutine, modifying the result of the FFT to yield the complex conjugate, and writing a subroutine that multiplies the FFTs of the initial sequence, a new program was generated that performs the required steps. The results are presented in Figure 7-28. It is left to the reader to perform the operation manually to check the results.

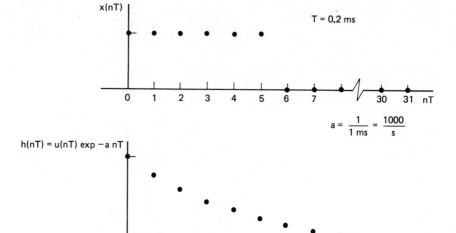

Figure 7–27 Sequences for Example 7–7.

```
1.000
1.819
2.489
3.038
3.488
3.856
3.157
2.585
2.117
1.733
1.418
1.161
0.951
0.778
0.637
0.522
0.428
0.350
0.286
0.234
0.191
0.156
0.127
0.104
0.085
0.069
0.052
0.037
0.025
0.015
0.007
0.000
```

Figure 7–28 Output of the convolution program.

Sec. 7.6   Application of the FFT

7.6.4 Discrete Correlation

As we did with the DFT, the cross- and autocorrelation values of two signals can be found. Using the approach of Section 5.7, Equations 5–20 through 5–25, we obtain the result for correlation.

Cross-correlation:
$$S_{12}(k) = S_1(k)S_2^*(k) \qquad (7\text{–}45\text{a})$$

Autocorrelation:.
$$S(k) = S_1(k)S_1^*(k) \qquad (7\text{–}45\text{b})$$

The FFT of each signal is calculated, the products are taken, and the IFFT is determined. The IFFT will yield the discrete correlation of the two signals. In either situation, the program that finds the FFT of the complex conjugate of a function is readily determined from the one provided in Appendix C by replacing the sign of the exponential term with its opposite. The same subroutine that performs the FFT also can be used for the IFFT by using the complex conjugate of the data, $S^*(k)$ rather than $S(k)$.

7.6.5 Digital Filter

Using the results of Chapter 6 (e.g., the procedure of Parks and McClellan), we can generate the impulse response, $h(nT)$, of a digital filter. These samples are used in the FFT program to obtain the discrete frequency response, $H(k)$. Multiplying these values by $X(k)$, the FFT of the input signal, we obtain the FFT of the sampled output, $Y(k)$. Using the IFFT, we then find the filtered output, $y(nT)$.

7.6.6 Efficient Use of the FFT

If we have two real sequences $g(nT)$ and $h(nT)$, both of length N, we can find the FFTs simultaneously by forming a complex sequence $a(nT) = g(nT) + jh(nT)$. Using this sequence as the input to the FFT routine, we obtain its FFT, $A(k)$ which is $G(k) + jH(k)$. Next, the sequence $a^*(nT) = g(nT) - jh(nT)$ is formed and its FFT is determined. This is given by

$$A^*(N - k) = G(k) - jH(k) \qquad (7\text{–}46)$$

$G(k)$ and $H(k)$ are extracted from $A(k)$ and $A^*(N - k)$ by forming Equations 7–47a and 7–47b.

$$G(k) = \frac{A(k) + A^*(N-k)}{2} \qquad (7\text{-}47a)$$

$$H(k) = \frac{A(k) - A^*(N-k)}{2j} \qquad (7\text{-}47b)$$

For lengthy sequences, the extra amount of arithmetic required is worth the increase in efficiency.

Efficient filtering can be performed if we are working with real time signals. If they are of equal length, such as adjacent blocks in the overlap-add technique, then a new complex sequence can be formed from them. This new sequence is convolved with the filter impulse response resulting in the convolution of each section. If we are given the real, N-length sequences $a(nT)$ and $b(nT)$ and form

$$d(nT) = a(nT) + jb(nT) \qquad (7\text{-}48a)$$

and convolve it with $h(nT)$, we have

$$h(nT)*d(nT) = h(nT)*a(nT) + jh(nT)*b(nT) \qquad (7\text{-}48b)$$

Using the FFT, we obtain

$$D(k) = \text{FFT}\{d(nT)\} \qquad (7\text{-}49a)$$

$$H(k) = \text{FFT}\{h(nT)\} \qquad (7\text{-}49b)$$

$$Y(k) = D(k)H(k) \qquad (7\text{-}49c)$$

$$\text{IFFT}\{Y(k)\} = h(nT)*d(nT) = y(nT) \qquad (7\text{-}49d)$$

The real part of $y(nT)$ is the convolution of $a(nT)$ with $h(nT)$ while its imaginary part is the convolution of $b(nT)$ with $h(nT)$.

The advantage of this procedure is a faster convolution of long sequences with a shorter filter sequence. Also, either two separate signals can be filtered simultaneously or we can have one signal convolved with two different filters, that is, where $a(nT)$ and $b(nT)$ represent the impulse responses of two filters such as a LP and a BP filter.

7.7 SUMMARY

This chapter has introduced the reader to the DFT and the FFT techniques for processing sampled signals.

PROBLEMS

1. Find the DFT of the sequence shown. $N = 8$.

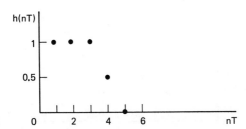

Figure P7-1

2. Find the DFT of the sequence shown. $N = 8$.

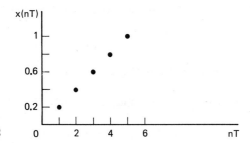

Figure P7-2

3. Using convolution in the sampled time domain, find $y(nT)$ for the sequences of problems 1 and 2.

4. Using the DFTs of Problems 1 and 2, find $Y(k)$.

5. Using the IDFT and the results of Problem 4, find $y(nT)$. Compare your results with that of Problem 3.

6. Repeat Problems 1 and 2 for $N = 16$.

7. Using the FFT program of Appendix C, repeat Problems 1 and 2.

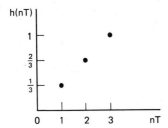

Figure P7-8

8. Find the eight-point DFT of the sequence shown.

9. Using the butterfly diagram, find the FFT for the sequence of Problem 1.

10. For the sequence shown, find $H(k)$ using the DFT and the FFT. Determine the number of computations in each case and compare them.

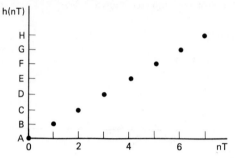

Figure P7–10

11. For a 16-point sequence, what would be the savings in calculations for the FFT over the DFT of Problem 10? What are the savings of a 128-point FFT over a DFT?

REFERENCES

BRIGHAM, O. E. *The Fast Fourier Transform.* Englewood Cliffs, N.J.: Prentice-Hall, 1974.

COOLEY, J. W. and TUKEY, J. W. "An Algorithm for the Machine Computation of Complex Fourier Series." *Math. Comp.,* 19, Apr., 1965: 297–301.

CONTE, S. D., and DEBOOR, C. *Elementary Numerical Analysis: An Algorithmic Approach.* New York: McGraw-Hill, 1980.

TRETTER, S. A. *Introduction to Discrete-Time Signal Processing.* New York, Wiley, 1976.

PROCESSOR INTEGRATED CIRCUITS

8

As of this point, the basic techniques and nomenclature of DSP have been introduced. It is now time to look at some of the hardware that is available to process the signal. These consist of integrated circuits (ICs) that will accept either the analog signal or its sample values in binary format and, after processing them, provide the result, also in binary format.

One of the main goals of a processor IC is to perform all the required operations in as short a time as possible, allowing the sampling to be performed at higher frequencies. This last fact allows wider-bandwidth signals (e.g., TV) to be processed using DSP techniques. We shall look at several ICs and the operations they perform on the signal, together with some of their applications.

8.1 AMD 29500 FAMILY

The AMD 29500 family of ICs, used for digital signal array processing, has been designed from the ground up with high-speed processing in mind. This is brought about by microprogramming, pipelining, and the use of parallel operations. In addition, the ICs are fabricated using a proprietary version of oxide-isolated bipolar technology in which an oxide-isolation region, instead of reverse-biased diffused junctions separates active devices. The end result of this technology is a device that is smaller, uses less power, and due to reduced capacitance, is faster. To bring about a further reduction of delays, emitter-coupled logic (ECL) is used for the internal circuitry while maintaining TTL circuitry for I/O. Figure 8–1 shows the block diagram of the signal processor. We will concentrate on the 29500 devices that are used in this processor.

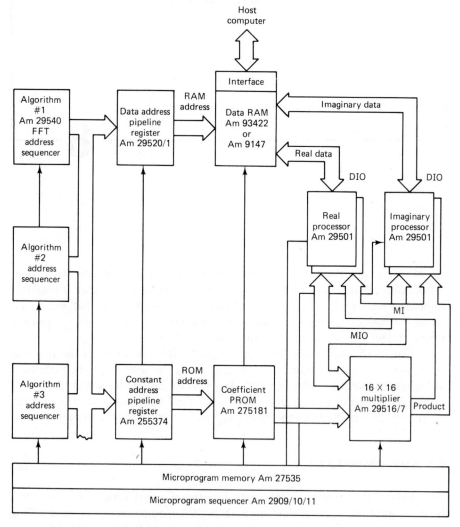

Figure 8–1 AMD 29500 signal processor. (Copyright © 1982. Advanced Micro Devices, Inc. Reprinted with permission of copyright owner. All rights reserved.)

Sec. 8.1   AMD 29500 Family

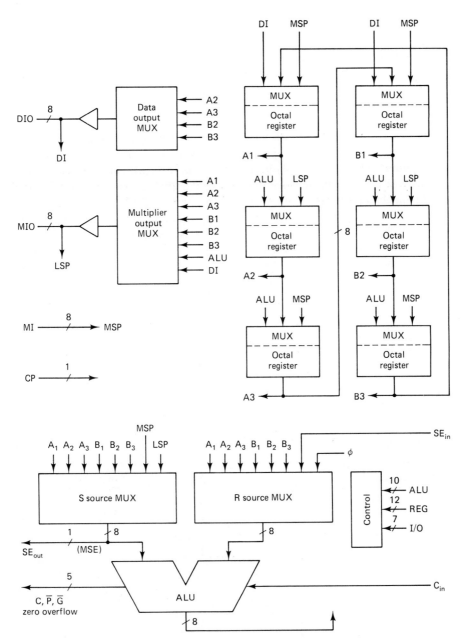

Figure 8–2 AMD 29501 microprogrammable signal processor. (Copyright © 1982. Advanced Micro Devices, Inc. Reprinted with permission of copyright owner. All rights reserved.)

8.1.1 Am 29501: Multiport Pipelined Processor

In a pipelined architecture, several instructions are executed simultaneously rather than sequentially. In the sequential mode, the arithmetic operations on data fetched from memory are performed one after the other. When one instruction is completed, the next one is started and executed. To speed up this process, registers are used to hold intermediate results during the execution of an instruction. This allows the following instruction to be initiated before the present one has been completed. In this manner, there is an increase in throughput, which brings about a substantial decrease in processing time. Figure 8–2 is a simplified block diagram of the 29501. The six registers used in the pipelining operation are shown in the upper right-hand corner. Figure 8–3 shows a nonpipelined organization versus the pipelined one. As can be seen for the pipeline architecture, each unit is being used at all times with many computations occurring at once rather than having to wait for the preceding stage to be completed.

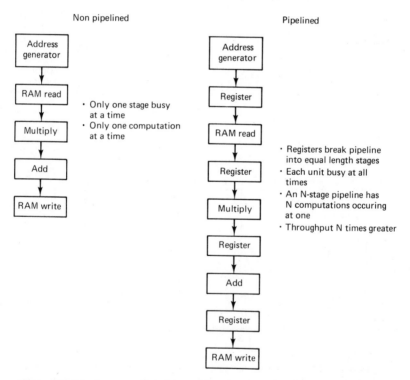

Figure 8–3 Pipelined vs. nonpipelined sequence. (Copyright © 1982. Advanced Micro Devices, Inc. Reprinted with permission of copyright owner. All rights reserved.)

8.1.2 Am 29516/517 16 x 16-Bit Parallel Multipliers

These units will accept two 16-bit binary numbers and produce a 32-bit product. The input registers are actually 17 bits in length, with the additional bit indicating whether the data are in 2's-complement or unsigned-number format. Figure 8-4 is the block diagram of the Am 29516.

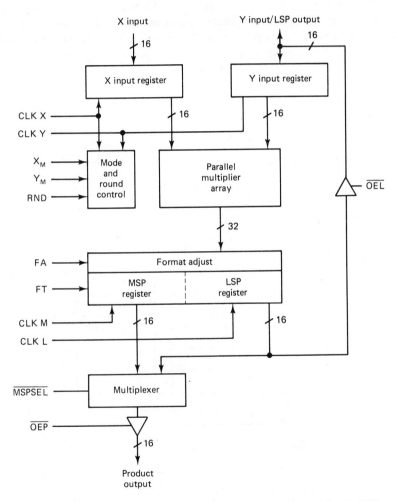

Figure 8-4 AMD 29516 16 × 16 bit parallel multipliers. (Copyright © 1982. Advanced Micro Devices, Inc. Reprinted with permission of copyright owner. All rights reserved.)

8.1.3 Am 29520/521 Multilevel Pipeline Registers

These units will provide temporary address or data storage. They are also capable of holding, transferring, and loading instructions. Figure 8-5 is a block diagram of the Am 29520. The structure allows for its use as a dual two-level pipeline or a single four-level pipeline.

8.1.4 Am 29540 Programmable FFT Address Sequences

This unit is capable of generating all the memory address necessary to perform the repetitive butterfly operations of the FFT. The data will be stored in the data RAM while the coefficients are stored in a ROM. The unit can be programmed to produce most common FFT address sequences by specifying:

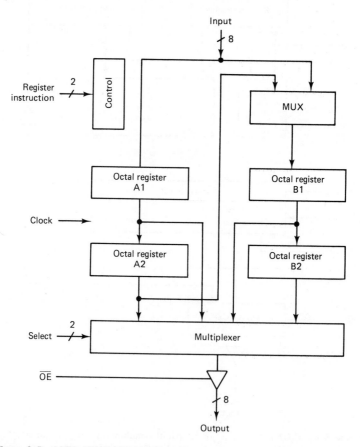

Figure 8-5 AMD 29520 Multilevel pipeline registers. (Copyright © 1982. Advanced Micro Devices, Inc. Reprinted with permission of copyright owner. All rights reserved.)

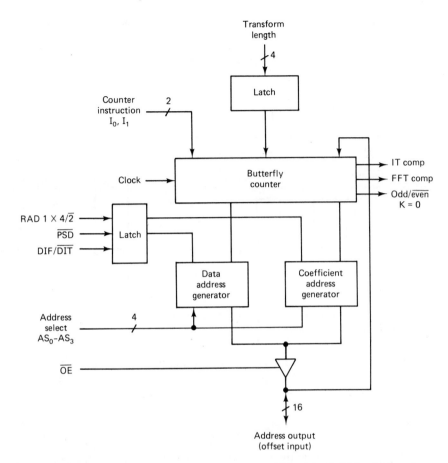

Figure 8-6 AMD 29540 FFT address generator. (Copyright © 1982. Advanced Micro Devices, Inc. Reprinted with permission of copyright owner. All rights reserved.)

radix-2 or radix-4; bit reversed input or output; and decimation in time or frequency. In addition, non-bit-reversing and real-valued input transforms are available. Figure 8-6 is the block diagram of the Am 29540. The 29500 family is a digital processor group and requires a digital input while providing a digital output. Additional circuitry is required to accept and provide analog signals.

8.2 INTEL 2920 SIGNAL PROCESSOR

Unlike the Am 29500 processor, the Intel 2920 is a single-chip IC that accepts an analog signal at one of four inputs and provides an analog output at one of eight output terminals. The output is from a digital-to-analog converter

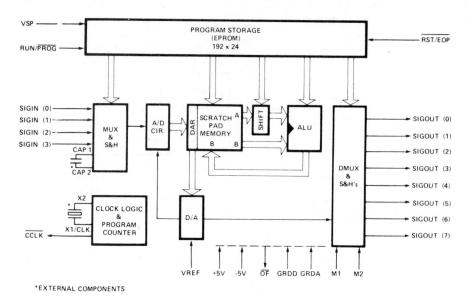

Figure 8-7 Block diagram of the Intel 2920 processor. (Reprinted with permission of Intel Corp., Santa Clara, CA.)

(DAC) and although continuous, has many jumps in the waveform values, giving rise to unwanted high-frequency components. These must be eliminated by use of an analog LP filter. This is termed the reconstruction filter and is discussed in Chapter 9.

Figure 8-7 is a block diagram of the Intel 2920 signal processor. The chip contains a 192-word x 24-bit erasable programmable read-only memory (EPROM), which contains the set of instructions that performs the processing. In addition, there is a two-port, 40-word x 25-bit RAM. For discussion, the unit is divided into a digital section and an analog section.

8.2.1 Analog Section

In these units, the desired analog signal is selected and converted to digital form. After processing, it is converted back to analog and then presented at one of eight outputs. Figure 8-8 shows the block diagram of the 2920 analog section.

Input multiplexer. Based on a 5-bit code from the EPROM, one of the four inputs will be selected for conversion.

Track-and-hold. An external capacitor is used in the TH section and is shared by all four inputs. The 5-bit code connects one of the four input channels to the capacitor for sampling. Its value is typically between

Sec. 8.2 Intel 2920 Signal Processor

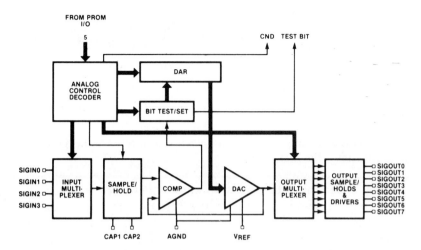

Figure 8–8 Block diagram of the Intel 2920 analog section. (Reprinted with permission of Intel Corp., Santa Clara, CA.)

100 and 1000 pF, depending on the clock rate. Once the acquisition time is determined, the user can determine how many instructions are required to sample an input.

ADC. Using the principles of successive approximation, the analog signal selected is converted to digital form using the comparator, bit test/set, DAR, and the DAC. This conversion is under the control of the instruction set contained in the EPROM. When the conversion is completed, the DAR contains the 9-bit number in 2's-complement form.

DAC. When an output command is received, the 9 bits of the DAR are converted to 1's-complement form and applied to the DAC. An external reference voltage between 1 and 2 V is required, causing the LSB to have a weight of $V_{REF}/256$. The resulting analog signal is then applied to the next unit.

Output multiplexer. Based on the instruction from the EPROM, one of the eight outputs will be selected.

Output track-and-hold and drivers. A TH circuit is used at the output to reduce the pulse width of the DAC output. This is shown in Figure 8–9. The effect of the reduced pulse width (i.e., $\tau/T < 1$) is to reduce the gain roll-off within the signal bandwidth. If this action is not taken, considerable distortion will appear in the output. The reader should refer to Section 3.5 to see the effect of sampling pulse width on the output spectrum.

Figure 8–9 Block diagram of the output circuit.

8.2.2 Digital Section

The operations that process the samples are performed in the digital section. Figure 8–10 shows the digital section.

RAM. Data are stored in a 40-word x 25-bit RAM which has two ports, each independently controlled by the EPROM. The A port can provide only data placed in storage, whereas the B port can receive and provide data. In addition, 16 constants are stored in this area and can be accessed through the A port. These constants range from $-8/8$ to $+7/8$ and are used in the various algorithms set up for processing the signal.

Also located in this region is the DAR, which is used to hold either the data from the ADC or the value to be provided to the DAC. Once the ADC conversion is completed, 16-LSB ones are concatenated to the DAR value and the result is then placed in the RAM. The RAM address associated with the DAR is 40, with 0–39 being associated with storage. Both A and B ports have access to this data, the A port on a "read-only" basis, while the B port is a read/write terminal.

Scaler. The scaler multiplies the output of the A port by a range of from 2^2 down to 2^{-13} upon a 4-bit command from the EPROM. The multiplication is accomplished by either a left or right shift of the value provided by port A and has the effect of multiplying the value by the desired power of 2.

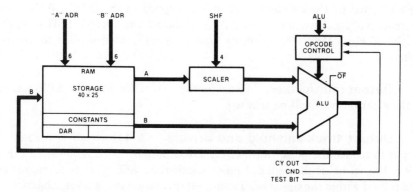

Figure 8–10 Block diagram of the Intel 2920 digital section. (Reprinted with permission of Intel Corp., Santa Clara, CA.)

Sec. 8.3   NEC μP7720   **287**

Arithmetic logic unit. The arithmetic logic unit (ALU) receives the scaled A-port value and the direct B-port value and performs one of several arithmetic or logic operations on them. The result is stored in the RAM through the B port, in address 0–39. If data are to be placed at the output, they are stored at address 40, where the nine MSBs are then passed to the DAR and thence to the DAC.

The ALU instruction set is a 3-bit code allowing for eight arithmetic/logic instructions. There are eight additional operations that are conditional based on the value of a tested bit. The full instruction set is presented in the following section.

8.2.3 Instruction Word

Figure 8–11 shows the instruction word and the corresponding mnemonic codes.

Program format. The first 3 bits (MSBs) provide the eight instructions for the ALU; the next 12 bits provide the destination and source addresses in the RAM; the next 4 bits provide the amount by which the A-port output is to be shifted, and the last 5 bits indicate the analog instruction.

Instruction mnemonics. The instruction set for the digital and analog sections is shown in Figure 8–11 together with the operations. The set is in assembly language format and must be converted to machine language format (binary or object code). The representation of the object code is in the form of a six-figure hexadecimal number.

8.3 NEC μP7720

This IC signal processor is fabricated using NMOS technology and uses an architecture that enables it to perform many calculations very rapidly. The μP7720 accepts and provides digital format data, requiring a DAC and ADC to process an analog signal. Figure 8–12 is the block diagram of the μP7720.

The technology and architecture allow the unit to handle sampling rates up to 10 kHz, making it readily applicable for voice-band applications, which typically include frequencies from 300 to 3000 kHz. The blocks making up the unit are considered in the following sections.

8.3.1 Multiplier

This is a unit that will multiply two 16 x 16-bit numbers in 2's-complement form and provide a 31-bit output. The memory has a pair of input registers (K, L) that hold the two operands, and a pair of registers (M, N) that hold

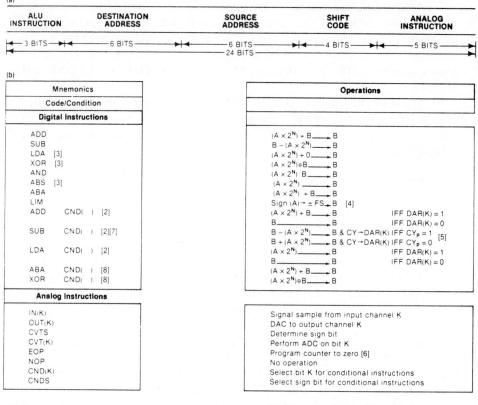

Figure 8–11 Intel 2920 instruction set and word. (Reprinted with permission of Intel Corp., Santa Clara, CA.)

the product. This product is available for operations at the next cycle as the M, N registers are directly connected to the ALU through a multiplexer.

8.3.2 Memory

This consists of three separate units, each performing a separate function.

Instruction ROM. The μP7720 can store up to 512 instructions of 23 bits each in the instruction ROM. It is addressed by a 9-bit program counter

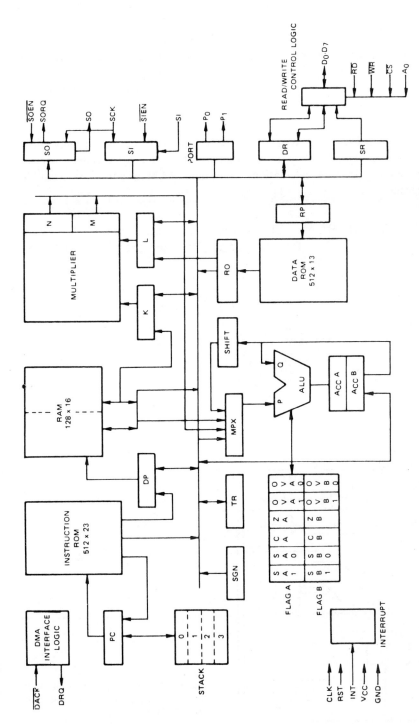

Figure 8-12 Block diagram of NEC μP7720 processor. (Reprinted with permission of NEC Electronics, Inc., Mountain View, CA.)

(PC) which is part of a stack. The stack configuration is a 4 x 9 LIFO register that allows nested interrupts and subroutines (no more than four), as it can store the subroutine/interrupt return address. As the instructions are 23 bits wide, they are all single instructions and can be obtained in a single fetch, reducing processing time.

Data RAM. This is a 128 x 16-bit storage unit that usually holds variable data: I/O information, intermediate results, and other variables required of the signal processing calculation. The data RAM is addressed from the DP register.

Coefficients ROM. This unit is usually used to provide data tables, conversion tables, and coefficients used in the signal processing calculation. Its organization is 512 x 13 bits.

```
NOP                    Increment Acc
OR/AND/XOR             Complement Acc
Subtract               1-bit Right Shift
ADD                    1-bit Left Shift
Subtract with borrow   2-bit Left Shift
ADD with Carry         4-bit Left Shift
Decrement Acc          8-bit Exchange
```

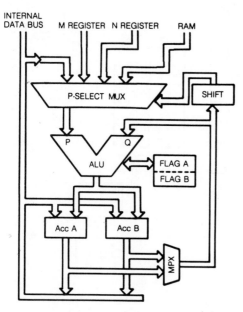

Figure 8–13 ALU and peripherals block diagram. (Reprinted with permission of NEC Electronics, Inc., Mountain View, CA.)

8.3.3 ALU

The ALU is a 16-bit, 2's-complement unit capable of executing 16 distinct operations on any internal register. These are listed in Figure 8–13. The output of the ALU can be stored in one of two accumulators, the choice being dependent on which one was used as a source of data. The use of dual accumulators decreases processing time further by allowing overlapping of data movement and calculations.

8.3.4 I/O

This is divided into two sections: serial for interfacing with analog-to-digital devices and parallel for interfacing with an external processor such as the Intel 80xx series.

8.4 TRW 10xx FAMILY

Among the devices fabricated by TRW are the TDC1008, 1009, and 1010. These are, respectively, 8-, 12-, 16-bit multiplier-accumulators (MACs). These devices will accept two n-bit numbers, x_{in} and y_{in}, and provide their product in either 2's-complement or unsigned-magnitude format. In addition, the product register's contents can be added to or subtracted from the next product, if desired. Figure 8–14 is the combined block diagram of the TDC 10xx LSI

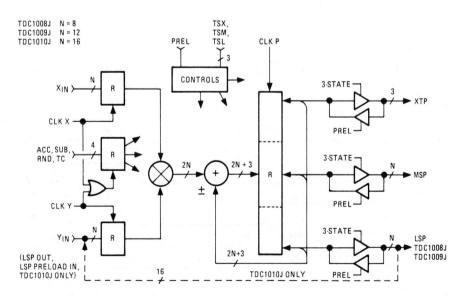

Figure 8–14 Block diagram of TRW TDC10xx MAC. (Reprinted with permission of TRW Inc., La Jolla, CA.)

OPERATION DESCRIPTION
(FOR THE TDC1008J, THE TDC1009J, AND THE TDC1010J)

Data Input

Data inputs are loaded into the X register and Y register at the rising edge of CLK X and CLK Y, respectively. Note: the LSP (P_0 to P_{15}) outputs are timeshared with the Y data inputs (Y_0 to Y_{15}) on the TDC 1010J.

Data Output

The product is divided into Least Significant Product (LSP), Most Significant Product (MSP), and Extended Product (XTP). The product generated is loaded into the output registers at the rising edge of CLK P.

Preload Data (PD)

Data is applied externally to output pins, to initialize output register to a given value at the rising edge of CLK P. (See Preload Control) Note: the PD LSP word (PD_0 to PD_{15}) inputs are timeshared with the Y data inputs on the TDC1010J.

Two's Complement Control (TC)

When TC is high, the inputs are two's complement numbers. When TC is low, the inputs are unsigned magnitude numbers. The TC signal is loaded into the TC register at the rising edge of (CLK X OR CLK Y). The TC signal must be valid over the same period that the input data is valid.

Round Control (RND)

When RND is high, a "1" is added to the most significant bit of the LSP in the multiplier array to round up the product in MSP and XTP rather than truncate it. The RND signal is loaded into the RND register at the rising edge of (CLK X OR CLK Y). The RND signal must be valid over the same period that the input data is valid.

Accumulation Control (ACC)

When ACC is high, the contents of the output registers are added to the next product generated and their sum is stored back into the output registers at the rising edge of the next CLK P. When ACC is low, multiplication without accumulation is performed and the next product generated will be stored into the output registers directly. The ACC signal is loaded into the ACC register at the rising edge of (CLK X OR CLK Y). The ACC signal must be valid over the same period that the input data is valid.

Subtraction Control (SUB)

When ACC and SUB are both high, the contents of the output register are subtracted from the next product generated and the difference is stored back into the output registers at the rising edge of the next CLK P. When ACC is high and SUB is low, addition instead of subtraction is performed. The SUB signal is loaded into the SUB register at the rising edge of (CLK X OR CLK Y). The SUB signal must be valid over the same period that the input data is valid. When ACC is low, SUB is a "Don't Care" pin.

Preload Control (PREL)

All output buffers are at high impedance (output disabled) when PREL is high. When TSL, TSM, or TSX is also high, the initial contents of their corresponding output register can be preset to the preload data applied to the output pins at the rising edge of CLK P. If TSL, TSM, or TSX is low while PREL is high, the contents of the respective output register remain unchanged while the output drivers remain at high impedance. PREL, TSL, TSM, TSX must be valid over the same period that the preload input data is valid.

Three State Least, Most, and Extended Control (TSL, TSM, TSX)

The LSP, MSP, or XTP output buffers are at high impedance (output disabled) when TSL, TSM, or TSX is high, respectively. These are direct, nonregistered control signals. The output drivers are enabled when TSL, TSM, or TSX is low, and PREL is low.

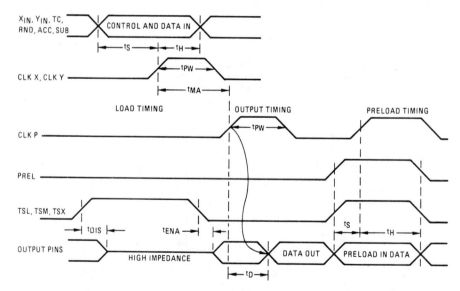

Figure 8-15 TRW TDC 10xx operation description (a) and timing diagram (b). (Reprinted with permission of TRW Inc., La Jolla, CA.)

Sec. 8.5 TMS32010 Digital Signal Processor

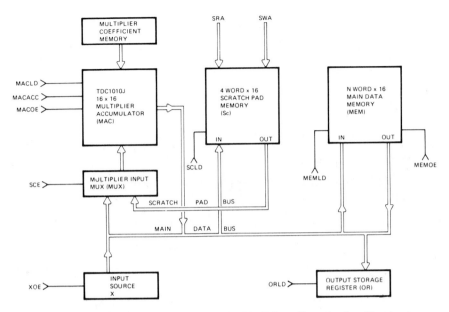

Figure 8-16 Use of the TRW TDC1010J MAC in a filter circuit. (Reprinted with permission of TRW Inc., La Jolla, CA.)

MACs. The "operation description" provided by the TRW data sheet is included together with the timing diagram in Figure 8-15. This device will take the contents of the two input registers, provide their product, and depending on the control signal, place it in the output register so that the data can be accessed.

A typical use of the TDC1010J is in the construction of a canonical, second-order, recursive filter. This is shown in Figure 8-16. The MAC receives the input signal to be filtered and, from a memory, the multiplier coefficients. The 4-word x 16-bit scratch-pad memory is used for temporary storage of data, while the main memory provides temporary storage for the data to be processed.

This circuit is usually under the control of a microprocessor. A direct memory access (DMA) function will allow the CPU to transfer data from the main memory system to the main data memory of the filter. Once the transfer has taken place, the processor can go on to other operations, leaving the multiplication to the MAC.

8.5 TMS32010 DIGITAL SIGNAL PROCESSOR

This processor chip from Texas Instruments is a 16/32-bit microcomputer that can execute 5 million instructions per second. Figure 8-17 is the block diagram of the TMS320M10 device. The TMS320M10 differs from the TMS32010 by the addition of a 3-kilobyte on-chip-program ROM.

functional block diagram

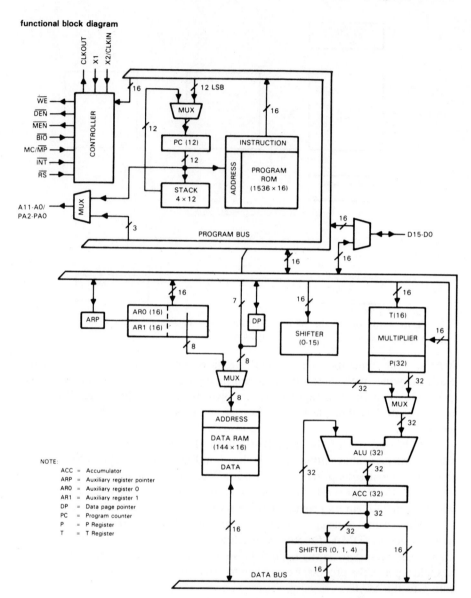

Figure 8–17 Chip architecture of the TMS320M10 processor. (Reproduced with permission of Texas Instruments.)

8.5.1 Chip Architecture

It has been found that separating the program memory locations from data memory locations results in an increase of operation speed, that is, the time it takes to perform an operation is reduced. This allows a complete overlap of instruction fetch and execution and is called *Harvard architecture*. TI has modified this form of architecture to allow transfers between program and data memory locations. They state that this modification ". . . allows transfers between program and data spaces, thereby increasing the flexibility of the device. This modification permits coefficients stored in program memory to be read into RAM, eliminating the need for a separate coefficient ROM."

32-bit ALU/accumulator. The ALU is 32-bits wide and will support double-precision arithmetic. The sources of the operands are either the data RAM or data contained in the instructions. The result of the ALU operation is stored in a 32-bit accumulator.

Shifters. The chip contains two shifters. The first can left-shift the data from 0 to 15 places; that is, the data can be multiplied by any power of two from 2^0 up to 2^{15}. This operation is performed prior to being fed to the accumulator as an operand. The second shifter will left-shift the upper half of the accumulator output by zero, one, or four places while it is being stored in the RAM.

16 × 16-bit parallel multiplier. Using hardware, the TMS32010 can perform a 2's-complement multiplication of two 16-bit numbers in a single 200-ns cycle. One number is stored in the T register while the other factor in the product comes from the data bus. The 32-bit product is stored in the P register. The source of the factors is either data, memory, or from an instruction word.

Input/output. Data are applied to the device through a 16-bit data bus. The address bus consists of 12 lines with the lower 3 bits used to select one of eight input ports or one of eight output ports.

8.5.2 Instruction Set

There are approximately 59 instructions that allow the TMS32010 to perform its functions. These are divided into (1) accumulator; (2) auxiliary register and data page pointer; (3) T register, P register, and multiply; (4) branch; (5) control; and (6) I/O and data memory instructions.

8.6 SUMMARY

In this chapter we have looked at some of the IC devices available for performing several of the operations used in DSP at the time this book was written. There is no doubt that advances in IC fabrication will lead to entirely new families of devices to perform all the functions of digital signal processing.

REFERENCES

2920 Analog Signal Processor Design Handbook. Santa Clara, Calif.: Intel Corporation, 1980.

TMS32010 User's Guide. Dallas: Texas Instruments, 1983.

μPD7220 Signal Processing Interface (SPI) Technical Manual. Natick, Mass.: NEC Electronics, 1983.

DIGITAL-TO-ANALOG CONVERTER

9

When the digital portion of the signal processing is complete, the output must be converted to analog format if it is to be utilized directly. Output devices such as the loudspeaker or cathode ray tube (CRT) require an analog signal at their input. The loudspeaker will convert the analog signal to sound energy while the CRT will convert it to light energy. To perform this conversion from digital to analog format, use is made of a digital-to-analog converter (DAC). This unit accepts an n-bit digital signal at its input and provides a voltage proportional to the value of the n-bit signal. As the conversion is made at a fixed rate and as the binary signal is quantized, the output will consist of a sequence of continuous, flat-topped pulses, connected by jumps. This is shown in Figure 9–1 where each level of output voltage corresponds to a particular binary signal at the input to the DAC. As this signal is quite distorted from the desired one due to the jumps caused by the changing binary input, some analog processing is required. This will be discussed later in this chapter. For now, we look at the devices used to perform the conversion from digital to analog format.

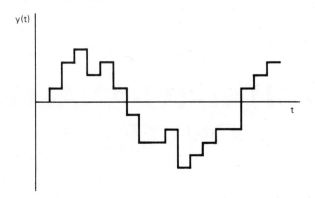

Figure 9-1 Analog output of the DAC.

9.1 DAC CIRCUITS

There are several circuits available for converting a binary signal to analog format. The two most widely used are presented here: the weighted current source and the R-$2R$ ladder network.

9.1.1 Weighted-Current Source DAC

The value of the current in a transistor is controlled by the value of a resistor in series with it and a constant reference voltage across the resistor. If the resistor value is changed by a power of 2, the current will change by the same factor, but in the opposite direction. The basic circuit of the weighted-current source DAC is shown in Figure 9-2.

Assuming that all transistors are identical, their base–emitter drops will be the same and the voltage across each resistor is V_{REF}. This yields a current in each resistor given by $V_{REF}/2^n R$, where n will correspond to a bit position. If the bit value connected to the emitter of the transistor by a diode is high (= 1), the diode is off and current will flow in the transistor; if the bit value is low (= 0), the diode is on and the current will be diverted from the transistor. In this manner, only transistors whose associated bit value is 1 will contribute to the sum at the node. To obtain the proper weighting for each bit, the resistor associated with the MSB has the smallest value, R, and each succeeding bit will have a resistor whose value is increased by a factor of 2 until the value of the LSB's resistor is 2^n times the value of R.

Depending on which bits are 1, the current through the transistors will sum at the node and be applied to the op-amp circuit. Since the current in the feedback resistor, R_F, is the same as the current sum at the node, and as the left side of R_F is at virtual ground, we have

$$V_{OUT} = -R_F I_{OUT} \qquad (9\text{--}1)$$

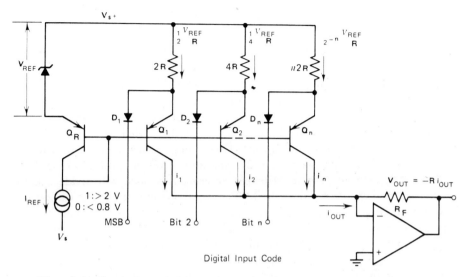

Figure 9–2 Circuit diagram of the weighted current source DAC. (Adapted from *A Users Handbook of D/A and A/D Converters,* E. R. Hnatek. Copyright © 1976, John Wiley and Sons, Inc. Reprinted by permission of John Wiley and Sons, Inc.)

Since I_{OUT} is the sum of the weighted currents, the output is given by

$$V_{\text{OUT}} = -R_F \sum I_n \quad (9\text{--}2)$$

For all values of n, I_n is determined by V_{REF}, the value of R, and the bit position to yield

$$V_{\text{OUT}} = -R_F \sum_{j=0}^{n-1} \frac{V_{\text{REF}} A_j}{2^j R} \quad (9\text{--}3)$$

where A_j is the value of the jth bit, 1 or 0. This is written as

$$V_{\text{OUT}} = -\frac{R_F V_{\text{REF}}}{R} \sum_{j=0}^{n-1} \frac{A_j}{2^j} \quad (9\text{--}4)$$

which shows the effect of the reference voltage on the output. If the reference voltage is external to the DAC, the device is termed a *multiplying DAC*.

Although this circuit realization of a DAC has a relatively low conversion time due to the fact that transistor current is switched, not turned on or off, problems can arise if n is large due to the resistor network which requires a range from R to $2^n R$. This wide range is difficult to obtain and shows drift effects with temperature swings: changes of resistance with temperature are not proportional, causing the ratio to vary.

9.1.2 R-2R Ladder Network

By using resistors whose ratio is either 1:1 or 2:1, the problem concerning the resistor values and their drift with temperature can be greatly reduced. Figure 9-3 shows the simplified circuit of the R-$2R$ ladder DAC. The position of each switch is determined by its associated bit value: if it is 0, the switch connects the resistor to ground; if it is 1, the resistor is connected to the reference. If we start on the right, the LSB point, R_A is in parallel with $R_B = R$ and when added to R_C yields $2R$ again. Continuing in this manner, we can see that each node will have a resistor $2R$ and an equivalent resistor, whose value is also $2R$, connected to it. Current entering this node will divide equally into each branch. Looking at the MSB node, the reference current will divide, with one-half going to each branch. As the current going to the right reaches the next MSB node, it will divide in half for each branch. This results in the resistor whose value is $2R$ receiving from left to right, $I_{\text{REF}}/2$, $I_{\text{REF}}/4$, ..., $I_{\text{REF}}/2^{n-1}$. Depending on the position of the switches, these currents will be summed at the input to the op amp. As before, the output voltage will be given as

$$V_{\text{OUT}} = -\frac{V_{\text{REF}} R_F}{R} \sum_{j=0}^{n-1} \frac{A_j}{2^j} \tag{9-5}$$

This circuit can be made to have a high level of accuracy over wide temperature variations by using lasers to fix the value of the resistors in a thin-film network. By choosing low values for R, the speed of the device's conversion can be increased.

Regardless of the circuit used to convert the digital signal, use is made

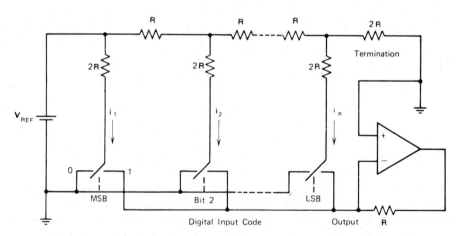

Figure 9-3 R-2R ladder network of the DAC. (Adapted from *A Users Handbook of D/A and A/D Converters*, E. R. Hnatek. Copyright © 1976, John Wiley and Sons, Inc. Reprinted by permission of John Wiley and Sons, Inc.)

Sec. 9.2   Commercial DACs   301

of a reference voltage. If it is internal, it is usually obtained from a zener diode in series with a forward-biased diode for temperature compensation. This provides the accuracy and stability required for the data conversion.

9.2 COMMERCIAL DACs

Although the circuits described above can be fabricated using discrete components, the problems that arise with matching resistors and temperature drift become quite formidable. Add to this the fact that many commercial devices are available, and it becomes clear that the purchased DAC is the choice. With this in mind, it is worthwhile to look at the devices presently available.

9.2.1 AD7520

The AD7520 is a CMOS 10-bit monolithic DAC that uses the R-$2R$ ladder network circuit. The switches are CMOS transistors whose geometry is chosen to provide on-resistance increases by a power of 2 for the first 6 bits, while the on-resistance remains constant for the last bits. With this technique, the linearity of the conversion is achieved, and over fairly wide temperature ranges. The value for R in the ladder network is chosen as 10 kΩ and is obtained by depositing a silicon–chromium mixture in the CMOS die.

The user of the AD7520 must provide the external op amp and a reference voltage. In unipolar conversion (also called two-quadrant multiplication) the output is given by Equation 9–5. As the feedback resistor for the op amp (internally provided) is 10 kΩ, Equation 9–5 becomes

$$V_{OUT} = - V_{REF} \sum_{j=0}^{n-1} \frac{A_j}{2^j} \qquad (9\text{–}6)$$

As n is 10, this can be rewritten as

$$V_{OUT} = - V_{REF} \frac{N}{2^n} \qquad (9\text{–}7)$$

where N is the decimal value of the binary input. This implies that the range of V_{OUT} is from 0 V (all binary values 0) to $-1023/1024$ V (all binary values 1). A change by 1 LSB will correspond to $V_{REF}/1024$. The data sheets for the AD7520 are presented in Figure 9–4.

9.2.2 AD370/371

The AD370/371 is a low-power, 12-bit DAC that contains a precision, high-speed FET input op amp together with a low-drift reference. The 370 provides a bipolar output range (-10 to $+10$ V), while the 371 provides a

CMOS 10- & 12-Bit Monolithic Multiplying D/A Converters
AD7520, AD7521

FEATURES
AD7520: 10-Bit Resolution
AD7521: 12-Bit Resolution
Linearity: 8-, 9- and 10-Bit
Nonlinearity Tempco: 2ppm of FSR/°C
Low Power Dissipation: 20mW
Current Settling Time: 500ns
Feedthrough Error: 1/2LSB @ 100kHz
TTL/DTL/CMOS Compatible

Note: AD7533 is recommended for new 10-bit designs.
AD7541, AD7542 or AD7543 is recommended for new 12-bit designs.

AD7520, AD7521 FUNCTIONAL BLOCK DIAGRAM

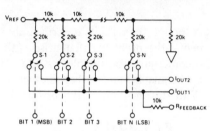

AD7520: N=10
AD7521: N=12
Logic: A switch is closed to I_{OUT1} for its digital input in a "HIGH" state.

GENERAL DESCRIPTION

The AD7520 (AD7521) is a low cost, monolithic 10-bit (12-bit) multiplying digital-to-analog converter packaged in a 16-pin (18-pin) DIP. The devices use advanced CMOS and thin wilm technologies providing up to 10-bit accuracy with TTL/DTL/CMOS compatibility.

The AD7520 (AD7521) operates from +5V to +15V supply and dissipates only 20mW, including the ladder network.

Typical AD7520 (AD7521) applications include: digital/analog multiplication, CRT character generation, programmable power supplies, digitally controlled gain circuits, etc.

PIN CONFIGURATIONS

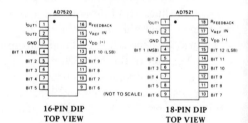

16-PIN DIP TOP VIEW   18-PIN DIP TOP VIEW

PACKAGE IDENTIFICATION[1]

Suffix D: Ceramic DIP Package
           AD7520: (D16B)
           AD7521: (D18B)

Suffix N: Plastic DIP Package
           AD7520 (N16B)
           AD7521 (N18B)

ORDERING INFORMATION

Nonlinearity	Temperature Range		
	0 to +70°C	-25°C to +85°C	-55°C to +125°C
0.2% (8-Bit)	AD7520JN AD7521JN	AD7520JD AD7521JD	AD7520SD AD7521SD
0.1% (9-Bit)	AD7520KN AD7521KN	AD7520KD AD7521KD	AD7520TD AD7521TD
0.05% (10-Bit)	AD7520LN AD7521LN	AD7520LD AD7521LD	AD7520UD AD7521UD

[1] See Section 20 for package outline information.

Figure 9–4 AD7520 data sheets. (Courtesy of Analog Devices, Two Technology Way, Norwood, MA.)

Sec. 9.2 Commercial DACs

SPECIFICATIONS (V_{DD} = +15, V_{REF} = +10V, T_A = +25°C unless otherwise noted)

PARAMETER	AD7520	AD7521	TEST CONDITIONS
DC ACCURACY[1]			
Resolution	10 Bits	12 Bits	
Nonlinearity (See Figure 5)			
	J, 0.2% of FSR max (8 Bit)	*	S,T,U: over −55°C to +125°C
	S, 0.2% of FSR max (8 Bit)	*	−10V ≤ V_{REF} ≤ +10V
	K, 0.1% of FSR max (9 Bit)	*	
	T, 0.1% of FSR max (9 Bit)	*	
	L, 0.05% of FSR max (10 Bit)	*	
	U, 0.05% of FSR max (10 Bit)	*	
Nonlinearity Tempco	2ppm of FSR/°C max	*	−10V ≤ V_{REF} ≤ +10V
Gain Error[2]	0.3% of FSR typ	*	−10V ≤ V_{REF} ≤ +10V
Gain Error Tempco[2]	10ppm of FSR/°C max	*	−10V ≤ V_{REF} ≤ +10V
Output Leakage Current			
(either output)	200nA max	*	Over specified temperature range
Power Supply Rejection	50ppm of FSR%/°C typ	*	
(See Figure 6)			
AC ACCURACY			To 0.05% of FSR
Output Current Settling Time	500ns typ	*	All digital inputs low to high
(See Figure 10)			and high to low
Feedthrough Error (See Figure 9)[4]	10mV p-p max		V_{REF} = 20V p-p, 100kHz
		*	All digital inputs low
REFERENCE INPUT			
Input Resistance[3]	5kΩ min	*	
	10kΩ typ	*	
	20kΩ max	*	
ANALOG OUTPUT			
Output Capacitance I_{OUT1}	120pF typ	*	All digital inputs high
(See Figure 8) I_{OUT2}	37pF typ	*	All digital inputs high
I_{OUT1}	37pF typ	*	All digital inputs low
I_{OUT2}	120pF typ	*	All digital inputs low
Output Noise (both outputs)	Equivalent to 10kΩ typ	*	
(See Figure 7)	Johnson noise		
DIGITAL INPUTS[5]			
Low State Threshold	0.8V max	*	Over specified temperature range
High State Threshold	2.4V min	*	Over specified temperature range
Input Current (low to high state)	1µA typ	*	Over specified temperature range
Input Coding	Binary	*	See Tables 1 & 2 under Applications
POWER REQUIREMENTS			
Power Supply Voltage Range	+5V to +15V	*	
I_{DD}	5nA typ	*	All digital inputs at GND
	2mA max	*	All digital inputs high or low
Total Dissipation (Including ladder)	20mW typ	*	

NOTES:
[1] Full scale range (FSR) is 10V for unipolar mode and ±10V for bipolar mode.
[2] Using the internal $R_{FEEDBACK}$
[3] Digital input levels should not go below ground or exceed the positive supply voltage, otherwise damage may occur.
[4] To minimize feedthrough with the ceramic package, the user must ground the metal lid. If the lid is not grounded, then the feedthrough is 10mV typical and 30mV maximum.
[5] Ladder and feedback resistor tempco is approximately −150ppm/°C.
Specifications subject to change without notice.

Figure 9-4 (cont.)

ABSOLUTE MAXIMUM RATINGS
(T_A = +25°C unless otherwise noted)

V_{DD} (to GND) . . . . +17V
V_{REF} (to GND). . . . . ±25V
Digital Input Voltage Range . . . . V_{DD} to GND
Output Voltage (Pin 1, Pin 2) . . . . -100mV to V_{DD}
Power Dissipation (package)
   up to +75°C. 450mW
   derates above +75°C by . . . . 6mW/°C
Operating Temperature
   JN, KN, LN Versions. . . . . 0 to +70°C
   JD, KD, LD Versions. . . . . -25°C to +85°C
   SD, TD, UD Versions. . . . . -55°C to +125°C
Storage Temperature . . . . -65°C to +150°C

CAUTION:
1. Do not apply voltages higher than V_{DD} or less than GND potential on any terminal except V_{REF}.
2. The digital control inputs are zener protected; however, permanent damage may occur on unconnected units under high energy electrostatic fields. Keep unused units in conductive foam at all times.

TYPICAL PERFORMANCE CURVES
T_A = +25°C, V_{DD} = +15V unless otherwise noted

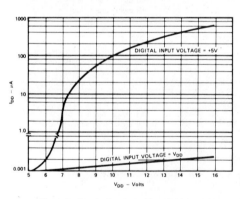

Figure 1. Supply Current vs. Supply Voltage

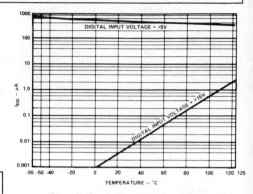

Figure 2. Supply Current vs. Temperature

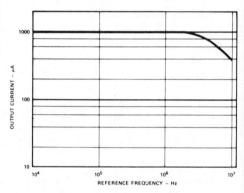

Figure 3. Output Current Bandwidth

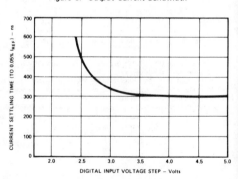

Figure 4. Output Current Settling Time vs. Digital Input Voltage

Figure 9–4 (cont.)

Sec. 9.2 Commercial DACs

TEST CIRCUITS

Note: The following test circuits apply for the AD7520. Similar circuits can be used for the AD7521.

DC PARAMETERS

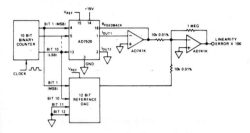

Figure 5. Nonlinearity

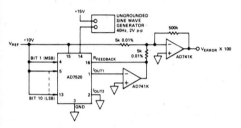

Figure 6. Power Supply Rejection

AC PARAMETERS

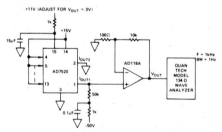

Figure 7. Noise

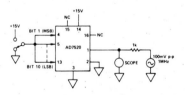

Figure 8. Output Capacitance

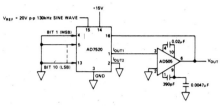

Figure 9. Feedthrough Error

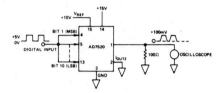

Figure 10. Output Current Settling Time

TERMINOLOGY

RESOLUTION: Value of the LSB. For example, a unipolar converter with n bits has a resolution of (2^{-n}) (V_{REF}). A bipolar converter of n bits has a resolution of $[2^{-(n-1)}]$ [V_{REF}]. Resolution in no way implies linearity.

SETTLING TIME: Time required for the output function of the DAC to settle to within 1/2 LSB for a given digital input stimulus, i.e., 0 to Full Scale.

GAIN: Ratio of the DAC's operational amplifier output voltage to the input voltage.

FEEDTHROUGH ERROR: Error caused by capacitive coupling from V_{REF} to output with all switches OFF.

OUTPUT CAPACITANCE: Capacity from I_{OUT1} and I_{OUT2} terminals to ground.

OUTPUT LEAKAGE CURRENT: Current which appears on I_{OUT1} terminal with all digital inputs LOW or on I_{OUT2} terminal when all inputs are HIGH.

Figure 9–4 (cont.)

CIRCUIT DESCRIPTION

GENERAL CIRCUIT INFORMATION

The AD7520 (AD7521), a 10-bit (12-bit) multiplying D/A converter, consists of a highly stable thin film R-2R ladder and ten (twelve) CMOS current switches on a monolithic chip. Most applications require the addition of only an output operational amplifier and a voltage or current reference.

The simplified D/A circuit is shown in Figure 11. An inverted R-2R ladder structure is used — that is, the binarily weighted currents are switched between the I_{OUT1} and I_{OUT2} bus lines, thus maintaining a constant current in each ladder leg independent of the switch state.

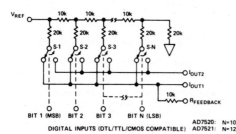

(Switches shown for Inputs "High")

Figure 11. AD7520 (AD7521) Functional Diagram

One of the CMOS current switches is shown in Figure 12. The geometries of devices 1, 2 and 3 are optimized to make the digital control inputs DTL/TTL/CMOS compatible over the full military temperature range. The input stage drives two inverters (devices 4, 5, 6 and 7) which in turn drive the two output N-channels. The "ON" resistances of the first six switches are binarily scaled so the voltage drop across each switch is the same. For example, switch-1 of Figure 12 was designed for an "ON" resistance of 20 ohms, switch-2 of 40 ohms and so on. For a 10V reference input, the current through switch 1 is 0.5mA, the current through switch 2 is 0.25mA, and so on, thus maintaining a constant 10mV drop across each switch. It is essential that each switch voltage drop be equal if the binarily weighted current division property of the ladder is to be maintained.

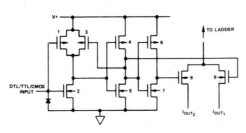

Figure 12. CMOS Switch

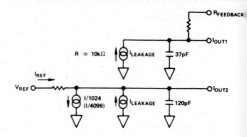

Figure 13. AD7520 (AD7521) Equivalent Circuit— All Digital Inputs Low

EQUIVALENT CIRCUIT ANALYSIS

The equivalent circuits for all digital inputs high and all digital inputs low are shown in Figures 13 and 14. In Figure 13 with all digital inputs low, the reference current is switched to I_{OUT2}. The current source $I_{LEAKAGE}$ is composed of surface and junction leakages to the substrate while the $\frac{1}{1024}\left(\frac{1}{4096}\right)$ current source represents a constant 1-bit current drain through the termination resistor on the R-2R ladder. The "ON" capacitance of the output N channel switch is 120pF as shown on the I_{OUT2} terminal. The "OFF" switch capacitance is 37pF, as shown on the I_{OUT1} terminal. Analysis of the circuit for all digital inputs high, as shown in Figure 14 is similar to Figure 13; however, the "ON" switches are now on terminal I_{OUT1}, hence the 120pF at that terminal.

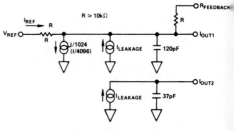

Figure 14. AD7520 (AD7521) Equivalent Circuit— All Digital Inputs High

Figure 9–4 *(cont.)*

APPLICATIONS

UNIPOLAR BINARY OPERATION

Figure 15 shows the circuit connections required for unipolar operation using the AD7520. Since V_{REF} can assume either positive or negative values, the circuit is also capable of 2-quadrant multiplication. The input code/output range table for unipolar binary operation is shown in Table 1.

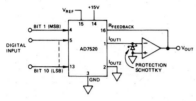

Figure 15. Unipolar Binary Operation (2-Quadrant Multiplication)

(Note: Protection Schottky not required with TRIFET output amplifier such as AD542 or AD544).

Zero Offset Adjustment

1. Tie all digital inputs to the AD7520 (AD7521) to GND potential.
2. Adjust the offset trimpot on the output operational amplifier for 0V ±1mV at V_{OUT}.

Gain Adjustment

1. Tie all digital inputs to the AD7520 (AD7521) to the +15V supply.
2. To increase V_{OUT}, place a resistor R in series with the amplifier output terminal and $R_{FEEDBACK}$ of the AD7520 (AD7521) (R = 0 to 500Ω)
3. To decrease V_{OUT}, place a resistor R in series with V_{REF}. (R = 0 to 500Ω)

DIGITAL INPUT	ANALOG OUTPUT
1 1 1 1 1 1 1 1 1 1	$-V_{REF} (1 - 2^{-10})$
1 0 0 0 0 0 0 0 0 1	$-V_{REF} (1/2 + 2^{-10})$
1 0 0 0 0 0 0 0 0 0	$\dfrac{-V_{REF}}{2}$
0 1 1 1 1 1 1 1 1 1	$-V_{REF} (1/2 - 2^{-10})$
0 0 0 0 0 0 0 0 0 1	$-V_{REF} (2^{-10})$
0 0 0 0 0 0 0 0 0 0	0

NOTE: 1 LSB = $2^{-10} V_{REF}$

Table 1. Code Table – Unipolar Binary Operation

BIPOLAR (OFFSET BINARY) OPERATION

Figure 16 illustrates the AD7520 connected for bipolar operation. Since the digital input can accept bipolar numbers and V_{REF} can accept a bipolar analog input, the circuit can perform a 4-quadrant multiplying function. Input coding is offset binary (modified 2's complement) as shown in Table 2.

DIGITAL INPUT	ANALOG OUTPUT
1 1 1 1 1 1 1 1 1 1	$-V_{REF} (1 - 2^{-9})$
1 0 0 0 0 0 0 0 0 1	$-V_{REF} (2^{-9})$
1 0 0 0 0 0 0 0 0 0	0
0 1 1 1 1 1 1 1 1 1	$V_{REF} (2^{-9})$
0 0 0 0 0 0 0 0 0 1	$V_{REF} (1 - 2^{-9})$
0 0 0 0 0 0 0 0 0 0	V_{REF}

NOTE: 1 LSB = $2^{-9} V_{REF}$

Table 2. Code Table – Bipolar (Offset Binary) Operation

When a switch's control input is a Logical "1", that switch's current is steered to I_{OUT1}, forcing the output of amplifier #1 to

$$V_{OUT} = - (I_{OUT1}) (10k)$$

where 10k is the value of the feedback resistor.

A Logical "0" on the control input steers the switch's current to I_{OUT2}, which is terminated into the summing junction of amplifier #2. Resistors R1 and R2 need not track the internal R-2R circuitry; however, they should closely match each other to insure that the voltage at amplifier #2's output will force a current into R2 which is equal in magnitude but opposite in polarity to the current at I_{OUT2}. This creates a push-pull effect which halves the resolution but doubles the output range for changes in the digital input.

With the MSB a Logic "1" and all other bits a Logic "0", a 1/2 LSB difference current exists between I_{OUT1} and I_{OUT2}, creating an offset of 1/2 LSB. To shift the transfer curve to zero, resistor R-9 is used to sum 1/2 LSB of current into the I_{OUT2} terminal.

Offset Adjustment

1. Make V_{REF} approximately +10V.
2. Tie all digital inputs to +15V (Logic "1").
3. Adjust amplifier #2 offset trimpot for 0V ±1mV at amplifier #2 output.
4. Tie MSB (Bit 1) to +15V, all other bits to ground.
5. Adjust amplifier #1 offset trimpot for 0V ±1mV at V_{OUT}.

Gain Adjustment

Gain adjustment is the same as for unipolar operation.

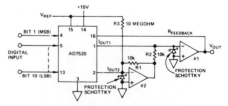

Figure 16. Bipolar Operation (4-Quadrant Multiplication)

(Note: Protection Schottky not required with TRIFET output amplifier such as AD542 or AD544).

Figure 9-4 (cont.)

DYNAMIC PERFORMANCE CHARACTERISTICS

The following circuits and associated waveforms illustrate the dynamic performance which can be expected using some commonly available IC amplifiers. All settling times are to 0.05% of 10V.

AD741J

Small Signal Bandwidth: 180kHz
Settling Time: 20µs

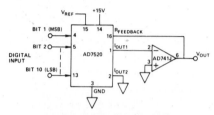

Figure 17. DAC Circuit Using AD741J

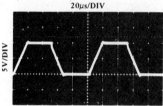

Figure 18. Output Waveform

AD518K

Small Signal Bandwidth: 1.0MHz
Settling Time: 6.0µs

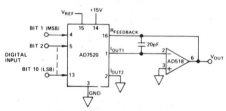

Figure 19. DAC Circuit Using AD518K

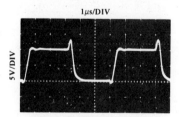

Figure 20. Output Waveform

AD505J

Small Signal Bandwidth: 1.0MHz
Settling Time: 2.5µs

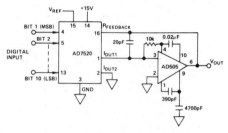

Figure 21. DAC Circuit Using AD505J

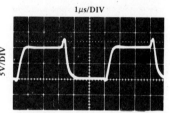

Figure 22. Output Waveform

AD509K

Small Signal Bandwidth: 1.6MHz
Settling Time: 2.0µs

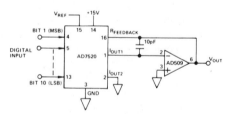

Figure 23. DAC Circuit Using AD509K

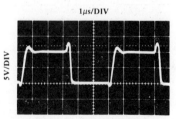

Figure 24. Output Waveform

Figure 9-4 (*cont.*)

ANALOG/DIGITAL DIVISION

With the AD7520 connected in its normal multiplying configuration as shown in Figure 15, the transfer function is

$$V_O = -V_{IN}\left(\frac{A_1}{2^1} + \frac{A_2}{2^2} + \frac{A_3}{2^3} + \cdots \frac{A_n}{2^n}\right)$$

where the coefficients A_x assume a value of 1 for an ON bit and 0 for an OFF bit.

By connecting the DAC in the feedback of an operational amplifier, as shown in Figure 25, the transfer function becomes

$$V_O = \left(\frac{-V_{IN}}{\frac{A_1}{2^1} + \frac{A_2}{2^2} + \frac{A_3}{2^3} + \cdots \frac{A_n}{2^n}}\right)$$

This is division of an analog variable (V_{IN}) by a digital word. With all bits off, the amplifier saturates to its bound, since division by zero isn't defined. With the LSB (Bit 10) ON, the gain is 1024. With all bits ON, the gain is 1 (±1 LSB).

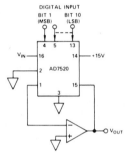

Figure 25. Analog/Digital Divider

Figure 9–4 (cont.)

unipolar range (0 to +10 V). Figure 9–5 presents the data sheets for the AD370/371 DAC.

9.2.3 BB DAC 71

This DAC is a 16-bit converter that provides an internal reference and op amp. In addition, it also allows the user to connect an external precision op amp to reduce the effect of lead and contact resistances on the output. This can be a problem with a 16-bit converter, as the value of the internal reference is (typically) 6.3 V. This implies that the weight of the LSB is $V_{REF}/2^{16}$ or 96 μV. If lead and contact resistance are not taken into consideration, the drops across them might exceed the weight of the LSB, causing errors in the output level. Figure 9–6 shows the data sheets for the BB DAC 71.

Complete, Low Power 12-Bit D/A Converter
AD370/AD371

FEATURES
Bipolar Voltage Output: AD370
Unipolar Voltage Output: AD371
Low Power: 150mW max
Linearity: ±1/2LSB, −55°C to +125°C (S Version)
TTL/CMOS Compatible
Compatible with Standard 18-Pin DAC Configurations
Hermetic 18-Pin DIP ("D" Package)
Factory Trimmed Gain and Offset: No External Adjustments Required
Monotonicity Guaranteed Over Specified Temperature Range

AD370/AD371 FUNCTIONAL BLOCK DIAGRAM

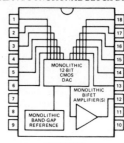

18-PIN DIP

PRODUCT DESCRIPTION
The AD370/AD371 is a complete 12-bit digital-to-analog converter fabricated with the most advanced monolithic and hybrid technologies. The design incorporates a low power monolithic CMOS DAC, precision high speed FET-input operational amplifiers and a low drift reference available in a hermetically sealed package. This innovative design results in significant performance advantages over conventional designs. The integral package-substrate combined with a lower chip count improves reliability over the standard low power hybrids of this type.

The converters come in two versions: AD370 with a bipolar output voltage range (−10V to +10V) and AD371 with a unipolar output voltage range (0 to +10V). Each device is internally laser trimmed for gain and offset to provide adjustment-free operation with only ±0.05% absolute error. The FET input operational amplifiers optimize the speed vs. power trade-off by settling to 1/2LSB from a full scale transition in 35μs with maximum total power dissipation of only 150mW. The low power monolithic CMOS DAC employs a current-switched silicon-chromium R-2R ladder to ensure that monotonicity is maintained over the full temperature range.

The AD370/AD371 "K" and "S" features ±1/2LSB maximum linearity error. Its rated temperature ranges are 0 to +70°C for the "J" and "K" versions and −55°C to +125°C for the "S" version.

PRODUCT HIGHLIGHTS
1. The AD370/AD371 replaces other devices of this type with significant increases in performance.
2. Reduced power consumption requirements (150mW max) result in improved stability and shorter warm-up time.
3. The precision output amplifiers and CMOS DAC have been optimized to settle within 1/2LSB for a full scale transition in 35μs.
4. Reduced chip count and integral package-subsrate improve reliability.
5. System performance upgrading is possible without redesign.
6. Internally laser trimmed—no gain or offset adjustments are required for specified accuracy.
7. The device is available in a hermetically-sealed ceramic 18 lead dual-in-line package. Processing to MIL-STD-883 Class B is available.
8. The AD370/AD371 is a second-source for 18-pin 12-bit DACs of the same configuration.

Figure 9–5 AD370 data sheets. (Courtesy of Analog Devices, Two Technology Way, Norwood, MA.)

Sec. 9.2   Commercial DACs   311

SPECIFICATIONS (typical at T_A = +25°C, V_S = ±15 Volts unless otherwise noted)

Model	AD370J	AD370K	AD371J	AD371K	AD370S[1]	AD371S[1]	Units
RANGE	-10 to +10	*	0 to +10	**	*	**	Volts
CODE	OCBI	*	CBI	**	*	**	
LINEARITY ERROR							
+25°C	±1	±1/2	±1	±1/2	±1/2	±1/2	LSB[2] max
$T_{min} - T_{max}$	±1	±1/2	±1	±1/2	±1/2	±1/2	LSB[2] max
ABSOLUTE ACCURACY							
+25°C	±0.05	*	*	*	*	*	% of FSR[3] max
$T_{min} - T_{max}$	±0.2	*	*	*	±0.3	±0.3	% of FSR[3] max
OFFSET ERROR							
+25°C	±5	*	±1	**	*	**	mV max
FULL SCALE SETTLING TIME							
TO ±1/2 LSB	25(35 max)	*	*	*	*	*	µs
INTERNAL REFERENCE	+10.0	*	*	*	*	*	Volts
DIGITAL INPUTS							
V_{INH}	2.0	*	*	*	*	*	Volts min
V_{INL}	0.8	*	*	*	*	*	Volts max
INPUT LEAKAGE CURRENT	±1.0	*	*	*	*	*	µA
INPUT CAPACITANCE	8	*	*	*	*	*	pF
POWER SUPPLY REJECTION RATIO							
+15V Supply	0.01	*	*	*	*	*	% FSR[3] /% V_S max
-15V Supply	0.01	*	*	*	*	*	% FSR[3] /% V_S max
POWER SUPPLY CURRENTS							
+15V Supply	3.5(5 max)	*	*	*	*	*	mA max
-15V Supply	2.5(4 max)	*	*	*	*	*	mA max
POWER DISSIPATION	105(150 max)	*	*	*	*	*	mW
TEMPERATURE RANGE	0 to +70	*	*	*	-55 to +125	***	°C

NOTES
[1] Also available to MIL-STD-883, Level B.
[2] LSB: Least Significant Bit
[3] FSR: Full Scale Range
*Specifications same as AD370J.
**Specifications same as AD371J.
***Specifications same as AD370S.

Specifications subject to change without notice.

Figure 9-5 (cont.)

ABSOLUTE MAXIMUM RATINGS
(T_A = +25°C unless otherwise noted)

V_{DD} (to GND) .... +17V
V_{EE} (to GND) .... −17V
Digital Input Voltage Range .... V_{DD} to GND
Storage Temperature .... −65°C to +150°C

CAUTION − ELECTROSTATIC SENSITIVE DEVICES

The digital control inputs are zener protected; however permanent damage may occur on unconnected units under high energy electrostatic fields. Keep unused units in conductive foam at all times.

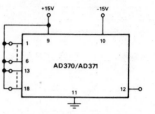

Figure 3. Burn-In Circuit

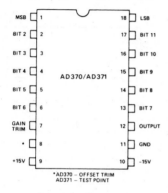

Figure 1. Pin Designations

DIGITAL INPUT	NOMINAL ANALOG OUTPUT
1 1 1 1 1 1 1 1 1 1 1 1	0
1 0 0 0 0 0 0 0 0 0 0 0	4.9975 Volts
0 1 1 1 1 1 1 1 1 1 1 1	5.0000 Volts
0 0 0 0 0 0 0 0 0 0 0 0	9.9975 Volts

Table 1. Code Table for the AD371 (CBI)

DIGITAL INPUT	NOMINAL ANALOG OUTPUT
1 1 1 1 1 1 1 1 1 1 1 1	−10.000 Volts
1 0 0 0 0 0 0 0 0 0 0 1	−0.0097 Volt
1 0 0 0 0 0 0 0 0 0 0 0	−0.0048 Volt
0 1 1 1 1 1 1 1 1 1 1 1	0
0 0 0 0 0 0 0 0 0 0 0 0	9.9952 Volts

Table 2. Code Table for the AD370 (OCBI)

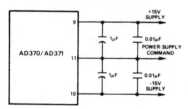

Figure 2. Power Supply Decoupling

Figure 9–5 (*cont.*)

ACCURACY

Accuracy error of a D/A converter is defined as the difference between the analog output that is expected when a given digital code is applied and the output that is actually measured with that code applied to the converter. Accuracy error can be caused by gain error, zero error or linearity error. The initial accuracy of the AD370/AD371 is trimmed to within 0.05% of full scale by laser trimming the gain and zero errors. Of the error specifications, the linearity error specification is the most important since it cannot be corrected by the user. The linearity error of the AD370/AD371 is specified over its entire temperature range. This means that the analog output will not vary by more than ±1/2LSB maximum from an ideal straight line drawn between the end points (inputs all "1s" and all "0s") over the specified operating temperature range of 0 to +70°C for the "K" version and -55°C to +125°C for the "S" version.

The absolute accuracy of the AD370/AD371 has been guaranteed to ±0.05% of full scale by internal factory trim of the gain and offset. External gain and offset adjustment terminals have been made available to allow fine adjustment to the ±0.012% accuracy level. The measurement system used to calibrate the output should be capable of stable resolution of 1/4LSB in the regions of zero and full scale. The adjustment procedure, described below, should be carefully followed to assure optimum converter performance.

The proper connections for the offset and gain adjustments are shown in Figure 4. For the AD371 full-scale calibration apply a digital input of all "1s" and adjust the gain potentiometer to +9.9975 volts (see Table 1).

The offset adjustment of the AD370 is made at the half-scale code. Adjust the offset potentiometer until 0.000V is obtained on the output. The full-scale adjustment is made at the negative endpoint or a code of all "1s". Adjust the gain potentiometer until -10.000 volts is obtained on the output.

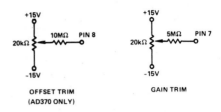

Figure 4. Optional External Trims

SETTLING TIME

Settling time for the AD370/AD371 is the total time required for the output to settle within ±1/2LSB band around its final value after a change in input (including slew time). The settling time specification is given for a full scale step which is 20V for the AD370 and 10V for the AD371.

IMPROVED SECOND SOURCE

The substrate design of the AD370/AD371 provides for complete pin-for-pin compatibility with other 18-pin DACs,; Hybrid Systems Corp. DAC340, DAC350 series and Micro Networks Corp. MN360, MN370, MN3200 series 18-pin 12-bit digital-to-analog converters all share the same pin configuration except for pin 7 and pin 8 (see Table 3). The AD370/AD371 is a superior direct replacement for these devices where the function of pins 7 and 8 allow. The versatility designed into the AD370/AD371 allows the function of pin 7 and pin 8 to be configured to exactly second source each of the other units. Information on other second source devices with 4 quadrant multiplying capability is available from Analog Devices.

Analog Devices	Hybrid Systems		Micro Networks		
AD370KD	DAC346C-12BPG		MN360	MN370	MN3211
AD371KD	DAC346C-12UP		MN362	MN371	MN3210
AD370SD	DAC347LPC-12G	DAC356C-12	MN360H	MN370H	
AD370SD/883B	DAC347LPS-12G	DAC356B-12			
AD371SD	DAC347LPC-12U	DAC356LPC-12	MN362H	MN371H	
AD371SD/883B	DAC347LPB-12U	DAC356LPB-12			

Table 3. Cross Reference

AD370/AD371 ORDERING GUIDE

Model	Package	Package Style[1]	Linearity	Output Voltage Range	Operating Temperature Range
AD370JN	Polymer Seal	HY18A	1LSB	-10V to +10V	0 to +70°C
AD370JD	Hermetic	HY18A	1LSB	-10V to +10V	0 to +70°C
AD371JN	Polymer Seal	HY18A	1LSB	0 to +10V	0 to +70°C
AD371JD	Hermetic	HY18A	1LSB	0 to +10V	0 to +70°C
AD370KN	Polymer Seal	HY18A	1/2LSB	-10V to +10V	0 to +70°C
AD370KD	Hermetic	HY18A	1/2LSB	-10V to +10V	0 to +70°C
AD371KN	Polymer Seal	HY18A	1/2LSB	0 to +10V	0 to +70°C
AD371KD	Hermetic	HY18A	1/2LSB	0 to +10V	0 to +70°C
AD370SD	Hermetic	HY18A	1/2LSB	-10V to +10V	-55°C to +125°C
AD370SD/883B	Hermetic	HY18A	1/2LSB	-10V to +10V	-55°C to +125°C
AD371SD	Hermetic	HY18A	1/2LSB	0 to +10V	-55°C to +125°C
AD371SD/883B	Hermetic	HY18A	1/2LSB	0 to +10V	-55°C to +125°C

[1] See Section 20 for package outline information.

Figure 9-5 (cont.)

DAC71

High Resolution
16-BIT DIGITAL-TO-ANALOG CONVERTER

FEATURES

- 16-BIT, 4-DIGIT RESOLUTION
- ±0.003% MAXIMUM NONLINEARITY
- LOW DRIFT ±7ppm/°C, (TYPICAL)
- CURRENT AND VOLTAGE MODELS
- LOW COST

DESCRIPTION

The DAC71 is a high quality 16-bit hybrid IC D/A converter available in a 24-pin dual-in-line ceramic package.

The DAC71 with internal reference and optional output amplifier offers a maximum linearity error of ±0.003% of FSR at room temperature and a maximum gain drift of ±15ppm/°C over a temperature range of 0°C to +70°C.

Three basic models accept complementary 16-bit binary or complementary 4-digit BCD TTL-compatible input codes.

Packaged within the DAC71 are fast-settling switches and stable laser-trimmed thin-film resistors that let you select output voltages 0 to +10V (CSB and CCD) or ±10V (COB) and output currents of ±1mA or 0 to -2mA. Input power is ±15VDC and +5VDC.

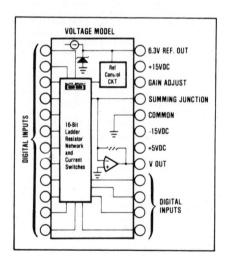

International Airport Industrial Park - P.O. Box 11400 - Tucson, Arizona 85734 - Tel. (602) 746-1111 - Twx: 910-952-1111 - Cable: BBRCORP - Telex: 66-6491

PDS-430

Figure 9–6 BBDAC71 data sheets. (Reprinted with permission of Burr-Brown Corporation, Tucson, AZ.)

SPECIFICATIONS

ELECTRICAL
Typical at $T_A = 25°C$ and rated power supplies unless otherwise noted.

MODEL	DAC71			UNITS
	MIN	TYP	MAX	
INPUT				
DIGITAL INPUT				
Resolution, CCD		4		Digits
CSB, COB		16		Bits
Logic Levels TTL-Compatible [1]				
Logical "1" at $+40\mu A$	+2.4		+5.5	VDC
Logical "0" at $-1.6mA$	0		+0.4	VDC
TRANSFER CHARACTERISTICS				
ACCURACY				
Linearity Error at 25°C, CCD			±0.005	% of FSR[2]
COB, CSB			±0.003	% of FSR
Gain Error,[3] Voltage		±0.01	±0.1	%
Current		±0.05	±0.25	%
Offset Error,[3] Voltage, Unipolar		±0.1	±2	mV
Voltage, Bipolar			±5	mV
Current, Unipolar			±1	μA
Current, Bipolar			±5	μA
Monotonicity Temperature Range 14 bits	0		+50	°C
DRIFT Over specified temp. range				
Total Bipolar Drift includes gain,				
offset, and linearity drift.[4] Voltage		±7	±15	ppm of FSR/°C
Current		±15	±50	ppm of FSR/°C
Total Error over Temperature Range[5]				
Voltage, Unipolar			±0.083	% of FSR
Bipolar			±0.071	% of FSR
Current, Unipolar			±0.23	% of FSR
Bipolar			±0.23	% of FSR
Gain, Voltage			±15	ppm/°C
Current			±45	ppm/°C
Offset				
Voltage, Unipolar		±1	±2	ppm of FSR/°C
Bipolar			±10	ppm of FSR/°C
Current, Unipolar			±1	ppm of FSR/°C
Bipolar			±40	ppm of FSR/°C
Differential Linearity over Temperature			±2	ppm of FSR/°C
Linearity Error over Temperature			±2	ppm of FSR/°C
SETTLING TIME				
Voltage Models to ±0.003% of FSR				
Output: 20V Step		5	10	μsec
1LSB Step[6]		3	5	μsec
Slew Rate		20		V/μsec
Current Models to ±0.003% of FSR				
Output: 2mA step 10Ω to 100Ω Load			1	μsec
1kΩ Load			3	μsec
Switching Transient		500		mV
OUTPUT				
ANALOG OUTPUT				
Voltage Models				
Ranges - CSB, CCD		0 to +10		V
COB		±10		V
Output Current	±5			mA
Output Impedance DC		0.05		Ω
Short Circuit Duration		Indefinite to Common		
Current Models				
Ranges - CSB, CCD		0 to -2		mA
COB		±1		mA
Output Impedance - Unipolar		15		kΩ
Bipolar		4.4		kΩ
Compliance		±2.5		V
INTERNAL REFERENCE VOLTAGE	6.0	6.3	6.6	V
Maximum External Current[7]		±200		μA
Temperature Coefficient of Drift		±10		ppm/°C

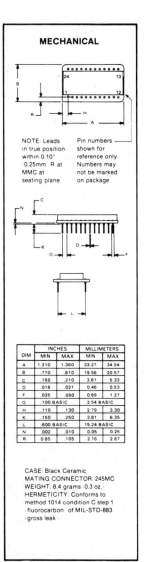

MECHANICAL

NOTE: Leads in true position within 0.10" 0.25mm R at MMC at seating plane.

Pin numbers shown for reference only. Numbers may not be marked on package.

DIM	INCHES		MILLIMETERS	
	MIN	MAX	MIN	MAX
A	1.310	1.360	33.27	34.54
B	.770	.810	19.56	20.57
C	.150	.210	3.81	5.33
D	.018	.021	0.46	0.53
F	.035	.050	0.89	1.27
G	.100 BASIC		2.54 BASIC	
H	.110	.130	2.79	3.30
K	.150	.250	3.81	6.35
L	.600 BASIC		15.24 BASIC	
N	.002	.010	0.05	0.25
R	0.85	.105	2.16	2.67

CASE: Black Ceramic
MATING CONNECTOR: 245MC
WEIGHT: 8.4 grams (0.3 oz.)
HERMETICITY: Conforms to method 1014 condition C step 1 fluorocarbon of MIL-STD-883 gross leak.

Figure 9–6 (*cont.*)

MODEL	DAC71			UNITS
	MIN	TYP	MAX	
POWER SUPPLY SENSITIVITY				
Unipolar Offset				
±15VDC		±0.0001		% of FSR/%Vs
+15VDC		±0.0001		% of FSR/%Vs
Bipolar Offset				
±15VDC		±0.0004		% of FSR/%Vs
+5VDC		±0.0001		% of FSR/%Vs
Gain				
±15VDC		±0.001		% of FSR/%Vs
+5VDC		±0.0005		% of FSR/%Vs
POWER SUPPLY REQUIREMENTS				
Voltage	±14.5, +4.75	±15, +5	±15, +5.25	VDC
Supply Drain, ±15VDC no load		±25	±35	mA
+5VDC logic supply		+20	+35	mA
TEMPERATURE RANGE				
Specification	0		+70	°C
Operating double above Drift Specs	-25		+85	°C
Storage	-55		+100	°C

NOTES:
1. Adding external CMOS hex buffers CD4009A will provide 15VDC CMOS input compatibility. The percent change in output ΔV_O as logic 0 varies from 0.0V to +0.4V and logic 1 changes from +2.4V to +5.0V on all inputs is less than 0.006% of FSR.
2. FSR means Full Scale Range and is 20V for ±10V range, 10V for ±5V range, etc.
3. Adjustable to zero with external trim potentiometer.
4. See "Computing Total Accuracy over Temperature".
5. With gain and offset errors adjusted to zero at 25°C.
6. LSB is for 14-bit resolution.
7. Maximum with no degradation of specifications.

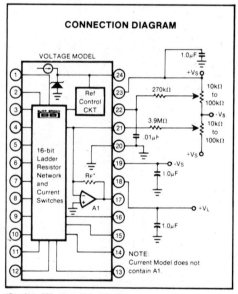

PIN ASSIGNMENTS

I Models	Pin No.	V Models
MSB Bit 1	1	Bit 1 MSB
Bit 2	2	Bit 2
Bit 3	3	Bit 3
Bit 4	4	Bit 4
Bit 5	5	Bit 5
Bit 6	6	Bit 6
Bit 7	7	Bit 7
Bit 8	8	Bit 8
Bit 9	9	Bit 9
Bit 10	10	Bit 10
Bit 11	11	Bit 11
Bit 12	12	Bit 12
Bit 13	13	Bit 13
Bit 14	14	Bit 14
Bit 15	15	Bit 15
LSB Bit 16	16	Bit 16 LSB
R_F	17	V_{OUT}
+5VDC	18	+5VDC
-15VDC	19	-15VDC
COMMON	20	COMMON
I_{OUT}	21	SUMMING JUNCTION
GAIN ADJUST	22	GAIN ADJUST
+15VDC	23	+15VDC
6.3V REF. OUT	24	6.3V REF. OUT

*R_F = 5kΩ CSB, 10kΩ COB, 8kΩ CCD

Figure 9–6 *(cont.)*

DISCUSSION OF SPECIFICATIONS

DIGITAL INPUT CODES

The DAC71 accepts complementary digital input codes in either binary (CSB, COB) or decimal (CCD) format. The COB model may be connected by the user for either complementary offset binary (COB) or complementary two's complement (CTC) codes (see Table I).

TABLE I. Digital Input Codes.

		DIGITAL INPUT CODES		
		CSB	COB	CTC*
CSB, COB MODELS		Compl Straight Binary	Compl Offset Binary	Compl Two's Complement
	MSB　　　LSB			
	All bits ON　0000 000	+Full Scale	+Full Scale	-1LSB
	Mid Scale　0111 111	+1/2 Full Scale	Zero	-Full Scale
	All bits OFF　1111 111	Zero	-Full Scale	Zero
	1000 000	Mid Scale -1LSB	-1LSB	+Full Scale
CCD MODELS		CCD		*Invert the MSB of the COB code with an external inverter to obtain CTC code
		Complementary Coded Decimal 4 Digits		
	F.S. bits ON　0110 0110	+Full Scale		
	All Bits OFF　1111 1111	Zero		

ACCURACY

LINEARITY

This specification describes one of the truest measures of D/A converter accuracy. As defined it means that the analog output will not vary by more than ±0.003% max (CSB, COB) or ±0.005% max (CCD) from a straight line drawn through the end points (all bits ON and all bits OFF) at +25°C.

DIFFERENTIAL LINEARITY

Differential linearity error of a D/A converter is the deviation from an ideal 1LSB voltage change from one adjacent output state to the next. A differential linearity error specification of ±1/2LSB means that the output voltage step sizes can be anywhere from 1/2LSB to 3/2LSB when the input changes from one adjacent input stage to the next.

MONOTONICITY

Monotonicity over 0°C to +50°C is guaranteed. This insures that the analog output will increase or remain the same for increasing 14-bit input digital codes.

DRIFT

Gain Drift is a measure of the change in the full scale range output over temperature expressed in parts per million per °C (see Figure 1). Gain Drift is established by: 1) testing the end point differences for each DAC71 model at +25°C and the appropriate specification temperature extremes; 2) calculating the gain error with respect to the +25°C value; and 3) dividing by the temperature change. This is expressed in ppm/°C.

Offset Drift is a measure of the actual change in output with all "1"s on the input over the specified temperature range.

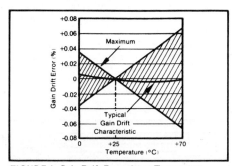

FIGURE 1. Gain Drift Error (%) vs Temperature.

The maximum change in offset is referenced to the offset at +25°C and is divided by the temperature range. This drift is expressed in parts per million of full scale range per °C (ppm of FSR/°C).

SETTLING TIME

Settling time for each DAC71 model is the total time (including slew time) required for the output to settle within an error band around its final value after a change in input (see Figure 2).

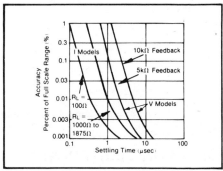

FIGURE 2. Full Scale Range Settling Time vs Accuracy.

VOLTAGE OUTPUT MODELS

Settling times are specified to ±0.003% of FSR; one for maximum full scale range changes of 20V and one for a 1LSB change. The 1LSB change is measured at the major carry (0111...11 to 1000...00), the point at which the worst-case settling time occurs.

CURRENT OUTPUT MODELS

Two settling time are specified to ±0.003% of FSR. Each is given for current models connected with two different resistive loads: 10Ω to 100Ω and 1000Ω.

COMPLIANCE

Compliance voltage is the maximum voltage swing

Figure 9–6　(cont.)

allowed on the output of the current models while maintaining specified accuracy. The typical compliance voltage of all current output models is ±2.5V and maximum safe voltage swing permitted without damage is ±5V.

POWER SUPPLY SENSITIVITY

Power supply sensitivity is a measure of the effect of a power supply change on the D/A converter output. It is defined as a percent of FSR per percent of change in either the positive, negative, or logic supplies about the nominal power supply voltages (see Figure 3).

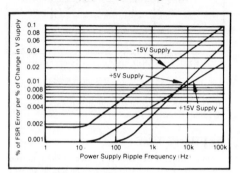

FIGURE 3. Power Supply Rejection vs Power Supply Ripple Frequency.

REFERENCE SUPPLY

All DAC71 models are supplied with an internal +6.3V reference voltage supply. This reference voltage (pin 24) has a tolerance of ±5% and is connected internally for specified operation. The zener is selected for a Gain Drift of typically ±3ppm/°C and is burned-in for a total of 168 hours for guaranteed reliability. This reference may also be used externally but the current drain is limited to 200μA. An external buffer amplifier is recommended if the DAC71 internal reference is used externally in order to provide a constant load to the reference supply output.

OPERATING INSTRUCTIONS

POWER SUPPLY CONNECTIONS

For optimum performance and noise rejection, power supply decoupling capacitors should be added as shown in the Connection Diagram. These capacitors (1μF tantalum or electrolytic recommended) should be located close to the DAC71. Electrolytic capacitors, if used, should be paralleled with 0.01μF ceramic capacitors for best high frequency performance.

EXTERNAL OFFSET AND GAIN ADJUSTMENT

Offset and gain may be trimmed by installing external offset and gain potentiometers. Connect these potentiometers as shown in the Connection Diagram and adjust as described below. TCR of the potentiometers should be 100ppm/°C or less. The 3.9MΩ and 510kΩ resistors (20% carbon or better) should be located close to the DAC71 to prevent noise pickup. If it is not convenient to use these high-value resistors, an equivalent "T" network, as shown in Figure 4, may be substituted in place of the 3.9MΩ. A 0.001μF to 0.01μF ceramic capacitor should be connected from Gain Adjust (pin 22) to common to prevent noise pickup. Refer to Figures 5 and 6 for relationship of offset and gain adjustments to unipolar and bipolar D/A converters.

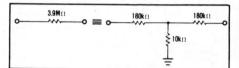

FIGURE 4. Equivalent Resistances.

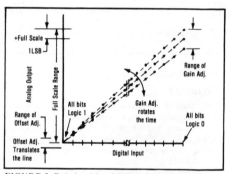

FIGURE 5. Relationship of Offset and Gain Adjustments for a Unipolar D/A Converter.

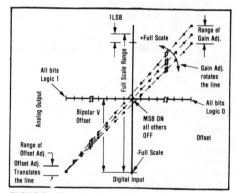

FIGURE 6. Relationship of Offset and Gain Adjustments for a Bipolar D/A Converter.

OFFSET ADJUSTMENT

For unipolar (CSB, CCD) configurations, apply the digital input code that should produce zero potential

Figure 9–6 (cont.)

Sec. 9.2 Commercial DACs

output and adjust the offset potentiometer for zero output.

For bipolar (COB, CTC) configurations, apply the digital input code that should produce the maximum negative output voltage. The COB model is internally connected for a 20V FSR range where the maximum negative output voltage is -10V. See Table II for corresponding codes and the Connection Diagram for offset adjustment connections. Offset adjust should be made prior to gain adjust.

TABLE II. Digital Input and Analog Output Relationships.

DIGITAL INPUT CODE	OUTPUT CODE			
	VOLTAGE		CURRENT	
	16-Bit Resolution	14-Bit Resolution	16-Bit Resolution	14-Bit Resolution
Complementary Unipolar Straight Binary CSB 0 to +10V or 0 to -2mA*				
One LSB	-153μV	+610μV	0.031μA	0.122μA
All Bits ON 00 00	+9.99985V	+9.99939V	-1.99997mA	-1.99988mA
All Bits OFF 11 11	Zero	Zero	Zero	Zero
Complementary Bipolar Offset Binary COB ±10V or ±1mA*				
One LSB	-305μV	-1.22mV	0.031μA	0.122μA
All Bits ON 00 00	+9.99969V	+9.99878V	-0.99997mA	-0.99988mA
All Bits OFF 11 11	-10.0000V	-10.0000V	-1.0000mA	-1.0000mA
Complementary Binary Coded Decimal CCD 0 to +10V or 0 to -1.25mA		4-Digit Resolution		4-Digit Resolution
One LSB	-1.0mV	N/A	0.125μA	N/A
Full Scale 0110 0110	-9.999V		-1.24987mA	
All Bits OFF 1111 1111	Zero		Zero	

* To obtain values for other binary CBI ranges 0 to +5V range. divide by 2. +10V range by 2. ±5V range, divide -10V range by 2, ±2.5V range, divide -10V range by 4

GAIN ADJUSTMENT

For either unipolar or bipolar configurations, apply the digital input that should give the maximum positive output voltage. Adjust the gain potentiometer for this positive full scale voltage. See Table II for positive full scale voltages and the Connection Diagram for gain adjustment connections.

INSTALLATION CONSIDERATIONS

The DAC71 is laser-trimmed to 14-bit linearity. The design of the device makes the 16-bit resolution available on binary units: If 16-bit resolution is not required, bit 15 (pin 15) and bit 16 (pin 16) should be connected to +5VDC through a single 1kΩ resistor.

Due to the extremely-high resolution and linearity of the DAC71, system design problems such as grounding and contact resistance become very important. For a 16-bit converter with a +10V full scale range, 1LSB is 153μV. With a load current of 5mA, series wiring and connector resistance of only 30mΩ, the output will be in error by 1LSB. To understand what this means in terms of a system layout, the impedance of #23 wire is about 0.021Ω/ft. Neglecting contact resistance, less than 6 inches of wire will produce a 1LSB error in the analog output voltage! Although the problems involved seem enormous, care in the installation planning can minimize the potential causes of error.

Figure 7 shows the connection diagram for a voltage output DAC71. Lead and contact resistances are represented by R_1 through R_5. As long as the load resistance (R_L) is constant, R_2 simply introduces a gain error than can be removed during initial calibration. R_3 is part of R_L if the output voltage is sensed at Common (pin 20) and therefore introduces no error. If R_L is variable then R_2 should be less than $R_{Lmin}/2^{16}$ to reduce voltage drops due to wiring to less than 1LSB. For example, if R_{Lmin} is 5kΩ then R_2 should be less than 0.08Ω. R_L should be located as close as possible to the DAC71 for optimum performance.

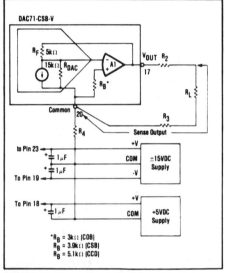

FIGURE 7. Output Circuit for Voltage Models.

Figures 8 and 9 show two methods of connecting current model DAC71's with external precision output op amps. By sensing the output voltage at the load resistor (i.e., by connecting R_F to the output of A1 at R_1) the effect of R_1 and R_2 is greatly reduced. R_1 will cause a gain error but is independent of the value of R_1 and can be eliminated during initial calibration. The effect of R_2 is negligible because it is inside the feedback loop of the output op amp and is therefore greatly reduced by the loop gain. If the output cannot be sensed at Common (pin 20), then the differential output circuit shown in Figure 9 is recommended. In this circuit the output voltage is sensed at the load common and not at the DAC common as in the previous circuits. The value of R_6 and R_7 must be adjusted for maximum common-mode rejection at R_1. Note that if R_1 is negligible the circuit of Figure 9 can be reduced to the one shown in Figure 8 because $R_{11} = (R_7 + R_5) \| R_6$. In all three circuits the effect of R_4 is negligible.

Figure 9-6   (cont.)

320   Chap. 9   Digital-to-Analog Converter

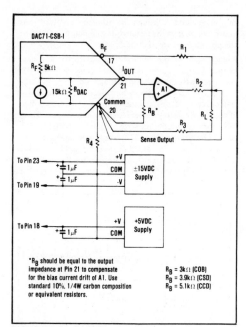

FIGURE 8. Preferred External Op Amp Configuration.

The DAC71 and the wiring to its connectors should be located to provide optimum isolation from sources of RFI and EMI. The key word in elimination of RF radiation or pickup is loop area. Therefore, signal leads and their return conductors should be kept close together. This reduces the external magnetic field along with any radiation. Also, if a signal lead and its return conductor are wired close together they present a small flux-capture cross section for any external field. This reduces radiation pickup in the circuit.

NOTE: It is recommended that the digital input lines of the DAC71 be driven from inverters or buffers of TTL input registers to obtain specified accuracy.

DRIVING A RESISTIVE LOAD UNIPOLAR

A load resistance, R_L, with the current output model connected as shown in Figure 10, will generate a voltage range, V_{OUT}, determined by:

$$V_{OUT} = -2mA[(15k\Omega \times R_L)/(15k\Omega + R_L)]$$
Where R_L max = $1.36k\Omega$
and V_{OUT} max = $-2.5V$

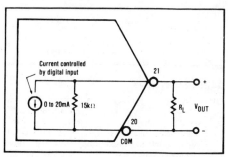

FIGURE 10. Equivalent Circuit DAC71-CSB-I Connected for Unipolar Voltage Output with Resistive Load.

Add an external low T.C. (<10ppm/°C) resistor (R_L) as shown in Figure 11 to obtain a 0 to -2V full scale output voltage range for CCD input codes.

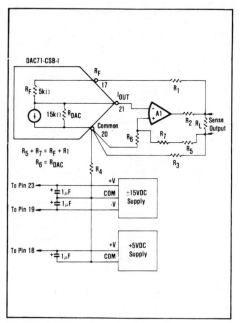

FIGURE 9. Differential Sensing Output Op Amp Configuration.

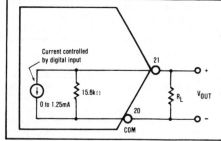

FIGURE 11. DAC71-CCD-I Connected for Voltage Output with Resistive Load.

Figure 9–6   (*cont.*)

Sec. 9.2 Commercial DACs

$V_{OUT} = -1.25\text{mA}[(15.6\text{k}\Omega \times R_L)/(15.6\text{k}\Omega + R_L)]$
Where R_L max = $1.79\text{k}\Omega$
and V_{OUT} max = -2.01V

DRIVING A RESISTIVE LOAD BIPOLAR

The equivalent output circuit for a bipolar output voltage range is shown in Figure 12. V_{OUT} is determined by:

$V_{OUT} = \pm 1\text{mA}[(4.44\text{k}\Omega \times R_L)/(4.44\text{k}\Omega + R_L)]$
Where R_L max = $5.72\text{k}\Omega$
and V_{OUT} max = $\pm 2.5\text{V}$

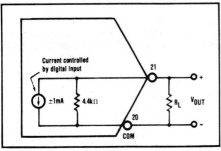

FIGURE 12. DAC71-COB-I Connected for Bipolar Output Voltage with Resistive Load.

APPLICATIONS

DRIVING AN EXTERNAL OP AMP WITH CURRENT OUTPUT DAC

The DAC71-(CSB, COB, CCD)-I are current output devices and will drive the summing junction of an op amp to produce an output voltage (see Figure 13). The op amp output voltage is:

$V_{OUT} = -I_{OUT} R_F$

where I_{OUT} is the DAC71 output current and R_F is the feedback resistor. Use of the internal feedback resistor (pin 17) is required to obtain specified gain accuracy and low gain drift.

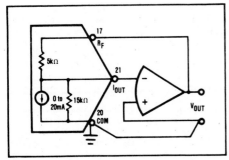

FIGURE 13. External Op Amp Using Internal Feedback Resistors.

The DAC71 can be scaled for any desired voltage range with an external feedback resistor, but at the expense of increased drifts of up to $\pm 25\text{ppm}/^{\circ}\text{C}$. The resistors in the DAC71 are chosen for ratio tracking of $\pm 1\text{ppm}/^{\circ}\text{C}$ and not absolute TCR (which may be as high at $\pm 25\text{ppm}/^{\circ}\text{C}$.)

An alternative method of scaling the output voltage of the DAC71 and preserving the low gain drift is shown in Figure 14.

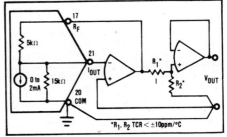

FIGURE 14. External Op Amp Using Internal and External Feedback Resistors to Maintain Low Gain Drift.

OUTPUTS LARGER THAN 20-VOLT RANGE

For output voltage ranges larger than $\pm 10\text{V}$, a high voltage op amp may be employed with an external feedback resistor. Use I_{OUT} values of $\pm 1\text{mA}$ for bipolar voltage ranges and -2mA for unipolar voltage ranges (see Figure 15). Use protection diodes when a high voltage op amp is used.

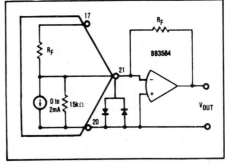

FIGURE 15. External Op Amp Using External Feedback Resistors.

COMPUTING TOTAL ACCURACY OVER TEMPERATURE

The accuracy drift with temperature of a DAC71 consists of three primary components: gain drift, unipolar or bipolar offset drift, and linearity drift. To obtain the

Figure 9–6 *(cont.)*

worst-case accuracy drift, most users would assume that all drift errors are random and would simply add them algebraically. However, the worst-case accuracy drift for a DAC71 operating in the bipolar mode is about one-half of the algebraic sum of the individual drift errors.

To explain this fact, it is necessary to consider the unipolar and bipolar modes of operation separately.

In the unipolar mode of operation, offset drift ($\pm$1ppm/°C) is due primarily to voltage offset drift of the output op amp and, to a lesser extent, to the leakage current through the quad current switches. Gain drift consists of several components: 1) $\pm$5ppm/°C due to ratio drift of current switch V_{BE} to the reference transistor, 2)$\pm$10ppm/°C due to the zener reference and, 3) $\pm$2ppm/°C linearity drift due to ratio drift of current weighting resistors and V_{BE} of the quad current switches. The sum of these three components, $\pm$17ppm/°C, is the maximum gain drift.

Because the parameters described could all drift in the same direction, the worst-case accuracy drift in the unipolar mode is simply the sum of the components, or $\pm$18ppm/°C.

In the bipolar mode the major portion of gain drift is due to the zener reference. The gain and offset drifts caused by reference drift are always in opposite directions. Therefore, the accuracy drift will be the difference rather than the sum of these drifts.

First, consider the effect of reference variations on offset drift. Figure 16 shows a simplified circuit diagram of a DAC71-COB-V with all bits off. The current switch leakage current is negligible, so

$V_{-FULL\ SCALE} = (-R_F/R_{BPO}) \cdot V_{REF}$
$= (-10k\Omega/6.3k\Omega) \cdot 6.3V = -10V$

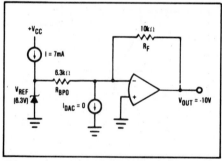

FIGURE 16. Simplified Diagram of DAC71-COB-V with "All Bits Off" ($\pm$10V Range).

This equation shows that if V_{REF} increases, the output voltage will decrease and vice versa. If the V_{REF} drift is +10ppm/°C, this equivalent to (+10ppm/°C) x (+6.3V) = +63μV/°C. This will result in a voltage drift at the amplifier output of

$\Delta V_{-FS}/\Delta T = -(R_F/R_{BPO}) \cdot (\Delta V_{REF}/\Delta T)$
$= -(10k\Omega/6.3k\Omega) \cdot (63\mu V/°C) = -100\mu V/°C$

Since the DAC71-COB-V is operating in the $\pm$10V range this equivalent to (-100μV/°C) $\div$ (20V range) = -5ppm of FSR/°C.

Now consider the effect of reference changes on gain drift. When all of the bits are turned on it can be shown that:

$\Delta V_{+FULL\ SCALE}/\Delta T = +(R_F/R_{BPO}) \cdot (\Delta V_{REF}/\Delta T)$
$= +(10k\Omega/6.3k\Omega) \cdot (63\mu V/°C) = +100\mu V/°C.$

and (+100μV/°C) 20V Range = +5ppm/°C of FSR.

This result indicates that the drift of the minus full scale voltage will be equal in magnitude to, and in the opposite direction of, the drift of the plus full scale voltage and that zener reference variations have virtually no effect on the zero point (see Figure 17). This equation also indicates that the gain drift is equal to the V_{REF} drift in ppm/°C, and the magnitude of the minus full scale drift and plus full scale drift is equal to one-half of the V_{REF} drift.

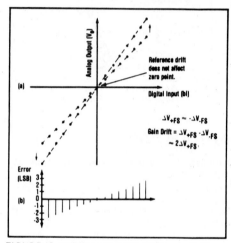

FIGURE 17. (a) Effect of a Positive Reference Drift on the Ideal D/A Transfer Function; (b) Error Distribution Due to Reference Voltage Drift in a DAC71.

Using this relationship, the worst-case accuracy drift for a DAC71-COB-V can be computed. The maximum TCR of the zener reference is $\pm$10ppm/°C. The gain drift due to the reference then is also $\pm$10ppm/°C. The full scale drift and bipolar offset drift are each half that amount or $\pm$5ppm/°C. The maximum gain and offset drifts of the DAC71, exclusive of the reference, are $\pm$5 and $\pm$3ppm/°C respectively. Adding this to the full scale drift due to the reference plus the linearity drift of $\pm$2ppm/°C gives a worst-case total accuracy drift of $\pm$15ppm/°C. (Random drifts, which these are, can be in the same direction so they add directly.) This is much less than the total drift obtained by simply adding the maximum gain, bipolar offset, and linearity drifts ($\pm$27ppm/°C). The maximum

Figure 9–6 (cont.)

zero point drift is equal to one-half of the gain drift exclusive of the reference plus the offset drift exclusive of the reference, or ±5.5ppm of FSR/°C.

The DAC71 is specified over a 0°C to +70°C temperature range giving a maximum excursion from room temperature (+25°C) of 45°C. Assuming that gain and offset errors have been adjusted to zero at room temperature,

total worst-case accuracy error
= Linearity error + Accuracy drift x ΔT
= ±0.003% + ±15ppm/°C (45°) (100)
= ±0.07%

total worst-case bipolar zero point error
= Bipolar zero drift x ΔT
= ±5.5ppm of FSR% (45°C) (100)
= ±0.025%

ORDERING INFORMATION

MODEL	INPUT CODE
CURRENT MODELS	
DAC71-COB-I	Complementary Offset Binary
DAC71-CSB-I	Complementary Straight Binary
DAC71-CCD-I	Complementary Coded Decimal
VOLTAGE MODELS	
DAC71-COB-V	Complementary Offset Binary
DAC71-CSB-V	Complementary Straight Binary
DAC71-CCD-V	Complementary Coded Decimal

Figure 9–6 (*cont.*)

9.2.4 Summary

This section has presented only a small sampling of the available DACs. Many more exist and from other manufacturers in addition to the ones listed. Each manufacturer, in addition to the data sheets for the device, provides detailed examples as to typical applications and circuit connections together with definitions of the terminology used in the data sheets. Several publications are available from the manufacturers which discuss these devices in detail.

9.3 DAC TERMINOLOGY

As with the ADC, several terms are used to describe the operation of the devices. These will help the user to interpret the data sheets; however, it should not be assumed that similar terms used on data sheets from different manufacturers are identically defined. The user should verify the interpretation of the terms with each supplier.

9.3.1 Linearity

If the LSB of the input changes, we want to be sure that the analog output will change by the equivalent amount, $V_{REF}/2^n$, where n is the number of bits. The ideal DA relationship, plotted in Figure 9-7, shows the ideal I/O relationship between a 4-bit input and its resulting analog output. The linearity is usually given in terms of the LSB, indicating the maximum amount of deviation that we can expect in the analog output for a given digital input. Typical values are 1 LSB or $\frac{1}{2}$ LSB.

9.3.2 Settling Time

This is the time that elapses from a binary input change to the time that the analog output reaches and remains within $\pm\frac{1}{2}$ LSB of its final value. This can also be termed the *conversion speed* of the unit. When the binary input changes, whether by one bit or many, the analog output must rapidly

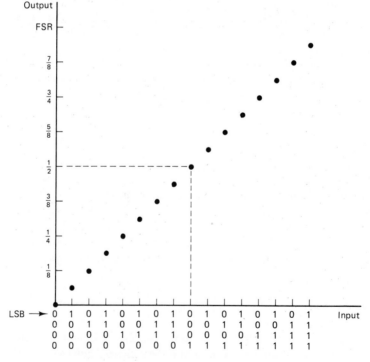

Figure 9-7 Ideal I/O relationship of a 4-bit DAC.

change to this new value. It is this rapid change that gives rise to the jumps between output values shown in Figure 9–1. These jumps represent many unwanted high frequency components in the output. An analog LP filter is required at the output to reduce this high frequency noise component.

9.3.3 Resolution

A converter that provides 2^n distinct and different analog output values corresponding to the set of n-bit binary words is said to have a *resolution* of n bits.

9.4 DAC OUTPUT

Prior to its being used, the DAC output requires some processing to put it in a form that will allow it to be used by output transducers. The defects in the waveform, which give rise to a distorted output, are glitches, jumps, and lengthy conversion time.

9.4.1 Glitches

When the input to a DAC changes by 1 LSB and that change causes a transition from many ones to many zeros (e.g., in going from 01111 . . . 11 to 10000 . . . 00), many internal switches are changing state. As is usually the case, the time for each transition is different, causing the output to go to zero or full scale before settling in at the value indicated by the new binary input. As these glitches are high, narrow pulses, they introduce many high-frequency components into the output.

The effect of these glitches can be eliminated by using a track-and-hold (TH) circuit at the output. When the input is about to change, the TH circuit is put into the hold state. When the DAC output has settled and all glitches have been terminated, the TH circuit is put into its track mode, where it can now attain the new DAC output value. Several of the available commercial DACs have included an internal TH circuit for this purpose and are termed "deglitched."

9.4.2 Reconstruction Filter

It was mentioned earlier that the DAC output has many jumps in its waveform. These are caused by the quantized values of the input and appear as unwanted high-frequency components in the output spectrum. To eliminate these components, the DAC output is passed through an analog LP filter, which is termed a *reconstruction filter*. Its design should provide a sufficiently high roll-off at the cutoff frequency to remove all unwanted high-frequency components.

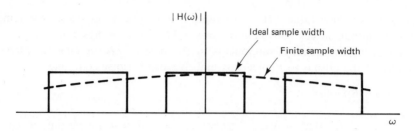

Figure 9–8 Effect of the sample pulse width on the spectrum of the signal.

9.4.3 Conversion-Time Effect

When practical sampling was investigated in Chapter 3, it was shown that the frequency content of the samples had a $(\sin x)/x$ factor multiplying the amplitude of each component, where x involved the sampling width. As the sample width decreased, the amount of roll-off of the $(\sin x)/x$ factor increased. This reduction in frequency-component amplitude as the sample width decreased is a form of distortion, resulting in a nonideal output signal. Figure 9–8 shows the results of ideal and nonideal sampling. As seen in the case of ideal sampling, there is no $(\sin x)/x$ term multiplying the output frequency content; however, in the nonideal sampling examples, the $(\sin x)/x$ term effect can be observed. This distortion can be reduced by passing the DAC output through a track-and-hold circuit such that its hold time is short compared with the conversion time of the DAC. This is shown in Figure 9–9. By keeping the TH output at the level of the DAC output for a relatively short time, the roll-off of the $(\sin x)/x$ term is reduced, resulting in less attenuation of the output frequency components at the high end of the spectrum.

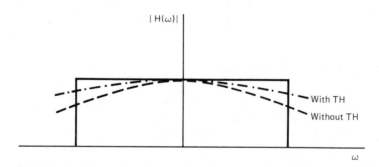

Figure 9–9 Effect of using the TH circuit on the output signal spectrum.

9.5 OUTPUT PROCESSING

Once the DAC output is passed through all TH and filter circuits, it is in analog form and ready for any final analog processing, such as amplification. When this is completed, the signal can be applied to an output transducer, such as a loudspeaker or cathode-ray tube.

9.6 SUMMARY

This chapter has presented some of the basic circuits used in converting a digital signal to analog form. Some commercial realizations of these circuits were also presented together with the terminology used in describing the characteristics. As this chapter was meant only to be an introduction to the topic, the reader is urged to review the References at the end of the chapter for a complete look at DACs.

REFERENCES

HNATEK, E. R. *A User's Handbook of D/A and A/D Converters.* New York: Wiley, 1976.

SHEINGOLD, D. H., ED. *Analog-Digital Conversion Notes.* Norwood, Mass.: Analog Devices, 1980.

ANALOG FILTER CHARACTERISTICS

appendix A

Analog filter theory is quite well developed and is presented in many texts. This appendix is a summation of some of the writings on filter theory which are listed in the References at the end. The purpose of including analog filter theory in this text is to provide readers with some information to which they can refer with regard to the filter topics discussed in Chapter 6. This appendix can also be used as a refresher for those who have not worked recently with analog filters.

A.1 IDEAL TRANSFER FUNCTION

The transfer function of an ideal low-pass (LP) filter is given in Figure A–1. The cutoff radian frequency, ω_c, is normalized to unity and the magnitude squared value of the transfer function in the passband is also chosen as unity (0% attenuation). For frequencies greater than cutoff, the amplitude of the transfer function is zero (100% attenuation). In addition, the phase shift in the passband is a linear function of the frequency.

Sec. A.2 Realizable Approximation

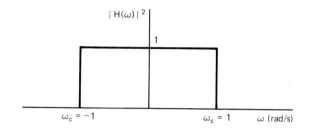

Figure A–1 Transfer function of an ideal LP filter.

A.2 REALIZABLE APPROXIMATION

If the inverse Fourier transform (IFT) of the ideal transfer function is obtained, it can be seen that the time-domain impulse response is noncausal. This is shown in Figure A–2. As the impulse response is not zero for $t < 0$ s, the ideal filter cannot be realized.

To be able to construct a filter that is realizable, the ideal conditions must be weakened, allowing for some deviation in the values associated with the pass- and stopbands. Figure A–3 shows the magnitude response versus frequency of a nonideal LP filter. The amplitude in the passband is not restricted to unity, but allowed to have some variation; the amplitude in the stopband which begins at ω_s is not restricted to zero, but also allowed to have some

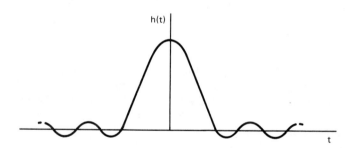

Figure A–2 Impulse response of an ideal filter.

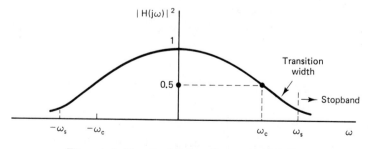

Figure A–3 Transfer function of a practical LP filter.

variation. In addition, the transition from the passband to the stopband does not take place at a single frequency, ω_c, but can take place over a range of frequencies. This range is termed the *transition region*.

A.3 ERROR

The design procedure introduces the concept of error. This is the difference between the ideal and actual transfer functions and is shown in Figure A-4. The error is defined in Equation A-1.

$$\text{error} = \epsilon(\omega) = \text{actual}(\omega) - \text{ideal}(\omega) \qquad (A-1)$$

The object of filter design is to minimize the error in either the passband, the stopband, or both.

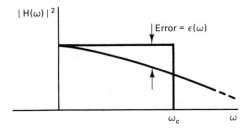

Figure A-4 Error between the ideal and practical transfer function.

A.4 APPROACHES

There are two approaches to minimizing the error in the design of a filter: the first attempts to minimize the error at a single point, while the second attempts to minimize the maximum value of the error over the passband.

A.4.1 Maximally Flat

In this approach to obtaining an approximation to the ideal passband characteristic, the actual transfer function is made to agree with the ideal one at a single point, ω_1. This is shown in Figure A-5. The first- and several higher-order derivatives of the practical transfer function will be zero at ω_1,

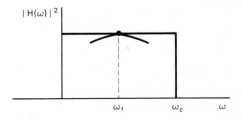

Figure A-5 Maximally flat agreement at ω_1.

Sec. A.5   Basic Equations   331

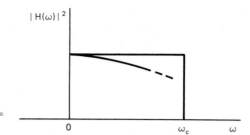

Figure A–6 Maximally flat curve at $\omega = 0$.

giving rise to the term *maximally flat*. As we are working with the LP filter, the frequency at which the agreement is to occur is chosen at the origin and shown in Figure A–6. $\epsilon(\omega) = 0$ at only one frequency, the origin.

A.4.2 Mini-max

The second approach for approximating the ideal transfer function is one that requires the maximum value of the error in the passband to be a minimum, hence the name *mini-max approach*. This is shown in Figure A–7, where the transfer function varies between 1 and $1/(1 + \epsilon^2)$. In addition, the choice can be made to (1) minimize the error in the stopband, or (2) simultaneously minimize the error in the stop- and passbands.

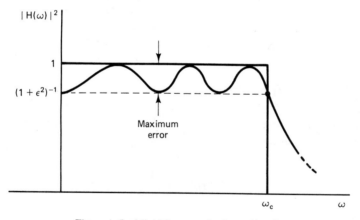

Figure A–7 Mini-Max error in the passband.

A.5 BASIC EQUATIONS

The magnitude squared transfer function is an even function of the radian frequency and, as a result, can be represented as a rational fraction of two polynomials in ω. This is given in Equation A–2.

$$|H(\omega)|^2 = \frac{N(\omega^2)}{D(\omega^2)} \qquad (A-2)$$

where $H(\omega)$ is obtained from $H(s)$ evaluated on the $j\omega$ axis. In order for this equation to approximate the ideal transfer function, it must show a value close to unity in the passband and a value close to zero in the stopband. To obtain this characteristic, we rewrite Equation A–2 as

$$|H(\omega)|^2 = \frac{N(\omega^2)}{N(\omega^2) + P(\omega^2)} \tag{A-3}$$

If we choose the functions $N(\omega^2)$ and $P(\omega^2)$ such that

$$N(\omega^2) \gg P(\omega^2), \quad \omega < \omega_c \tag{A-4a}$$

$$N(\omega^2) \ll P(\omega^2), \quad \omega > \omega_c \tag{A-4b}$$

we will have the desired approximation to the ideal transfer function.

We can express the magnitude squared value of the transfer function as

$$|H(j\omega)|^2 = H(j\omega)\overline{H(j\omega)} = H(j\omega)H(-j\omega)$$

where the last equality is true if the coefficients of $H(s)$ are real. Setting $j\omega = s$, we have

$$|H(j\omega)|^2 = H(j\omega)H(-j\omega) = H(s)H(-s)|_{s=j\omega} \tag{A-5}$$

To ensure that the filter is stable, we will associate all the poles that occur in the left-hand plane with $H(s)$, which will then be the desired transfer function of the filter. As all the poles in the right-hand plane are associated with $H(-s)$, this ensures that the resulting filter design will be stable.

A.6 BUTTERWORTH APPROXIMATIONS

The first approximation equation is obtained by using the Butterworth polynomials, $N(\omega^2) = 1$, $P(\omega^2) = (\omega^2)^n$, in Equation A-3. This is given in Equation A–6.

$$|H(\omega)|^2 = \frac{1}{1 + \omega^{2n}} \tag{A-6a}$$

A plot of the Butterworth approximation for several values of n is given in Figure A–8. As can be seen, the larger n, the closer the Butterworth curves will approximate the ideal transfer characteristic; that is, the error goes to zero as n goes to infinity. If the cutoff frequency is not at $\omega_c = 1$, Equation A–6a is expressed as

$$|H(\omega)|^2 = \frac{1}{1 + \left(\dfrac{\omega}{\omega_c}\right)^{2n}} \tag{A-6b}$$

Sec. A.7 Chebychev Approximations

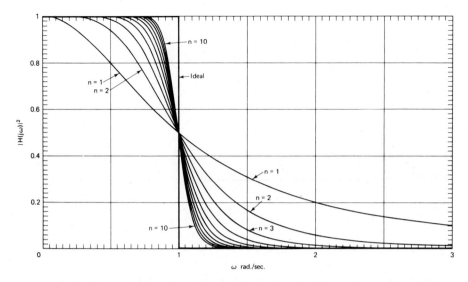

Figure A-8 Plot of Butterworth polynomials used as the filter transfer function. (From *Analog and Digital Filters: Design and Realization*, H. Y-F Lam © 1979. Reprinted by permission of Prentice-Hall, Inc., Englewood Cliffs, N.J.)

Several characteristics of the Butterworth approximation are presented:

1. The function is maximally flat at the origin.
2. All curves pass through the value one-half at $\omega = \omega_c$.
3. The curves are monotonically decreasing.
4. The roll-off of the transfer-function squared magnitude is at a rate of $20n$ dB/decade.

The transfer function of the approximation is obtained by replacing ω with s/j in the Butterworth approximation, Equation A-6. This yields $H(s)H(-s)$ as shown in Equation A-7.

$$|H(\omega)|^2 \Big|_{\omega=s/j} = H(s)H(-s) = \frac{1}{1+\omega^{2n}} \Big|_{\omega=s/j}$$
$$= \frac{1}{1+(-1)^n s^{2n}} \tag{A-7}$$

A.7 CHEBYCHEV APPROXIMATIONS

As we did with the Butterworth approximation, an equation is presented that yields an approximation to the ideal transfer function. This equation uses the Chebychev polynomials to obtain the desired values.

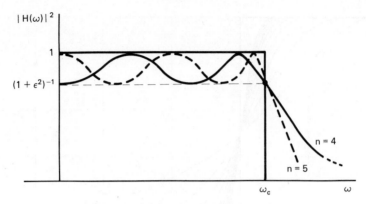

Figure A-9 Chebychev approximation for $n = 4, 5$.

$$|H(\omega)|^2 = \frac{1}{1 + \epsilon^2 C_n^2(\omega)} \tag{A-8}$$

where $C_n(\omega)$ is the nth order Chebychev polynomial and ϵ is a measure of the allowed error in the passband. Figure A-9 shows the Chebychev approximation for n equal to 4 and 5. Table A-1 is a listing of the Chebychev polynomials for $n = 0, 1, \ldots, 4$. Note that the value of the Chebychev polynomial is 1 when $\omega = 1$ for any value of n. Chebychev polynomials of any order can be obtained by the recursive relation of Equation A-9.

$$C_{n+1}(\omega) = 2\omega C_n(\omega) - C_{n-1}(\omega), \qquad n = 1, 2, 3, \ldots \tag{A-9}$$

As we know that $C_0(\omega)$ is 1, and $C_1(\omega) = \omega$, all the higher-order polynomials can be generated from Equation A-9. Figure A-10 is a plot of several of the Chebychev polynomials for n equals 1 through 4.

The larger the value of n, the closer the Chebychev approximation is to the ideal transfer function. Some properties of the Chebychev filter are:

1. For $|\omega| < 1$, $|H(\omega)|^2$ will vary between $1/(1 + \epsilon^2)$ and 1.
2. For $|\omega| > 1$, $|H(\omega)|^2$ decreases monotonically toward zero with a rolloff rate of $20n$ dB/decade.
3. All curves pass through $1/(1 + \epsilon^2)$ at the cutoff frequency.
4. $|H(0)| = 1$ for odd values of n while $|H(0)| = (1 + \epsilon^2)^{-1/2}$ for even values.

TABLE A-1 Chebychev Polynomials for $n = 0$ to 4

$C_0(\omega) = \cos 0 = 1$
$C_1(\omega) = \cos(\cos^{-1} \omega) = \omega$
$C_2(\omega) = \cos(2 \cos^{-1} \omega) = 2\omega^2 - 1$
$C_3(\omega) = \cos(3 \cos^{-1} \omega) = -3\omega + 4\omega^3$
$C_4(\omega) = \cos(4 \cos^{-1} \omega) = 1 - 8\omega^2 + 8\omega^4$

Sec. A.8 Filter Poles

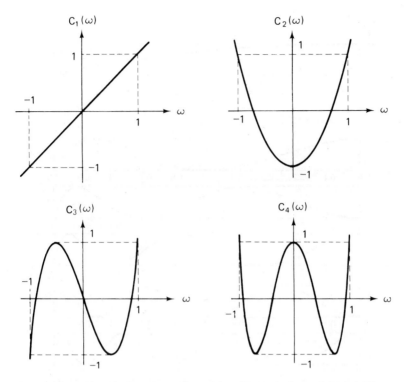

Figure A–10 Plot of Chebychev polynomials. (From *Analog and Digital Filters: Design and Realization,* H. Y-F Lam. Copyright © 1979, Prentice-Hall, Inc. Reprinted by permission of Prentice-Hall, Inc.)

Figure A–11 shows the passband ripples of a Chebychev filter for n equals 1 through 10. The transfer function of the filter is obtained by replacing ω with s/j in the Chebychev polynomial, Equation A–8.

A.8 FILTER POLES

Using the transfer function equations, the (stable) poles for the filters can be obtained.

A.8.1 Butterworth Poles

Using Equation A–6, evaluated at $\omega = s/j$, we see that all poles can be found from the solution of

$$1 + (-1)^n s^{2n} = 0 \tag{A–10}$$

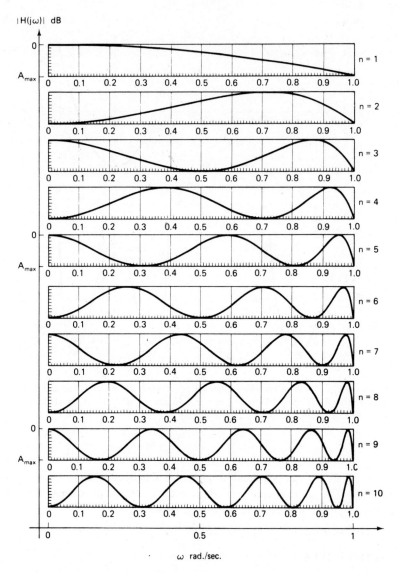

Figure A–11 Chebychev passband ripples. (From *Analog and Digital Filters: Design and Realization*, H. Y-F Lam. Copyright © 1979, Prentice-Hall, Inc. Reprinted by permission of Prentice-Hall, Inc.)

This allows us to express $H(s)$ as

$$H(s) = \prod_{k=1}^{n} \frac{1}{s - s_k} \tag{A-11a}$$

where

Sec. A.8 Filter Poles

$$s_k = \sigma_k + j\omega_k \tag{A-11b}$$

The solutions of Equation A-10 that yield the left-hand-plane poles of $H(s)$ are

$$s_k = \sigma_k + j\omega_k \tag{A-12a}$$
$$= -\sin\theta_k + j\cos\theta_k \tag{A-12b}$$

where

$$\theta_k = \frac{2k-1}{2n}\pi, \quad k = 1, 2, 3, \ldots, n \tag{A-12c}$$

This indicates that the poles of $H(s)$ lie on a circle of unit radius as shown in Figure A-12 for n equals 4 and 5. Note that the (unstable) poles in the right-hand plane are associated with $H(-s)$ and are not considered further.

For values of ω_c other than 1 rad/s, the equations of the poles are identical to Equation A-12, but multiplied by ω_c. As seen from Figure A-12, the poles of the Butterworth filter are complex conjugates. If we take a pair of factors in Equation A-11 and replace the roots with their complex conjugate values, we obtain

$$H(s) = \prod_{k=1}^{n/2} \frac{1}{s^2 + (2\sin\theta_k)s + 1} \tag{A-13a}$$

for even n and

$$H(s) = \frac{1}{s+1} \prod_{k=1}^{(n-1)/2} \frac{1}{s^2 + (2\sin\theta_k)s + 1} \tag{A-13b}$$

for odd n.

A.8.2 Chebychev Poles

Due to the use of Chebychev polynomials in the transfer function equations, the determination of the poles is more involved. The poles are the solution of the denominator of Equation A-8.

$$1 + \epsilon^2 C_n^2(\omega) = 1 + \epsilon^2 \cos^2(n \cos^{-1}\omega) = 0 \tag{A-14}$$

As we are interested only in those poles that lie in the left-hand plane, they are found to be $s_k = \sigma_k + j\omega_k$, where

$$\sigma_k = -\sinh\left(\frac{1}{n}\sinh^{-1}\frac{1}{\epsilon}\right)\sin\left(\frac{2k-1}{2n}\pi\right) \tag{A-15a}$$

$$\omega_k = \cosh\left(\frac{1}{n}\sinh^{-1}\frac{1}{\epsilon}\right)\cos\left(\frac{2k-1}{2n}\pi\right) \tag{A-15b}$$

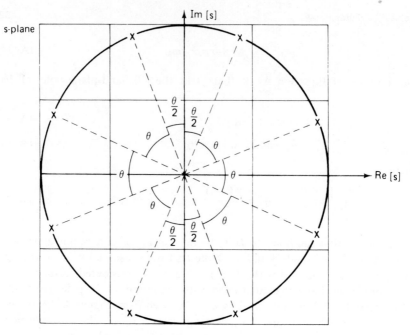

(a) n = even. No pole on real axis. In this case n = 4, θ = 45°

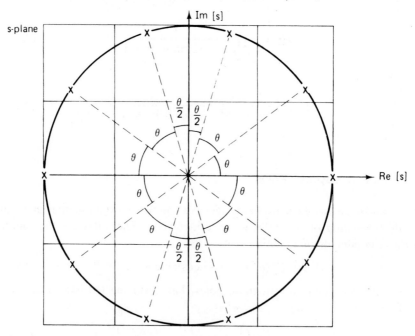

(b) n = odd. A pole as at s = −1. In this case n = 5, θ = 36°

Figure A–12 Pole diagram for $n = 4, 5$. (From *Analog and Digital Filters: Design and Realization*, H. Y-F Lam. Copyright © 1979, Prentice-Hall, Inc. Reprinted by permission of Prentice-Hall, Inc.)

Sec. A.8 Filter Poles

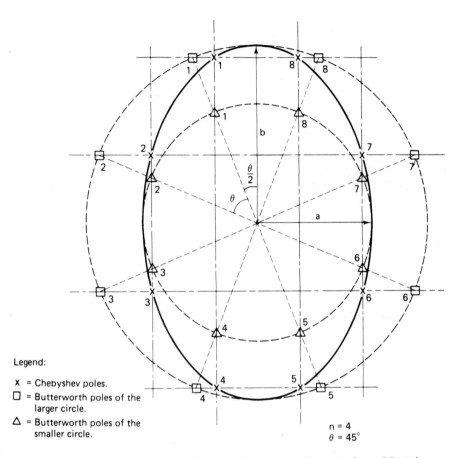

Legend:
× = Chebyshev poles.
□ = Butterworth poles of the larger circle.
△ = Butterworth poles of the smaller circle.

$n = 4$
$\theta = 45°$

Figure A–13 Pole diagram for Chebychev filter, $n = 4$. (From *Analog and Digital Filters: Design and Realization,* H. Y-F Lam. Copyright © 1979, Prentice-Hall, Inc. Reprinted by permission of Prentice-Hall, Inc.)

where $k = 1, 2, \ldots, n$. The poles of the Chebychev filter equation are found to lie on an ellipse as shown in Figure A–13. Butterworth poles for n equals 4, but with different cutoff frequencies are also shown for comparison.

As with the Butterworth filters, the pole locations for other than $\omega_c = 1$ rad/s are the values of Equation A–15 multiplied by the value of ω_c. The transfer function of the Chebychev filter can be written as

$$H(s) = \prod_{k=1}^{n} \frac{1}{s - s_k} \qquad (A\text{–}16)$$

where $k = 1, 2, \ldots, n$.

A.9 DESIGN PROCEDURES

Once the pass- and stopband frequencies and characteristics have been chosen, the design becomes one of choosing the proper value of n. This will provide sufficient gain rolloff to ensure that for frequencies greater than f_c, the gain has been reduced to the desired values. Knowledge of n allows us to determine the pole locations and the desired transfer function.

For a Chebychev design, the ripple parameter, ϵ, must also be stated: The smaller ϵ, the smaller the ripple width. The usual specification that is provided is not ϵ but the maximum value that the passband ripple attenuation can attain. This is given in Equation A–17.

$$A_{max} \text{ (dB)} = -10 \log \frac{1}{1+\epsilon^2} \tag{A-17}$$

which allows us to determine ϵ by

$$\epsilon = \sqrt{10^{(A/10)} - 1} \tag{A-18}$$

The design procedures are used to find the transfer function for two filters in the examples that follow.

Example A–1

A Butterworth filter is required such that its magnitude squared gain is greater than 0.95 when ω is 10 rad/s, and less then 0.02 when ω is greater than 30 rad/s. What is the filter design if ω_c is 15 rad/s?

Solution Using the equation for the Butterworth transfer function for the first condition, we have

$$|H(\omega)|^2 = \frac{1}{1+(\omega/\omega_c)^{2n}} \geq 0.95$$
$$= \frac{1}{1+(10/15)^{2n}} \geq 0.95 \tag{A-6b}$$

The solution of this equation yields a value of 4 for n.

To satisfy the stopband requirements, we have

$$|H(\omega)|^2 = \frac{1}{1+(\omega/\omega_c)^{2n}} \leq 0.02$$
$$= \frac{1}{1+(30/15)^{2n}} \leq 0.02 \tag{A-6b}$$

This yields 3 for a value of n. To ensure that both conditions are met, the larger value is chosen. The poles of $H(s)$ are found to be

$$s_1 = 15 \frac{\text{rad}}{\text{s}} (-0.3827 + j0.9239)$$

Sec. A.9 Design Procedures

$$s_2 = 15\,\frac{\text{rad}}{\text{s}}(-0.9239 + j0.3827)$$

$$s_3 = 15\,\frac{\text{rad}}{\text{s}}(-0.9239 - j0.3827) \quad\quad (A-12)$$

$$s_4 = 15\,\frac{\text{rad}}{\text{s}}(-0.3827 - j0.9239)$$

and the transfer function is

$$H(s) = \prod_{k=1}^{n}\frac{1}{s-s_k} = \frac{1}{(s-s_1)(s-s_2)(s-s_3)(s-s_4)}$$

$$= \frac{(15\text{ rad/s})^4}{s^4 + 39.197s^3 + 768.2s^2 + 8819.2s + 50{,}625} \quad\quad (A-11)$$

Example A-2

A Chebychev filter is desired with a maximum passband attenuation of 2.5 dB. If the square of the stopband gain must be down to 0.02 by $\omega = 50$ rad/s, what are the pole locations and the transfer function if $\omega_c = 20$ rad/s?

Solution The value of ϵ is determined from Equation A-18.

$$\epsilon = \sqrt{10^{(2.5/10)} - 1} = 0.8822 \quad\quad (A-18)$$

The order of the filter is obtained from Equation A-8 by equating it to the desired values of the passband at the specified frequencies. Using trial and error, we obtain

$$|H(\omega)|^2 = \frac{1}{1 + \epsilon^2 C_n^2(\omega/\omega_c)} \leq 0.02$$

$$= \frac{1}{1 + (0.7783)C_n^2(\omega/\omega_c)} \leq 0.02$$

$$\frac{1}{0.02} - 1 \leq (0.7783)C_n^2\left(\frac{50}{20}\right)$$

$$49 \leq (0.7783)C_n^2(2.5)$$

$n = 1: \quad 49 \leq (0.7783)(\omega)^2$
$\quad\quad\quad 49 \leq (0.7783)(2.5)^2 = 4.864$

$n = 2: \quad 49 \leq (0.7783)(2\omega^2 - 1)^2$
$\quad\quad\quad 49 \leq (0.7783)[2(2.5)^2 - 1]^2 = 132.25$

For $n = 2$, we see that all specifications will be met. The poles are determined using Equation A-15.

$$s_1 = 20\,\frac{\text{rad}}{\text{s}}(-0.3576 + j0.7924)$$

$$\quad\quad\quad\quad\quad\quad\quad\quad\quad\quad\quad\quad\quad\quad (A-15)$$

$$s^2 = 20\,\frac{\text{rad}}{\text{s}}(-0.3576 - j0.7924)$$

The transfer function is found using Equation A–16.

$$H(s) = \prod_{k=1}^{2} \frac{1}{s - s_k} = \frac{1}{(s - s_1)(s - s_2)}$$
$$= \frac{1}{s^2 + 14.31\,s + 302.3} \tag{A–16}$$

A.10 ADDITIONAL FILTERS

In addition to the Butterworth and Chebychev filters, there are two others that should be mentioned, the elliptic and Bessel filters.

A.10.1 Elliptic Filters

If it is desired to have both the stop- and passbands show mini-max characteristics simultaneously, the elliptic filter is chosen. The equation chosen must approximate the ideal filter in both bands. The transfer function is given by

$$|H(\omega)|^2 = \frac{1}{1 + \epsilon^2 R_n^2(\omega)}$$

where $R_n(\omega)$ is a polynomial in ω.

A.10.2 Bessel Filters

The filters discussed above attempted to approximate the ideal transfer function as closely as possible by judiciously choosing a function that would reduce the passband error to as small a value as practical. The Bessel filter design is based on the approximation of a linear phase shift in the passband, and the resulting transfer function equation yields an all-pole equation (i.e., the filter has no zeros). The equation that is generated involves Bessel functions; as these go beyond the scope of this text, the Bessel filter will not be discussed further.

A.11 SUMMARY

This appendix has presented a brief review of analog filter design to serve as a background for their use in the design of digital filters. The reader is strongly urged to refer to the texts listed in the References for a more in-depth coverage of the topic.

REFERENCES

BUDAK, A. *Passive and Active Network Analysis and Synthesis.* Boston: Houghton Mifflin, 1974.

LAM, H. Y.-F. *Analog and Digital Filters.* Englewood Cliffs, N.J.: Prentice-Hall, 1979.

TEMES, G. C., and MITRA, S. K., Eds. *Modern Filter Theory and Design.* New York: Wiley, 1973.

PASCAL PROGRAM FOR THE DFT

appendix B

{This program will find the DFT of an N-point sample, X(k), where

$$X(k) = \sum_{n=0}^{N-1} x(n)[\cos(2\pi kn/N) - j\sin(2\pi kn/N)],$$

x(n) are the input samples, n = 0, 1, ..., N - 1. The range of k is also 0, 1, ..., N - 1.

The computer will prompt the user to enter the number of sample points, N. Another prompt will ask for the data to be entered, one value per line.}

```
program dft(input,output,go);
```

{The arrays are defined here. hold will store the input sequence; xrk[k] and xik[k] will hold the real and imaginary parts of X(k); mag[k] and phi[k] will hold the magnitude and phase of X(k).

If it is desired to change the number of sample points, e.g., to zero fill the data, the upper limit of the array must be changed to the new value. For example, if a N-point DFT is required, the upper limit of ALL the arrays is set to N-1.}

```
const N = 256;
      pi = 3.14159265;
```

Appendix B Pascal Program for the DFT

```pascal
type hold = array[0..(N-1)]of real;
var xrk, xik, smlx, mag, phi: hold;
    k: integer;
    go: text;

{Loadarray is the procedure that reads in the sample points
 to be transformed.}

procedure loadarray(var smlx: hold; nn: integer; var gon: text);
var data, currentindex: integer;
    Y, N, character: char;
begin
writeln('Are you zero-filling this DFT?  Enter Y or N.');
readln(character);
if character = Y
   then begin
           writeln('How many points are data?');
           readln(data);
           writeln('Enter ',data:2,'-data points, one per line.');
           reset(gon);
           for currentindex:= 0 to (data-1)
               do readln(gon,smlx[currentindex]);
           for currentindex:= data to nn-1
               do smlx[currentindex]:= 0.0
           end {1}

   else begin {2}
           writeln('Enter ',N:2,'-data points, one per line.');
           for currentindex:= 0 to nn-1
               do readln(smlx[currentindex])
           end; {2}
end;

{compute is the procedure that calculates the transform values
 X(k).}

procedure compute(var smlx: hold;
                  var xxrk: hold;
                  var xxik: hold;
                  nn: integer);
var k, n: integer;
    temp1, temp2: real;
begin
for k:= 0 to nn-1
    do begin
       temp1:= 0.0;
       temp2:= 0.0;
       for n:= 0 to nn-1
           do begin
              temp1:= temp1 + smlx[n]*cos((6.28318531*k*n)/nn);
              temp2:= temp2 + smlx[n]*sin((6.28318531*k*n)/nn)
              end;
       XXRK[K]:= TEMP1/N;
       XXIK[K]:= -TEMP2/N
          end;
end;

{The procedure that calculates the magnitude of X(k) begins
 here.}

procedure magnitude(var mmag: hold;
                    var xxrk: hold;
                    var xxik: hold;
                    nn: integer);
var k: integer;
begin
for k:= 0 to nn-1
    do begin
       mmag[k]:= sqrt(sqr(xxrk[k]) + sqr(xxik[k]));
       end;
end;
```

{The procedure that calculates the phase of X(k) begins here.}

```pascal
procedure phase(var pphi, xxrk, xxik: hold; nn: integer);
var k: integer;
begin {1}
for k:= 0 to nn-1
    do begin {2}
        if abs(xxrk[k]) < 0.0001
          then begin {3}
           if abs(xxik[k]) < 0.0001
             then pphi[k]:= 0.0
             else begin {4}
              if xxik[k] > 0.0
                then pphi[k]:= pi*0.5
                else pphi[k]:= -pi*0.5;
                  end; {4}
                end {3}
          else begin {5}
            pphi[k]:= arctan((xxik[k])/(xxrk[k]));
            if xxrk[k] < 0.0
              then begin {6}
               if xxik[k] = 0.0
                 then pphi[k]:= -pi
                 else begin {7}
                  if xxik[k] > 0.0
                    then pphi[k]:= pi+pphi[k]
                    else pphi1[k]:= -(pi-pphi[k]);
                       end; {7}
                   end; {6}
                end; {5}
      pphi[k]:= ((pphi[k]*360)/(2.0*pi))
        end; {2}
end; {1}
```

{The procedure that prints the data begins here.}

```pascal
procedure writedata(nn: integer;
                    smlx: hold; xrk: hold; xik: hold;
                    mag: hold; phi: hold);
var k: integer;
begin
writeln;
writeln;
writeln('  n    x(n)       k       XR(k)       XI(k)       Mag(k)      Phi(k)');
writeln('----------------------------------------------------------------------');
for k:= 0 to nn-1
    do begin
        writeln(k:3, smlx[k]:5:1, k:6, xrk[k]:11:4,
                xik[k]:11:4, mag[k]:11:4, phi[k]:11:2);
        end;
end;
```

{The procedure that graphs the values of the magnitude of the DFT begins here.}

```pascal
procedure graph(promag: hold; nn: integer);
var counter, k: integer;
    temp: real;
begin
writeln;
writeln('         The values of the magnitude of the DFT.     ');
writeln;
writeln('             0.2        0.4       0.6       0.8      1.0');
writeln('      ___|___|___|___|___|___|___|___|___|___|');
temp:= 0.0;
for k:= 0 to nn-1
    do begin
        if promag[k] > temp
          then temp:= promag[k]
        end;
```

Appendix B   Pascal Program for the DFT

```pascal
        temp:= 50.0/(temp);
        for k:= 0 to nn-1
            do begin
                write(k:3,'|');
                promag[k]:= (promag[k])*temp;
                for counter:= 5 to ((round(promag[k]))+4)
                    do begin
                        write(' ');
                        end;
                write('x');
                writeln
                end;
        end;

{The main program begins here.}

begin
writeln('*********NOTICE*********');
writeln;
writeln('This is a ', N:2, '-point DFT. If you wish another value,');
writeln('then you must change the value of N in the program.');
writeln;

{The procedure that reads in the sample points is called here.}

loadarray(smlx, N, go);

{The procedure that calculates the X(k) is called here.}

compute(smlx, xrk, xik, N);

{The procedure that calculates the magnitude of X(k) is
 called here.}

magnitude(mag, xrk, xik, N);

{The procedure that calculates the phase of X(k) is called
 here.}

phase(phi, xrk, xik, N);

{The procedure that prints the data is called here.}

writedata(N, smlx, xrk, xik, mag, phi);

{The procedure that graphs the magnitude of the DFT is
 called here.}

graph(mag, N)

end.

*********NOTICE*********

This is a 64-point DFT.  If you wish another value,
then you must change the value of N in the program.

Are you zero-filling this DFT?  Enter Y or N.
y

How many points are data?
8

Enter   8-data points, one per line.
1
1
1
1
1
1
1
1
```

n	x(n)	k	XR(k)	XI(k)	Mag(k)	Phi(k)
0	1.0	0	0.1270	0.0000	0.1270	0.00
1	1.0	1	0.1166	-0.0417	0.1238	-19.69
2	1.0	2	0.0885	-0.0726	0.1145	-39.38
3	1.0	3	0.0514	-0.0857	0.0999	-59.06
4	1.0	4	0.0159	-0.0798	0.0814	-78.75
5	1.0	5	-0.0089	-0.0597	0.0604	-98.44
6	1.0	6	-0.0182	-0.0341	0.0387	-118.13
7	1.0	7	-0.0134	-0.0121	0.0180	-137.81
8	0.0	8	0.0000	0.0000	0.0000	0.00
9	0.0	9	0.0142	0.0007	0.0142	2.81
10	0.0	10	0.0228	-0.0069	0.0238	-16.88
11	0.0	11	0.0229	-0.0170	0.0285	-36.56
12	0.0	12	0.0159	-0.0238	0.0286	-56.25
13	0.0	13	0.0060	-0.0239	0.0246	-75.94
14	0.0	14	-0.0017	-0.0176	0.0177	-95.63
15	0.0	15	-0.0039	-0.0082	0.0090	-115.31
16	0.0	16	0.0000	0.0000	0.0000	0.00
17	0.0	17	0.0074	0.0035	0.0082	25.31
18	0.0	18	0.0144	0.0014	0.0145	5.63
19	0.0	19	0.0177	-0.0044	0.0183	-14.06
20	0.0	20	0.0159	-0.0106	0.0191	-33.75
21	0.0	21	0.0102	-0.0137	0.0171	-53.44
22	0.0	22	0.0037	-0.0122	0.0127	-73.13
23	0.0	23	-0.0003	-0.0067	0.0067	-92.81
24	0.0	24	0.0000	0.0000	0.0000	0.00
25	0.0	25	0.0043	0.0048	0.0065	47.81
26	0.0	26	0.0103	0.0055	0.0117	28.12
27	0.0	27	0.0150	0.0022	0.0151	8.44
28	0.0	28	0.0159	-0.0032	0.0162	-11.25
29	0.0	29	0.0127	-0.0076	0.0148	-30.94
30	0.0	30	0.0072	-0.0087	0.0113	-50.62
31	0.0	31	0.0020	-0.0057	0.0061	-70.31
32	0.0	32	0.0000	0.0000	0.0000	0.00
33	0.0	33	0.0020	0.0057	0.0061	70.31
34	0.0	34	0.0072	0.0087	0.0113	50.62
35	0.0	35	0.0127	0.0076	0.0148	30.94
36	0.0	36	0.0159	0.0032	0.0162	11.25
37	0.0	37	0.0150	-0.0022	0.0151	-8.44
38	0.0	38	0.0103	-0.0055	0.0117	-28.13
39	0.0	39	0.0043	-0.0048	0.0065	-47.81
40	0.0	40	0.0000	0.0000	0.0000	0.00
41	0.0	41	-0.0003	0.0067	0.0067	92.81
42	0.0	42	0.0037	0.0122	0.0127	73.12
43	0.0	43	0.0102	0.0137	0.0171	53.44
44	0.0	44	0.0159	0.0106	0.0191	33.75
45	0.0	45	0.0177	0.0044	0.0183	14.06
46	0.0	46	0.0144	-0.0014	0.0145	-5.63
47	0.0	47	0.0074	-0.0035	0.0082	-25.31
48	0.0	48	0.0000	0.0000	0.0000	0.00
49	0.0	49	-0.0039	0.0082	0.0090	115.31
50	0.0	50	-0.0017	0.0176	0.0177	95.62
51	0.0	51	0.0060	0.0239	0.0246	75.94
52	0.0	52	0.0159	0.0238	0.0286	56.25
53	0.0	53	0.0229	0.0170	0.0285	36.56
54	0.0	54	0.0228	0.0069	0.0238	16.87
55	0.0	55	0.0142	-0.0007	0.0142	-2.81
56	0.0	56	0.0000	0.0000	0.0000	0.00
57	0.0	57	-0.0134	0.0121	0.0180	137.81
58	0.0	58	-0.0182	0.0341	0.0387	118.12
59	0.0	59	-0.0089	0.0597	0.0604	98.44
60	0.0	60	0.0159	0.0798	0.0814	78.75
61	0.0	61	0.0514	0.0857	0.0999	59.06
62	0.0	62	0.0885	0.0726	0.1145	39.37
63	0.0	63	0.1166	0.0417	0.1238	19.69

Appendix B Pascal Program for the DFT

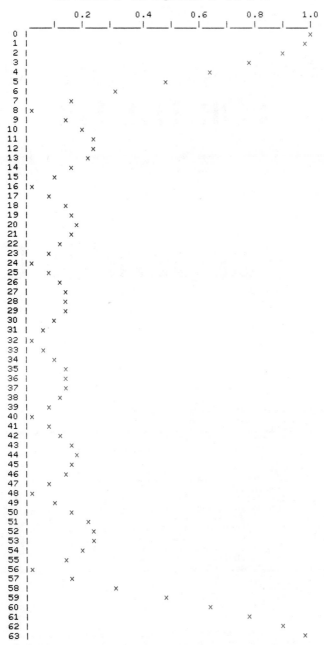

The values of the magnitude of the DFT.

FORTRAN PROGRAM FOR THE FFT

appendix C

```
C       THIS IS THE SUBROUTINE THAT CALCULATES THE FFT.
        SUBROUTINE FFTSTP(ZIN, AFTER, NOW, BEFORE, ZOUT)
        INTEGER AFTER,BEFORE,NOW,IA,IB,IN,J
        REAL ANGLE,RATIO,TWOPI,T
        COMPLEX ZIN(AFTER,BEFORE,NOW),ZOUT(AFTER,NOW,BEFORE)
        COMPLEX ARG,OMEGA,VALUE
        DATA TWOPI / 6.2831853071795864769 /
        T=NOW*AFTER
        ANGLE=TWOPI/T
        OMEGA=CMPLX(COS(ANGLE),-SIN(ANGLE))
        ARG=CMPLX(1.,0.)
        DO 100 J=1,NOW
            DO 90 IA=1,AFTER
                DO 80 IB=1,BEFORE
                    VALUE=ZIN(IA,IB,NOW)
                    DO 70 IN=(NOW-1),1,-1
   70                   VALUE=VALUE*ARG+ZIN(IA,IB,IN)
   80               ZOUT(IA,J,IB)=VALUE
   90       ARG=ARG*OMEGA
  100   CONTINUE
        RETURN
        END

C       THIS PROGRAM PERFORMS A FOURIER TRANSFORM ON DATA
C       INSERTED IN THE PROGRAM, OR ON DATA ENTERED UPON A
C       PROMPT FROM THE COMPUTOR.
C       N=16 IS THE NMBR OF POINTS OF THE FNCTN THAT IS SMPLD.
C       INZEE DTRMNES WHICH WRK ARRAY IS TO BE OPRATED ON.  THE
C       OTHER WORK ARRAY WILL HOLD THE RESULT OF THE FFT.
```

Appendix C Fortran Program for the FFT

```
      INTEGER INZEE,N,AFTER,BEFORE,NEXT,NEXTMX,NOW,PRIME(16)
      COMPLEX Z1(16),Z2(16)
      READ(5,80)(Z1(I),I=1,16)
   80 FORMAT(8F6.3)
      DATA PRIME/2,3,5,7,11,13,17,19,23,29,31,37,41,43,47,49/
      NEXTMX=16
      N=16
      INZEE=1
      AFTER=1
      BEFORE=16
      NEXT=1
   10 IF((BEFORE/PRIME(NEXT))*PRIME(NEXT)-BEFORE)1,2,2
    1 NEXT=NEXT+1
      GO TO 3
    2 NOW=PRIME(NEXT)
      BEFORE=BEFORE/PRIME(NEXT)
      GO TO 4
    3 IF(NEXT.LE.NEXTMX)GO TO 10
      NOW=BEFORE
      BEFORE=1
    4 IF(INZEE-1)6,5,6
    5 CALL FFTSTP(Z1, AFTER, NOW, BEFORE, Z2)
      GO TO 7
    6 CALL FFTSTP(Z2, AFTER, NOW, BEFORE, Z1)
    7 INZEE=3-INZEE
      IF(BEFORE-1)9,8,9
    9 AFTER=AFTER*NOW
      GO TO 10
    8 IF(INZEE-1)21,22,21
   21 WRITE(6,750)(Z2(I),I=1,16)
  750 FORMAT('Z2=',2F14.7)
      GO TO 11
   22 WRITE(6,751)(Z1(J),J=1,16)
  751 FORMAT('Z1=',2F14.7)
   11 STOP
      END
```

INDEX

A

Abscissa, 2
Absolutely integrable, 5
Accumulators, 291, 295
Accuracy, 63, 106, 107, 111
Acquisition time, 94, 285
Active filter, 69
Adaptive filter, 58
ADC, 106, 173, 285, 286, 287
Addition, 7
Admittance, 73
Advance signal, 9
Algorithm, 258
Aliasing, 62, 63, 88, 95, 102, 160, 173, 212, 253
All-pole equation, 342
All-zero system, 194
Alternation theorem, 231
ALU, 287, 288, 291, 295
AMD 29500, 277
Amplification, 67
Amplifiers, 76
Am 29516/517, 281
Am 29540, 282
Analog, 5, 106
Analog filters, 67, 201
Analog filter theory, 328
Analog processing, 67
Analog signal processing, 57, 59, 60
Analog signals, 58, 61, 64
Analog-to-digital converter (ADC), 62, 63, 106
Analog transfer function, 201
Analytic, 186
Antisymmetric, 219, 227
Aperiodic, 30
Aperiodic jitter, 93
Aperiodic time varying signal, 31
Aperture time, 90
Argument, 3, 8, 195
Arithmetic logic unit, 287
Attenuation, 328
Autocorrelation, 40, 44
Autocorrelation function, 40, 42, 43, 44, 245, 247
Autocorrelation integral, 46

B

Band edges, 232
Band limit, 62, 67
Band limited, 88, 101, 209, 212
Band limiting, 62, 72
Band reject, 232
Band reject filter, 68
Band width, 199
Bartlett, 225
Basis vectors, 27

Index

Batch effect, 60
Bessel filters, 342
Bessel functions, 342
Bilinear transform, 208
Bilinear transformation (BLT), 201, 202, 205, 206, 207, 237
Binary, 39
Bipass filter, 198
Bipolar, 63, 108
Biquad, 76
Biquad filter, 75
Bit reversed order, 261
Blackman, 225, 226
Buffer amplifier, 89
Butterfly flow graph, 268
Butterfly operations, 282
Butterfly relationship, 260
Butterworth, 73, 201, 206, 208, 209, 332, 339
Butterworth filters, 70

C

Canonical, 192, 193
Canonical form, 190, 192, 200
Canonical signal flow graphs, 268
Cartesian, 27
Cascade, 192
Cascade filter, 209
Cascade form, 190
Cathode ray tube (CRT), 297
Causal, 20, 70, 99, 151, 152, 164, 168, 181, 183, 186, 188, 218, 220, 245
Causality, 10
Causal sequence, 167, 168
Causal system, 175
Center frequency, 235
Chebychev, 73, 201, 208, 209, 244
Chebychev approximations, 230, 333
Chebychev filter, 70, 339
Chebychev poles, 337
Chebychev polynomials, 230, 334
Circular convolution, 250
CMOS, 301
CMOS technology, 118
Coefficients ROM, 290
Collection, 2
Common mode, 77
Common mode gain, 77, 78
Common mode rejection ratio (CMRR), 77, 78, 79
Common mode signal, 79
Communications, 45
Comparator, 115, 117
Complete set, 28
Complex conjugates, 207
Complex frequency, 48
Complex z plane, 161
Conductances, 73
Contact resistance, 390
Continuous, 1
Continuous time, 174
Continuous time function, 149
Continuous time system (CTS), 149
Conversion interval, 96
Conversion rate, 114
Conversion speed, 324

Conversion time, 114, 118, 299
Conversion time effect, 326
Convert, 118
Convolution, 1, 16, 25, 35, 36, 45, 69, 87, 149, 153, 160, 161, 168, 174, 175, 194, 250, 222, 223, 244, 245, 251, 274
Convolution equation, 17, 183, 245
Convolution integral, 18, 36, 70
Convolving, 59, 244
Cooley, J.W., 257, 262
Cooley-Tukey algorithm, 268
Correlation, 1, 39, 47, 63, 244, 254
Correlation function, 4
Correlation integral, 40
Counter, 117
CPU, 293
Cross correlation function, 40
Cutoff, 328
Cut off frequency, 70, 71, 201, 205, 234

D

DAC, 115, 284, 285, 286, 287, 297, 299
DAR, 286, 287
Data acquisition IC, 135
Data RAM, 282, 290
Data ready, 118
Data ready signal, 89
Decimation, 103
Decimation in frequency (DIF), 262
Decimation in time (DIT), 262, 297
Decomposition, 192
Deglitched, 325
Delay, 8, 183, 221
Delayed impulse, 150
Delay property, 167, 182
Delay section, 178
Delay unit, 177, 178
Delta function, 11, 70
Deviation, 230, 234
DFT, 168, 244, 245, 248, 249, 253, 255, 257, 258, 260
DIF, 262
Difference, 227
Difference equation, 47, 181, 184, 188, 189, 194, 207, 219, 221, 238
Differential amplifier, 77
Differential equation, 47, 174, 175
Differential gain, 77, 78
Differential linearity, 113
Differential mode, 77
Differential output voltage, 79
Digital computers, 176
Digital filters, 58, 196, 273
Digital filter transform function, 201
Digital processing system, 62
Digital processor, 106
Digital signal processing (DSP), 1, 61
Digital-to-analog converter (DAC), 62, 64, 115, 283, 297
Direct form: type 1, 189
Direct form: type 2, 189
Direct memory access (DMA), 293
Discontinuity, 22, 23
Discrete, 2
Discrete convolution, 154, 250, 271

Discrete convolution relation, 150
Discrete correlation, 273
Discrete Fourier transform, 244, 245, 248
Discrete impulse function, 150, 151, 152
Discrete sinusoidal wave, 150
Discrete system, 194
Discrete-time, 174, 181
Discrete-time function, 149
Discrete-time impulse response, 149, 151
Discrete-time signals, 150
Discrete-time systems, 149
Distortion, 204, 209
DIT, 262, 268
DMA, 293
Domain, 2, 26, 49, 149
Doppler, 45
Doppler filter bank, 46
Doppler frequency shift, 46
Dot product, 28
Double precision mathematics, 295
Drift, 68
Droop, 94
Dual-slope converter, 118
Dual-slope integrater, 115, 116
Dual-slope technique, 118

E

ECL, 278
EKG, 254
Electrical system, 1
Electrocardiogram (EKG), 62
Electroencephalogram (EEG), 62
Electromagnetic interference (EMI), 76
Ellipse, 339
Elliptic filters, 342
Emitter-coupled logic, 278
End-of-conversion (EOC), 106
Energy content, 43
Energy density spectrum, 43
Envelope distortion, 212
EPROM, 284, 285, 286
Equivalent, 6
Error, 107, 220, 226, 227, 229, 330, 332
Error terms, 173
Even function, 331
Extrema, 237
Extremal frequencies, 231, 234
Extremum, 231

F

Fast Fourier transform (FFT), 245, 257, 258
Feedthrough, 94
FET, 89, 253, 262, 268, 269, 270, 282
FET input op amp, 301
FFT, 245, 258, 260
FFT address sequences, 282
FFT algorithm, 262
Filter, 182, 189
Filtering, 67
Filter poles, 335
Finite, 218
Finite impulse response (FIR), 152, 220
Finite impulse sequence, 163
Finite width pulse train, 99

FIR, 152, 181, 182, 197, 254
FIR filter, 181, 212, 213, 214, 216, 218, 221, 223, 226, 227
First order filter, 205
First order term, 191
Flow graph diagram, 257, 261
Folding, 19
Fourier, 48, 69, 156
Fourier series, 26, 29, 30, 159, 217, 218, 226, 250
Fourier transform, 25, 26, 30, 39, 42, 157, 161, 166, 184, 195
Fourier transform properties, 34, 209
Frequency content, 60, 159, 166
Frequency domain, 35, 59, 60
Frequency impulse sequence, 163
Frequency response, 157, 158, 166, 184, 194, 195, 199, 205, 209, 211, 212, 216, 218, 247
Frequency response functions, 227
Frequency transformation, 234
FT pairs, 33
Function, 1, 2
Fundamental frequency, 28

G

Gain, 340
Gain error, 112
Gain roll-off, 72
Geometric progression, 165
Geometric series, 212, 223
Gibbs phenomenon, 223
Glitches, 325
Grid, 231
Grid density, 232
Guard band, 88

H

Hamming, 225, 226
Hanning, 225
Harvard architecture, 295
High frequency noise, 67
Hold command, 86, 89, 93

I

IC, 227
IC ADCs, 117
Ideal amplifier, 76
Ideal filter, 217, 218, 219
Ideal filter response, 226
Ideal frequency response, 216, 218
Ideal sampling, 86, 87, 89
Ideal sampling circuit, 87
Ideal transfer characteristics, 332
IDFT, 248, 249, 250, 251, 268, 269, 270
IFFT, 270, 273
IGMF filter, 74, 76
IIR, 152, 181, 201
IIR filter, 196
Implicit function, 3, 5, 180, 196
Impulse, 11, 14
Impulse invariance technique, 209
Impulse response, 14, 16, 17, 18, 22, 44, 69,

Index

151, 166, 180, 209, 212, 213, 216, 217, 218, 220, 230, 231, 232, 234, 244, 329
Impulse sampling, 149
Impulse train, 86
Impulse transfer function, 189
Independent variable, 2, 18
Infinite domain, 2
Infinite gain multiple feedback filter, 74, 76
Infinite impulse response (IIR), 186, 222
Inner product, 28
Input impedance, 76, 78
Input multiplexer, 284
Input section, 62
Instability, 187
Instruction, 176
Instruction fetch, 295
Instruction mnemonic, 287
Instruction ROM, 288
Instruction set, 295
Instruction word, 287
Instrumentation amplifier, 78
Integrated circuit (IC), 277
Integrator, 116, 117
Intel 2920, 283
Interference, 67
Interrupts, 290
Inverse, 211
Inverse DFT, 269
Inverse FFT, 270
Inverse Fourier transform (IFT), 329
Inverse Laplace transform, 211
Inverse z-transform, 187, 188, 189, 190
Inverting, 78
I/O, 291
Iteration, 230

K

Kernel, 27
Kirchoff, 47
KVL, 78

L

Laplace transfer function, 48, 182
Laplace transform, 27, 48, 69, 156, 157, 161, 209
Least significant bit, 107
LIFO, 290
Limiting process, 30, 223
Linear, 1, 70, 151, 152, 181
Linear convolution, 250, 251
Linear DT system, 182
Linearity, 1, 7, 324
Linear phase, 213
Linear phase response, 197
Linear phase shift, 212, 221, 342
Linear system output energy, 44
Linear systems, 1, 9, 44, 47
Linear time invariant system (LTIS), 10
Long sequence, 254
Loudspeaker, 297
Low pass digital filter, 204
Low pass (LP) filter, 59, 65, 67, 328
LP-to-BP filter, 235
LP-to-HP filter, 236

LP-to-LP filter, 234
LP-to-notch filter, 236
LSB, 107, 111, 112, 113, 173, 285, 298, 300, 301, 324

M

MAC, 291, 293
Magnitude, 28, 29
Magnitude squared, 328
Main-lobe width, 225
Matched filter, 45
Maximally flat, 330, 331, 333
Maximum valued pole, 188
McClellan, J., 226, 232, 244, 273
Memory, 176, 177, 288
Microprocessor chips, 63
Microprocessors, 176
Microprogramming, 277
Minimax, 226, 230, 231, 244, 331, 342
Minimum conversion time, 96
Modulating process, 97
Monolithic A/D converter, 118
Monotically decreasing, 333, 334
Most significant bit, 109
MSB, 30, 109, 115, 287, 298
Multiplex (MUX), 135, 288
Multiplication, 7
Multiplier, 177, 287
Multiplier accumulator (MAC), 291
Multiplying DAC, 299

N

n-bit binary signal, 106
n-bit register, 115
n-dimensional space, 28
NEC μP7720, 287
NMOS technology, 287
Noise, 64, 76
Noise rejection capability, 77
"No missing codes," 114
Noncausal, 10, 218, 329
Noninverting, 78
Nonlinearity, 112
Nonlinear phase shift, 197
Nonperiodic function, 30
Nonrecursive, 176, 183, 184
Nonrecursive filter, 221
Nonrecursive systems, 178, 179, 180, 181, 182
Nonzero pulse width, 95
Nonzero transition width, 221
Normalized frequency, 195
Notch, 232
Notch filter, 63, 68, 73
$(N + 1)$ tap, 194
n-point convolution, 253
n-point DFT, 259, 261
Nyquist's sampling frequency, 195
Nyquist's sampling rate theorem, 103
Nyquist's theorem, 88

O

Object code, 287
Offset errors, 112

Offset voltage, 60, 68
1's-complement, 285
Op amp, 60, 68, 73, 78, 79
Op amp drift, 74
Optimal, 226, 244
Optimal filter, 226, 232
Ordinate, 3
Orthogonal, 28
Orthogonality condition, 30
Orthonormal, 28
Orthonormal set, 46
Output impedance, 76–77
Output multiplexer, 284
Overlap, 18, 19, 20, 21, 23, 37, 38, 155, 209, 211, 223
Overlap add, 254
Overrange output, 109
Overshoot, 220, 221, 223, 224
Oxide isolated bipolar technology, 277

P

Parallel form, 192
Parallel form realization, 193
Parallel multipliers, 281, 295
Parameter, 9, 17, 223
Parks, T., 226, 232, 244, 273
Parsaval, 43
Partial fraction expansion, 187, 192, 212
Partial fraction sum, 211
Pascal, 249
Passband, 70, 201, 202, 212, 216, 328, 332
Passband ripple, 230, 335
Passband ripple attenuation, 340
Past output, 176, 196
Periodic, 28
Periodic sequence, 248
Phase shift, 197, 198, 328
Pipelined architectures, 280
Pipelined processors, 280
Pipelined registers, 282
Pipelining, 277
Poles, 162, 185, 186, 187, 188, 195, 197, 200, 209, 330, 332
Pole-zero diagram, 183
Positive time offset binary, 109
Practical sampling, 102
Practical sampling rate, 89
Practical sampling theorem, 96
Practical transfer function, 330
Present output, 176, 196
Prewarp, 208
Prewarping, 176, 204, 277
Processor, 63, 176, 277
Processor noise, 63
Program counter (PC), 288, 290
Program format, 287
Pseudo-random noise, 39

Q

Quantization, 107, 149
Quantization effect, 63
Quantized, 111
Quantizing, 173

R

Rabinier and Gold, 232
Radar signal, 245
Radar systems, 45
Radix-2, 283
Radix-4, 283
RAM, 284, 286, 287, 295
Range, 2, 26, 149
Range rates, 46
Ratio, 111
Rational fraction, 331
Rational function, 182
Realizable, 218, 221
Reconstructed filter, 284, 325
Rectangular form, 207
Recursive, 176, 179, 181, 184, 185, 196
Recursive digital filter, 203
Recursive relation, 334
Recursive system, 181
Region of convergence, 183, 186, 187, 188
Region of conversions, 164
Relative accuracy, 118
Residue theorem, 187
Resolution, 325
Rightsided sequence, 186
Ripple, 70, 220, 221, 223, 224, 226, 229
Ripple specifications, 229
Ripple width, 71
ROC, 184
Rolloff, 70, 333, 340
ROM, 282, 293, 295
Roundoff error, 271
R-2R ladder network, 293, 300
Rule, 2, 3, 26

S

Sample-and-hold (SH), 62
Sampled output, 87
Sample points, 226
Sampling, 63, 67
Sampling noise, 95
Sampling process, 63
Sampling rate, 87, 101, 149, 175, 209
Sampling rate variations, 95
Sampling width, 326
Scaler, 286
Scratchpad memory, 293
Searching parameter, 4, 245
Second order, 192
Second order recursive filter, 295
Sequence, 149
Set, 2
Settling time, 324
Shift invariance, 151
Shift invariant, 181
Shift invariant property, 152
Shift register, 177
Sifting integral, 16, 17
Sifting property, 14
Signal, 5
Signal processing, 58, 278
Signal-to-noise ratio (SNR), 77
Signal variation, 63
Smoothing, 194

Index

Software, 61
Source resistance, 78
Span, 28
Spectral analysis, 63, 253
Speed, 106
SQNR, 111
Stable, 162, 164, 183, 186, 188, 332
Stable system, 197, 202
Start conversion, 89
Step function, 150
Stopband, 70, 201, 202, 208, 216, 329, 332
Stopband attenuations, 226
Stopband characteristic, 71
Stopband ripple, 230
Subroutines, 290
Successive approximation ADs, 135
Successive approximations, 117
Superposition, 78
Susceptances, 73
Symmetric, 218, 219, 227
Symmetric impulse response, 215
Symmetry, 213
System, 1
System transfer, 49
System transfer function, 182, 188

T

TDC 100, 8, 291
Temperature compensation, 301
TH, 115, 117
TH circuit, 89, 149, 173
Thin film network, 300
Throughput, 280
Time delayed function, 8
Time domain, 60
Time invariance, 8, 9, 14, 157
Time invariant, 8, 70, 152
Time shifted, 8, 14
Time shifted delta function, 13, 15
Time shifted property
Time shifting property, 35
Time sifting property, 14, 35, 39
TMS32010, 293
Track-and-hold (TH), 63, 85, 86, 96
Track command, 89
Transducer, 62
Transfer characteristics, 199
Transfer function, 1, 48, 59, 69, 180, 198, 199, 206, 208, 328
Transform, 226, 156
Transformation, 235, 236, 237
Transform concepts, 156
Transition band, 201, 202, 220

Transition region, 330
Transition width, 223, 224, 226, 233
Transistor beta, 60
Transversal, 194
Tri-state buffers, 118
Truncated, 223
Truncated impulse response, 221
Truncating, 218, 219, 221
Truncation, 220
TRW, 291
TTL, 278
Tukey, J.W., 257, 262
Two quadrant multiplication, 301
2's-complement, 281, 291, 295

U

Unipolar, 63, 108
Unipolar conversion, 301
Unit circle, 166, 245, 247
Unit delay, 185
Unity-gain buffer amplifier, 89
Unrealizable system, 10
Unsigned magnitude format, 291
Unsigned-number format, 281

V

VCVS, 75
Vector, 27, 198
Vector concepts, 27
VLSI, 64
Voltage-controlled voltage source, 73

W

Warping, 204
Waveform generator, 59
Weighted current source, 298
Weighting function, 229, 233
Window functions, 223, 224, 225, 226
Windowing, 216, 221

Z

Zener diode, 301
Zero extended, 254
Zero filled, 255, 270
Zero filling, 171, 250
Zeros, 162, 185, 197, 200
z plane, 162, 189, 197
z transform, 15, 157, 161, 162, 163, 165, 166, 182, 185, 189, 194, 211, 218, 221, 245, 246, 247, 250, 252